Directed Self-assembly of Block Copolymers for Nano-manufacturing

Related titles

Modeling, characterization and production of nanomaterials: Electronics, photonics and energy applications
(ISBN 978-1-78242-228-0)

Magnetic Nano- and Microwires: Design, Synthesis, Properties and Applications
(ISBN 978-0-08-100164-6)

Robust Design of Microelectronic Assemblies Against Mechanical Shock, Temperature and Moisture
(ISBN 978-1-84569-528-6)

Woodhead Publishing Series in Electronic and Optical Materials: Number 83

Directed Self-assembly of Block Copolymers for Nano-manufacturing

Edited by

Roel Gronheid and Paul Nealey

ELSEVIER

AMSTERDAM • BOSTON • CAMBRIDGE • HEIDELBERG
LONDON • NEW YORK • OXFORD • PARIS • SAN DIEGO
SAN FRANCISCO • SINGAPORE • SYDNEY • TOKYO
Woodhead Publishing is an imprint of Elsevier

WP WOODHEAD PUBLISHING

Woodhead Publishing is an imprint of Elsevier
80 High Street, Sawston, Cambridge, CB22 3HJ, UK
225 Wyman Street, Waltham, MA 02451, USA
Langford Lane, Kidlington, OX5 1GB, UK

ISBN 978-0-08-100250-6 (print)
ISBN 978-0-08-100261-2 (online)

British Library Cataloguing in Publication Data
A catalogue record for this book is available from the British Library.

Library of Congress Control Number: 2015939626

For information on all Woodhead Publishing publications visit our website at http://store.elsevier.com/

Contents

List of contributors

K.G. Amir Tavakkoli Massachusetts Institute of Technology, Cambridge, MA, USA

J. Andres Torres Mentor Graphics Corporation, Wilsonville, OR, USA

Karl K. Berggren Massachusetts Institute of Technology, Cambridge, MA, USA

K.R. Carter University of Massachusetts, Amherst, Massachusetts, USA

J. Choi University of Massachusetts, Amherst, Massachusetts, USA

T.H. Coskun GLOBALFOUNDRIES Inc., Santa Clara, CA, USA

G.S.W. Craig University of Chicago, Chicago, IL, USA

W.B. Farnham E. I. duPont de Nemours and Company, Central Research and Development, Wilmington, DE, USA

G.H. Fredrickson University of California, Santa Barbara, CA, USA

V.V. Ginzburg The Dow Chemical Company, Midland, MI, USA

Y. Han State Key Laboratory of Polymer Physics and Chemistry, Changchun Institute of Applied Chemistry, Chinese Academy of Sciences, Changchun, P.R. China

P.D. Hustad Dow Electronic Materials, Marlborough, MA, USA

B. Kim University of California, Santa Barbara, CA, USA

R.J. Kline NIST, Gaithersburg, MD, USA

N. Laachi University of California, Santa Barbara, CA, USA

A. Latypov GLOBALFOUNDRIES Inc., Santa Clara, CA, USA

P.F. Nealey University of Chicago, Chicago, IL, USA

Samuel M. Nicaise Massachusetts Institute of Technology, Cambridge, MA, USA

T.P. Russell University of Massachusetts, Amherst, Massachusetts, USA

T. Segal-Peretz Division of Materials Science, Argonne, IL, USA

R. Seidel University of Chicago, Chicago, IL, USA

M.T. Sheehan DuPont Electronic Polymers, Gregory, TX, USA

H.S. Suh Division of Materials Science, Argonne, IL, USA

D.F. Sunday NIST, Gaithersburg, MD, USA

M. Takenaka Kyoto University, Kyoto, Japan

P. Trefonas Dow Electronic Materials, Marlborough, MA, USA

J.D. Weinhold The Dow Chemical Company, Freeport, TX, USA

L. Williamson University of Chicago, Chicago, IL, USA

H.-S.P. Wong Stanford University, Stanford, CA, USA

G. Wu University of Chicago, Chicago, IL, USA

S. Xiong Division of Materials Science, Argonne, IL, USA

H. Yi Stanford University, Stanford, CA, USA

H. Yoshida Hitachi Research Laboratory Hitachi Ltd., Ibaraki, Japan

X. Yu State Key Laboratory of Polymer Physics and Chemistry, Changchun Institute of Applied Chemistry, Chinese Academy of Sciences, Changchun, P.R. China

C. Zhou University of Chicago, Chicago, IL, USA

Woodhead Publishing Series in Electronic and Optical Materials

1 **Circuit analysis**
J. E. Whitehouse
2 **Signal processing in electronic communications: For engineers and mathematicians**
M. J. Chapman, D. P. Goodall and N. C. Steele
3 **Pattern recognition and image processing**
D. Luo
4 **Digital filters and signal processing in electronic engineering: Theory, applications, architecture, code**
S. M. Bozic and R. J. Chance
5 **Cable engineering for local area networks**
B. J. Elliott
6 **Designing a structured cabling system to ISO 11801: Cross-referenced to European CENELEC and American Standards**
Second edition
B. J. Elliott
7 **Microscopy techniques for materials science**
A. Clarke and C. Eberhardt
8 **Materials for energy conversion devices**
Edited by C. C. Sorrell, J. Nowotny and S. Sugihara
9 **Digital image processing: Mathematical and computational methods**
Second edition
J. M. Blackledge
10 **Nanolithography and patterning techniques in microelectronics**
Edited by D. Bucknall
11 **Digital signal processing: Mathematical and computational methods, software development and applications**
Second edition
J. M. Blackledge
12 **Handbook of advanced dielectric, piezoelectric and ferroelectric materials: Synthesis, properties and applications**
Edited by Z.-G. Ye
13 **Materials for fuel cells**
Edited by M. Gasik
14 **Solid-state hydrogen storage: Materials and chemistry**
Edited by G. Walker
15 **Laser cooling of solids**
S. V. Petrushkin and V. V. Samartsev
16 **Polymer electrolytes: Fundamentals and applications**
Edited by C. A. C. Sequeira and D. A. F. Santos
17 **Advanced piezoelectric materials: Science and technology**
Edited by K. Uchino

18 **Optical switches: Materials and design**
Edited by S. J. Chua and B. Li
19 **Advanced adhesives in electronics: Materials, properties and applications**
Edited by M. O. Alam and C. Bailey
20 **Thin film growth: Physics, materials science and applications**
Edited by Z. Cao
21 **Electromigration in thin films and electronic devices: Materials and reliability**
Edited by C.-U. Kim
22 ***In situ* characterization of thin film growth**
Edited by G. Koster and G. Rijnders
23 **Silicon-germanium (SiGe) nanostructures: Production, properties and applications in electronics**
Edited by Y. Shiraki and N. Usami
24 **High-temperature superconductors**
Edited by X. G. Qiu
25 **Introduction to the physics of nanoelectronics**
S. G. Tan and M. B. A. Jalil
26 **Printed films: Materials science and applications in sensors, electronics and photonics**
Edited by M. Prudenziati and J. Hormadaly
27 **Laser growth and processing of photonic devices**
Edited by N. A. Vainos
28 **Quantum optics with semiconductor nanostructures**
Edited by F. Jahnke
29 **Ultrasonic transducers: Materials and design for sensors, actuators and medical applications**
Edited by K. Nakamura
30 **Waste electrical and electronic equipment (WEEE) handbook**
Edited by V. Goodship and A. Stevels
31 **Applications of ATILA FEM software to smart materials: Case studies in designing devices**
Edited by K. Uchino and J.-C. Debus
32 **MEMS for automotive and aerospace applications**
Edited by M. Kraft and N. M. White
33 **Semiconductor lasers: Fundamentals and applications**
Edited by A. Baranov and E. Tournie
34 **Handbook of terahertz technology for imaging, sensing and communications**
Edited by D. Saeedkia
35 **Handbook of solid-state lasers: Materials, systems and applications**
Edited by B. Denker and E. Shklovsky
36 **Organic light-emitting diodes (OLEDs): Materials, devices and applications**
Edited by A. Buckley
37 **Lasers for medical applications: Diagnostics, therapy and surgery**
Edited by H. Jelínková
38 **Semiconductor gas sensors**
Edited by R. Jaaniso and O. K. Tan
39 **Handbook of organic materials for optical and (opto)electronic devices: Properties and applications**
Edited by O. Ostroverkhova
40 **Metallic films for electronic, optical and magnetic applications: Structure, processing and properties**
Edited by K. Barmak and K. Coffey
41 **Handbook of laser welding technologies**
Edited by S. Katayama
42 **Nanolithography: The art of fabricating nanoelectronic and nanophotonic devices and systems**
Edited by M. Feldman
43 **Laser spectroscopy for sensing: Fundamentals, techniques and applications**
Edited by M. Baudelet

44 **Chalcogenide glasses: Preparation, properties and applications**
Edited by J.-L. Adam and X. Zhang
45 **Handbook of MEMS for wireless and mobile applications**
Edited by D. Uttamchandani
46 **Subsea optics and imaging**
Edited by J. Watson and O. Zielinski
47 **Carbon nanotubes and graphene for photonic applications**
Edited by S. Yamashita, Y. Saito and J. H. Choi
48 **Optical biomimetics: Materials and applications**
Edited by M. Large
49 **Optical thin films and coatings**
Edited by A. Piegari and F. Flory
50 **Computer design of diffractive optics**
Edited by V. A. Soifer
51 **Smart sensors and MEMS: Intelligent devices and microsystems for industrial applications**
Edited by S. Nihtianov and A. Luque
52 **Fundamentals of femtosecond optics**
S. A. Kozlov and V. V. Samartsev
53 **Nanostructured semiconductor oxides for the next generation of electronics and functional devices: Properties and applications**
S. Zhuiykov
54 **Nitride semiconductor light-emitting diodes (LEDs): Materials, technologies and applications**
Edited by J. J. Huang, H. C. Kuo and S. C. Shen
55 **Sensor technologies for civil infrastructures**
Volume 1: Sensing hardware and data collection methods for performance assessment
Edited by M. Wang, J. Lynch and H. Sohn
56 **Sensor technologies for civil infrastructures**
Volume 2: Applications in structural health monitoring
Edited by M. Wang, J. Lynch and H. Sohn
57 **Graphene: Properties, preparation, characterisation and devices**
Edited by V. Skákalová and A. B. Kaiser
58 **Silicon-on-insulator (SOI) technology**
Edited by O. Kononchuk and B.-Y. Nguyen
59 **Biological identification: DNA amplification and sequencing, optical sensing, lab-on-chip and portable systems**
Edited by R. P. Schaudies
60 **High performance silicon imaging: Fundamentals and applications of CMOS and CCD sensors**
Edited by D. Durini
61 **Nanosensors for chemical and biological applications: Sensing with nanotubes, nanowires and nanoparticles**
Edited by K. C. Honeychurch
62 **Composite magnetoelectrics: Materials, structures, and applications**
G. Srinivasan, S. Priya and N. Sun
63 **Quantum information processing with diamond: Principles and applications**
Edited by S. Prawer and I. Aharonovich
64 **Advances in non-volatile memory and storage technology**
Edited by Y. Nishi
65 **Laser surface engineering: Processes and applications**
Edited by J. Lawrence, C. Dowding, D. Waugh and J. Griffiths
66 **Power ultrasonics: Applications of high-intensity ultrasound**
Edited by J. A. Gallego-Juárez and K. F. Graff
67 **Advances in delay-tolerant networks (DTNs): Architectures, routing and challenges**
Edited by J. J. P. C. Rodrigues
68 **Handbook of flexible organic electronics: Materials, manufacturing and applications**
Edited by S. Logothetidis

69 **Machine-to-machine (M2M) communications: Architecture, performance and applications**
Edited by C. Anton-Haro and M. Dohler
70 **Ecological design of smart home networks: Technologies, social impact and sustainability**
Edited by N. Saito and D. Menga
71 **Industrial tomography: Systems and applications**
Edited by M. Wang
72 **Vehicular communications and networks: Architectures, protocols, operation and deployment**
Edited by W. Chen
73 **Modeling, characterization and production of nanomaterials: Electronics, photonics and energy applications**
Edited by V. Tewary and Y. Zhang
74 **Reliability characterisation of electrical and electronic systems**
Edited by J. Swingler
75 **Handbook of industrial wireless sensor networks: Monitoring, control and automation**
Edited by R. Budampati and S. Kolavennu
76 **Epitaxial growth of complex metal oxides: Techniques, properties and applications**
Edited by G. Koster, G. Rijnders and M. Huijben
77 **Semiconductor nanowires: Materials, synthesis, characterization and applications**
Edited by J. Arbiol and Q. Xiong
78 **Superconductors in the Power Grid**
Edited by C. Rey
79 **Optofluidics, sensors and actuators in microstructured optical fibres**
Edited by S. Pissadakis
80 **Magnetic Nano- and Microwires: Design, Synthesis, Properties and Applications**
Edited by M. Vázquez
81 **Robust Design of Microelectronic Assemblies Against Mechanical Shock, Temperature and Moisture**
E-H. Wong and Y-W. Mai
82 **Biomimetic technologies**
Edited by T. D. Ngo
83 **Directed Self-assembly of Block Copolymers for Nano-manufacturing**
Edited by R. Gronheid and P. Nealey

Part One

Physics and chemistry of block copolymer (BCP) materials

Physics of block copolymers from bulk to thin films

1

H. Yoshida[*], *M. Takenaka*[†]
[*]Hitachi Research Laboratory Hitachi Ltd., Ibaraki, Japan; [†]Kyoto University, Kyoto, Japan

1.1 Introduction

Block copolymers are composed of more than two kinds of chemically distinct polymers linked by covalent bonds. Various kinds of block copolymers have been synthesized since Szwarc firstly synthesized a block copolymer by using living-anionic polymerization technique (Szwarc et al., 1956). Recently, the development of living-cationic polymerization (Miyamoto et al., 1984) and living radical polymerization techniques (Otsu et al., 1982) has enabled us to synthesize block copolymers, which cannot be synthesized with living anionic polymerization and thus have widened the field of their applications. Block copolymers have been utilized as thermoplastic elastomers, pressure sensitive hot-melt adhesives, drug delivery systems, and recently have been extensively studied for use in the field of nano-patterning, including next-generation lithography for semiconductors. The superior physical properties of block copolymers originate from their self-assembled structures in the length scale of 10–100 nm. The physics of the self-assembling of block copolymers is one of the most important issues of statistical physics of polymers (Strobl, 2007). Particularly, in lithographic application, in which a higher degree of control over the self-assembly process is required, it is important to understand the physics of block copolymers in depth.

This chapter is aimed to give the fundamentals of block copolymer self-assembly. First, we shall focus on the physics of block copolymers in their bulk state. We will describe the order–disorder transition and the phase diagram of block copolymers and then discuss the dynamics of disorder to order transitions. Second, we shall address the physics of block copolymers in thin films. We will discuss the effects of confined space and the surface free energy on the morphologies.

1.2 Order–disorder transition of block copolymers

Here, we focus on the order–disorder transition of the simplest AB diblock copolymer melts (Hashimoto, 1996; Bates and Fredrickson, 1990; Hamley, 1998). The order–disorder transition is a thermodynamic phase transition and is controlled by the competition between the energetics and the entropy. The order–disorder transition of the AB diblock copolymer depends on several parameters, including (i) the volume fraction of one block in a diblock copolymer (f), (ii) the Flory–Huggins interaction parameter between two block chains (χ), which represents the energy cost of

Directed Self-assembly of Block Copolymers for Nano-manufacturing. http://dx.doi.org/10.1016/B978-0-08-100250-6.00001-8

two spices mixed homogeneously and is inversely proportional to temperature, and (iii) the polymerization index or total number of monomers per a polymer chain of the diblock copolymer ($N = N_A + N_B$), where subscripts A and B represent the two distinct block chains and N_i is the total number of monomers in ith chain. The strength of the segregation power of diblock copolymer is expressed by the reduced parameter, χN. When χN does not exceed a critical value, A and B block chains are mixed in the molecular level (disordered state, Figure 1.1a). In contrast, when χN exceeds a critical value, A and B block chains segregate into A- and B-rich phases, respectively (ordered state, Figure 1.1b). Because the chemical junctions of A and B block chains are located at the interface between two domains, the phase separation occurs on a microscopic scale, not on a macroscopic scale (larger than 1 μm) as is the common case for the blend of A and B homopolymer chains. This phase separation is called microphase separation. The dimensions of domains formed by microphase separation are therefore in the order of the radius of the gyration, R_g, of each block chain: ~5–100 nm. Typical morphologies formed by microphase separation are lamellae (LAM), the Double Gyroid ($Ia\bar{3}d$), hexagonally packed cylinders (HEX), and spheres in a body-centered lattice (BCC) as shown in Figure 1.1c.

In the case of low molecular weight diblock copolymers, we can observe the order–disorder transition without degradation of the diblock copolymers. Figure 1.2 shows the transmission electron microscope (TEM) images of the disordered and ordered states of symmetric poly(styrene-*block*-isoprene) (PS-*b*-PI) (Sakamato and Hashimoto, 1995). In the disordered state, we can observe an isotropic bicontinuous pattern reflecting the frozen concentration fluctuations, while this particular PS-*b*-PI has a lamellar structure with long-range order in the ordered state.

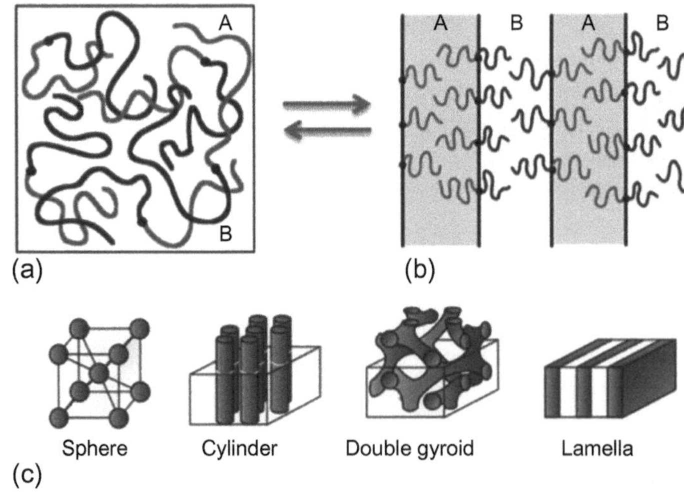

Figure 1.1 Schematic illustration of chain distributions in (a) disordered state and (b) ordered state of an A–B diblock copolymer. Typical domain morphologies are also presented in (c).

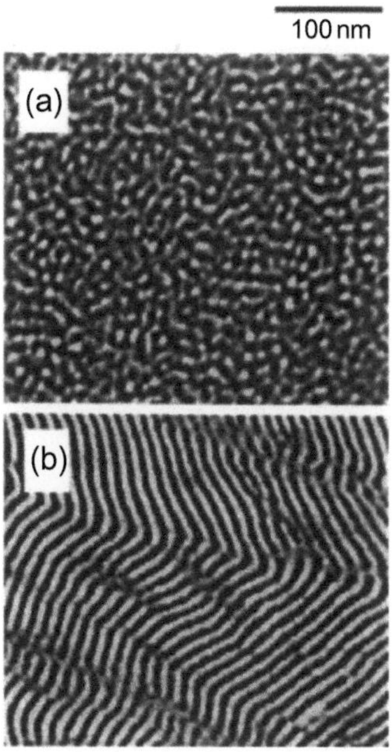

Figure 1.2 Transmission electron microscope images of (a) disordered state and (b) ordered state of symmetric poly(styrene-*block*-isoprene).

1.2.1 Disordered state

In the disordered state, A and B block chains are mixed in the molecular level and the thermal concentration fluctuations exists. The thermal concentration fluctuations are expressed by random phase approximation (RPA) (Leibler, 1980). Figure 1.3 shows structure factors of a diblock copolymer AB in disordered state with RPA (Mori et al., 1987). The structure factor is proportional to the scattering function and can be observed by using the scattering techniques such as small-angle X-ray scattering (SAXS) and small-angle neutron scattering. In Figure 1.3, we also show the structure factors of the corresponding A/B polymer blends. The structure factor of an AB diblock copolymer exhibits its peak at $q = q_m \sim 1/R_g \sim N^{-1/2}$ due to the connectivity between A and B polymers, where R_g is radius of gyration of an AB diblock copolymer. The connectivity reduces the concentration fluctuations at wavelength larger than R_g in comparison with the structure factor of the A/B blend. We can calculate the structure factor and spinodal point in various kinds of architectures of binary polymer systems including ABA triblock copolymers (Sakurai et al., 1992) and AB star polymers $(AB)_n$ (Hashimoto et al., 1989) with RPA. Effects of polydispersity and composition distribution (Mori et al., 1989), and asymmetry of monomer on the structure factor are also considered. We can also estimate the temperature dependence of χ by fitting the scattering intensity of binary block copolymers with the structure factors calculated by RPA (Mori et al., 1987).

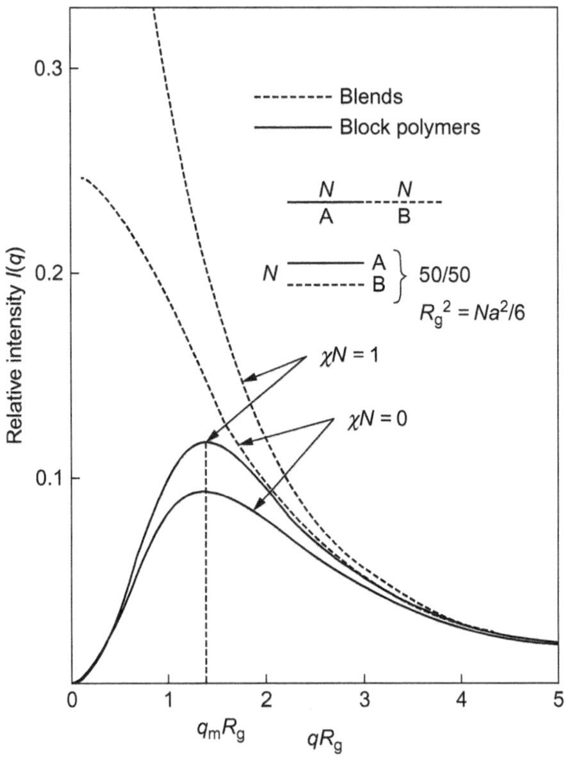

Figure 1.3 Structure factor calculated with random phase approximation for A–B diblock copolymer and A/B polymer blend at $\chi N=0$ and $\chi N=1$.
Reprinted with permission from Mori et al. (1987). Copyright 1987 American Chemical Society.

1.2.2 Weak segregation limit in ordered state

At the weak segregation limit in the vicinity of $\chi N=10$ of the ordered state, the segregation power is not strong enough to form a sharp phase boundary between the two phases, and the interface becomes diffuse. The phase diagram in the weak segregation limit has been predicted by Leibler (1980). Leibler identified stable phases in parameter space of χN on f by comparing the calculated Ginzburg–Landau free energy functional of the diblock copolymer for possible morphologies. The obtained phase diagram is displayed in Figure 1.4 (Hamley, 1998; Leibler, 1980; Fredrickson and Helfand, 1987). Around $f=0.5$, LAM are stable and the stable morphology changes to HEX and then to BCC as f decreases. At asymmetric volume fractions, the stable phase changes in the order of disorder (DIS), BCC, HEX, and LAM with increasing χN. Milner found the $Ia\bar{3}d$ phase between LAM and HEX phases (Milner and Olmsted, 1997), which is also predicted by the self-consistent field theory as discussed later. The order of phase transition becomes the second-order phase transition only at $f=0.5$ and χN at the order–disorder transition point $(\chi N)_c$ is 10.495.

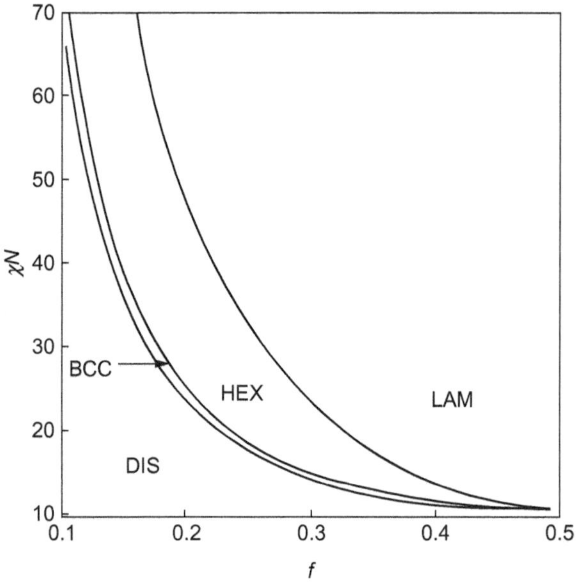

Figure 1.4 Phase diagram of A–B diblock copolymer obtained by mean-field theory without considering the effect of thermal fluctuations.
Reprinted with permission from Fredrickson and Helfand (1987). Copyright 1987 AIP Publishing LLC.

Figure 1.5 shows the phase diagram of low-molecular-weight AB diblock copolymers, including the effects of the thermal fluctuations. At near order–disorder transition point, the effects of the thermal fluctuations reduce the concentration fluctuations predicted by the mean-field theory (Fredrickson and Helfand, 1987; Schwahn et al., 1995), and the critical point of the phase diagram calculated with the mean-field theory becomes DIS (Milner and Olmsted, 1997). Thus, at asymmetric volume fraction, the direct phase transitions from DIS to LAM, HEX, and $Ia\bar{3}d$ phases are predicted, differing from the results of the mean-field theory. Fredrickson–Helfand (Fredrickson and Helfand, 1987) predicted $(\chi N)_c$ is higher than that estimated with the mean-field theory. For $f=0.5$, $(\chi N)_c$ is expressed by

$$\left(xN\right)_c = 10.495 + 41.022N^{-1/3} \tag{1.1}$$

The order of order–disorder transition becomes the weak first-order phase transition at $f=0.5$, and peak intensity of the structure factor has discontinuity between ordered and disordered states, while the second-order phase transition is predicted by the mean-field theory. Figure 1.6 shows the temperature dependence of the scattering function near order–disorder transition of PS-*b*-PI (Sakamato and Hashimoto, 1995). The temperature dependence of the scattered intensity shows discontinuity in the peak intensities between 98.2 and 100.2 °C, indicating that the order–disorder transition of AB diblock copolymers exhibits the first-order transition.

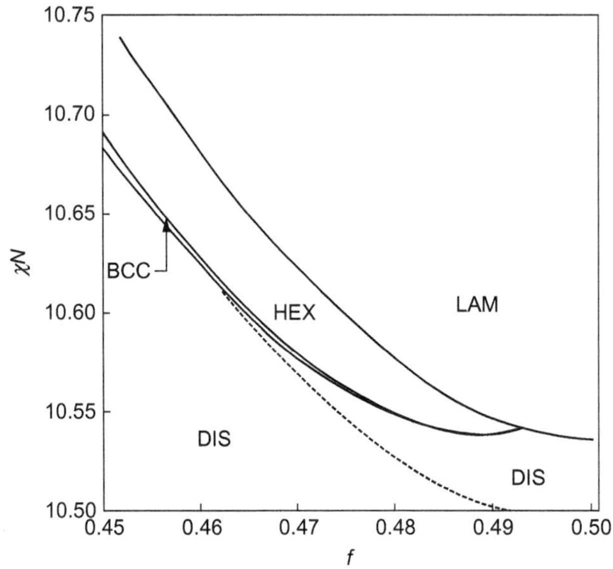

Figure 1.5 Phase diagram of A–B diblock copolymer, including the effects of thermal fluctuations for low molecular weight.
Reprinted with permission from Fredrickson and Helfand (1987). Copyright 1987 AIP Publishing LLC.

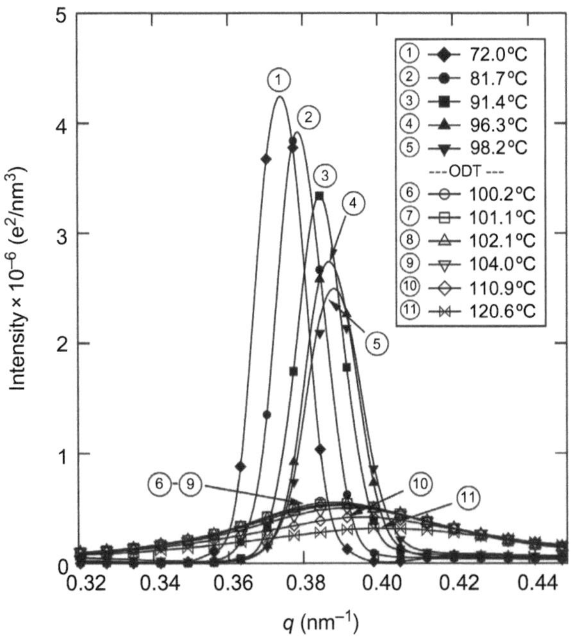

Figure 1.6 Small-angle X-ray scattering profiles of poly(styrene-*block*-isoprene) above and below ODT.
Reprinted with permission from Sakamato and Hashimoto (1995). Copyright 1995 American Chemical Society.

1.2.3 Strong segregation limit in ordered state

At $\chi N > 100$, or at the strong segregation limit, the thickness of the phase boundary becomes proportional to $a\chi^{1/2}$ (where a is the statistical segment length of diblock copolymer) and the volume fraction of A monomer in each phase is almost 1 or 0. Helfand et al. calculated the phase diagram in strong segregation limit and found the morphologies in strong segregation limit only depends on volume fraction f and is independent of χ or temperature (Helfand and Wasserman, 1976; Semenov, 1985; Ohta and Kawasaki, 1986). Figure 1.7 shows the phase diagram obtained experimentally by Hasegawa et al. for PS-*b*-PI. The experiment supports the theoretical results (Hasegawa et al., 1987).

The equilibrium domain spacing (*d*-spacing) of microdomains in the strong segregation limit is determined by the balance between the minimization of interfacial free energy and the maximization of conformational entropy of polymer chain under incompressible conditions. The following scaling law was obtained:

$$d \propto aN^{2/3}\chi^{1/6} \tag{1.2}$$

Figure 1.8 shows the molecular weight dependence of *d*-spacing of LAM for PS-*b*-PI in double logarithmic scale. The slope estimated from the results by small-angle X-ray scattering experiments becomes 2/3 (Hasegawa et al., 1987; Hashimoto et al., 1974, 1977, 1980a,b; Todo et al., 1977; Fujimura et al., 1981), which agrees with the theoretical results by Helfand and Wasserman (1976). Although experimental results have not been reported for χ-dependence of *d*-spacing, the following concentration

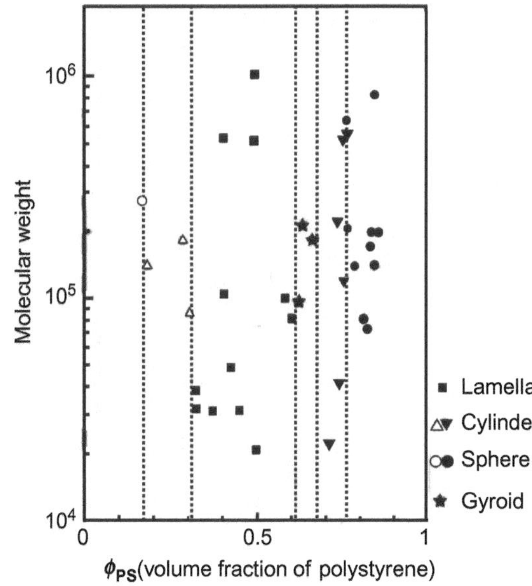

Figure 1.7 Phase diagram of poly(styrene-*block*-isoprene) at strong segregation region. Reprinted with permission from Hasegawa et al. (1987). Copyright 1987 American Chemical Society.

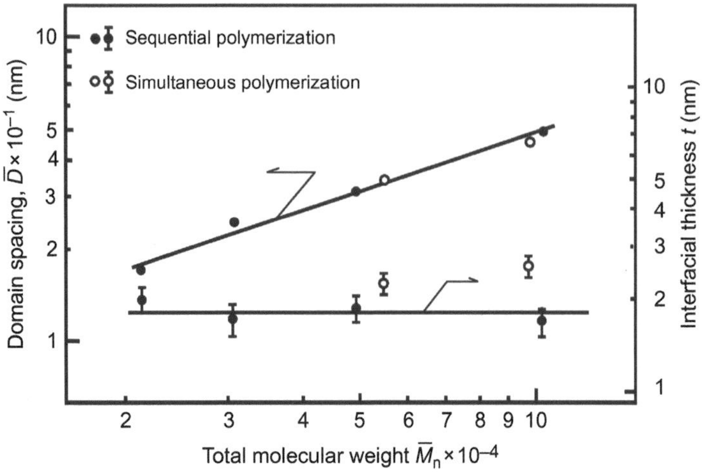

Figure 1.8 Molecular weight dependencies of domain spacing and interfacial thickness for poly(styrene-*block*-isoprene) at strong segregation region.
Reprinted with permission from Hashimoto et al. (1980a). Copyright 1995 American Chemical Society.

dependencies of *d*-spacing are reported for PS-*b*-PI dissolved into common good solvents in terms of the concentration of polymer ϕ_P (Shibayama et al., 1983; Hashimoto et al., 1983; Lodge et al., 2003),

$$d \propto \left(\phi_P / T\right)^{1/3} \tag{1.3}$$

$$d \propto \phi_P^{0.33} \chi^{0.25} \tag{1.4}$$

1.2.4 Phase diagram obtained by using self-consistent field theory

Matsen et al. constructed the phase diagram of the AB diblock copolymer from the weak segregation limit to the strong segregation limit by using self-consistent filed theory (SCFT) (Matsen and Schick, 1994a; Matsen and Bates, 1996). Figure 1.9 compares the phase diagram calculated by using SCFT for diblock copolymer melts (a) and that determined by the experimental results of PS-*b*-PI obtained by Khandpur et al. (1997) (b). In the weak segregation limit, the phase diagram obtained by SCFT results agrees with that by Leibler (1980). LAM phase around $f = 0.5$, and $Ia\bar{3}d$, HEX, and BCC phases appear as the volume fraction *f* is biased. In addition to those phases, closed-packed spheres (CPS) phase appears outside of BCC phase. In the strong segregation limit, $Ia\bar{3}d$ phase disappears while LAM, HEX, BCC, and CPS phases remain and depend only on the volume fraction.

As shown in Figure 1.9, although the phase diagram calculated with SCFT agrees with the experimental results qualitatively, there are some disagreements between them. In

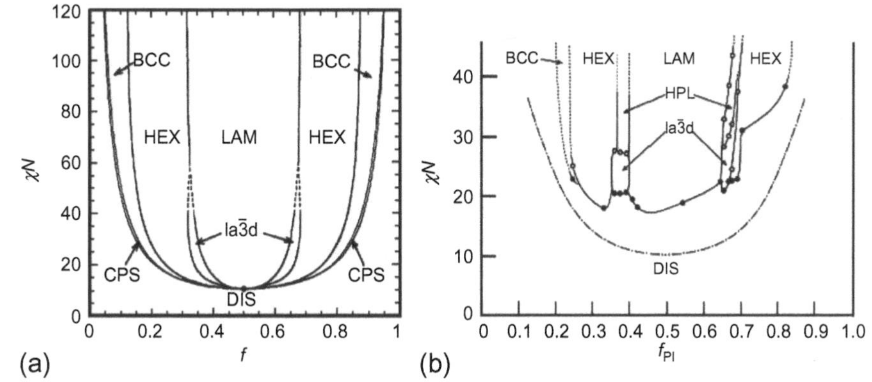

Figure 1.9 (a) Theoretical phase diagram calculated with self-consistent field theory by Matsen, and (b) experimental phase diagram of poly(styrene-*block*-isoprene) obtained by Khandpur et al.

the SCFT calculation, effects of the thermal concentration fluctuations are not included. Thus, the phase diagram by the SCFT exhibits the critical point at $\chi N=10.5$ for $f=0.5$ and the stable phase changes in the order of DIS, BCC, HEX, and to LAM with increasing χN near $f=0.5$. However, in the experiments, the order–disorder transition occurred around $\chi N=20$. Also, direct phase transitions from DIS to LAM, HEX, and $Ia3\bar{d}$ phases were found, and CPS phase was not found in the experimental. Moreover, the phase diagram, as determined by the experiments, is asymmetric in terms of volume fraction f. This asymmetry originates from the asymmetry in monomer volume and statistical segment length of PS and PI. The SCFT calculation, including the asymmetry in monomer volume and statistical segment length was proposed, which expressed the asymmetry in the phase diagram qualitatively (Bates et al., 1994). We can calculate phase diagram in various kinds of architectures of binary polymer systems, including ABA triblock copolymer (Matsen and Thompson, 1999) and AB star polymer $(AB)_n$ (Matsen and Schick, 1994b; Matsen, 2012). Recently, a fifth morphology, *Fddd*, which is an interconnected but orthotropic structure, has been found as an equilibrium phase between L and G phases (Takenaka et al., 2007; Kim et al., 2008a,b; Tyler and Morse, 2005). In Figure 1.10, the SAXS profile, the TEM image, and the model of *Fddd* are shown.

1.3 Morphologies of diblock copolymer/homopolymer mixtures

Binary mixtures of an AB diblock copolymer (AB) and a homopolymer A (HA) are often employed to tailor the domain spacing and the morphologies. In the mixture systems, an important parameter controlling the microdomain morphologies is $r_A=N_{HA}/N_A$, where N_{HA} is the degree of polymerization of HA (Hasegawa and Hashimoto, 1989; Tanaka et al., 1992).

The three mixing states of AB/HA mixtures are shown in Figure 1.11. In the case of $r_A \ll 1$ (b), HA is solubilized into the A phase of the AB microdomain structure

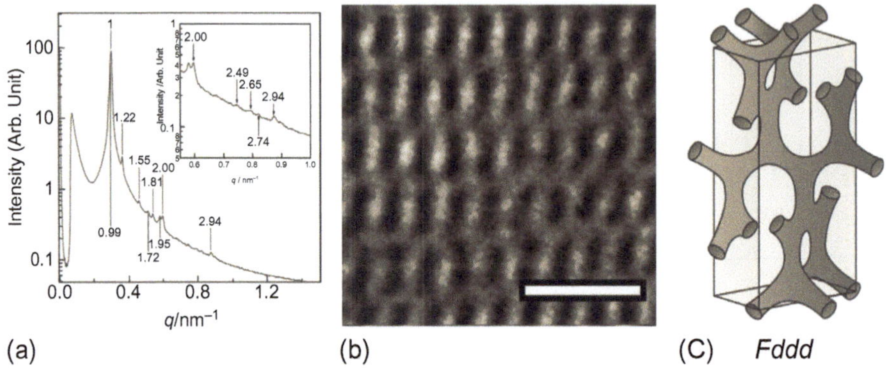

Figure 1.10 (a) Small-angle X-ray scattering profile and (b) transmission electron microscope image for *Fddd* structure in poly(styrene-*block*-isoprene). The white scale bar corresponds to 50nm. Schematic 3D image of *Fddd* is shown in (c).

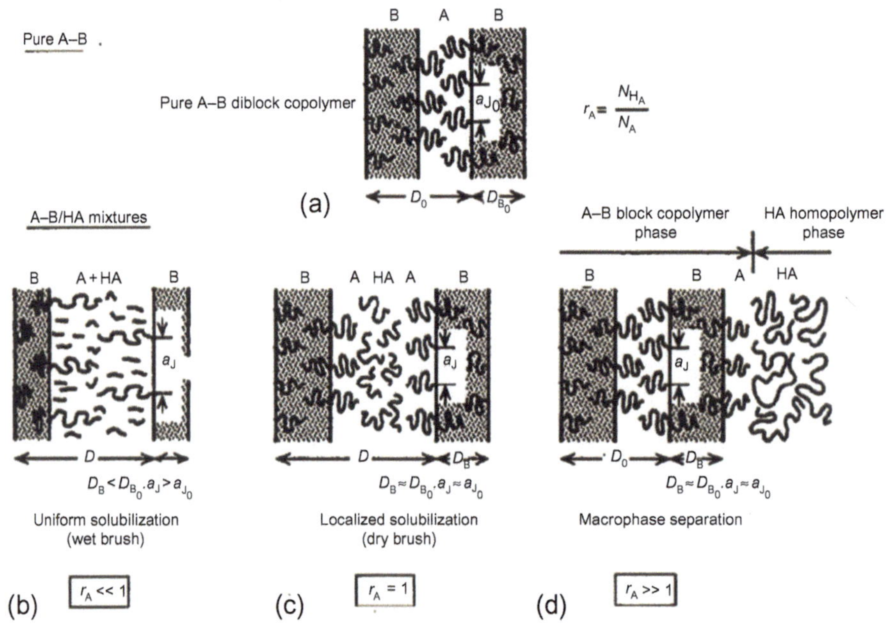

Figure 1.11 Possible microphase separated states of binary mixtures of AB diblock copolymer and A homopolymer (HA) with r_A. (a) Neat AB diblock copolymer with domains spacing D_0, thickness of B domain D_{B0}, and average distance between neighboring junction points a_{J0}. (b) Wet brush condition with $r_A \ll 1$. HA is uniformly solubilized in the A microdomains. Swollen HA expands D_0 and a_{J0} to D and a_J, respectively, and reduces D_{B0} to D_B. Change in the effective volume of A microdomain may induce morphological transition. (c) Dry brush condition with $r_A \approx 1$. Localized HA expands D_0 to D, while D_{B0} and a_{J0} are not affected by HA. (d) Macroscopic phase separation state at $r_A \gg 1$.
Reprinted with permission from Hasegawa and Hashimoto (1989). Copyright 1989 Pergamon Press plc.

uniformly and swells the A block chains. This situation results in expanding the average distance between neighboring junction points of AB block copolymer from a_{J0} to a_J, where a_{J0} and a_J are, respectively, the average distance between neighboring junction points of neat AB block copolymer and that of AB/HA mixtures. This case is called "wet brush." The swelling causes changes in the average volumes occupied per single A and B block chains and hence changes the morphology of the microdomain. Figure 1.12 shows the change in the morphology of PS-*b*-PI/Homo-PS mixtures in terms of the amount of Homo-PS in wet brush condition (Tanaka et al., 1992). While a neat PS-*b*-PI applied in this experiment exhibited lamellar structure, the morphology changed from lamellae to cylinder and to sphere with increasing mixing amount of Homo-PS.

Around $r_A \approx 1$ (Figure 1.11c), HA tends to be solubilized into the A phase, but HA does not significantly swell the A block chains. Thus, the average distance between neighboring junction points of AB almost does not change ($a_{J0} \approx a_J$). HA tends to be localized in the center regions of A microdomains. This state is called "dry brush." Under the "dry brush" condition, HA does not affect the morphology of AB microdomain significantly, while domain spacing increases with the amount of HA associating with the increase in the size of A microdomain. Figure 1.13 shows the TEM images of PS-*b*-PI/Homo-PS with $r_S = 1.2$ with increasing weight fraction of Homo-PS (Koizumi et al., 1992). The domain spacing increased while the thickness of PI lamellae remained constant, which is a typical behavior of the dry brush system.

When $r_A \gg 1$ (Figure 1.11d), AB and HA undergo macroscopic phase separation and AB forms microdomain structure in the AB-rich region.

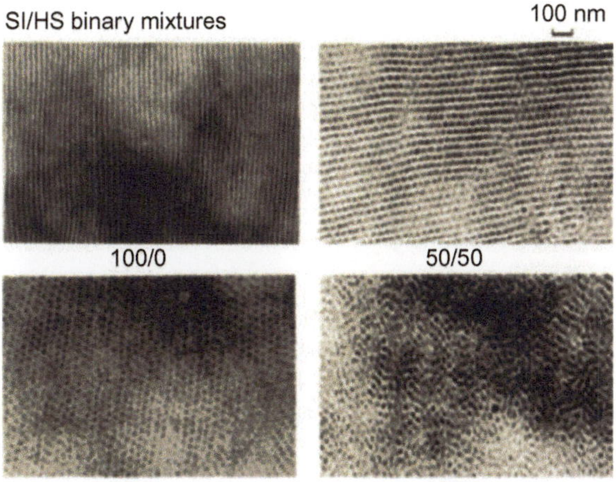

Figure 1.12 Transmission electron microscope images of PS-*b*-PI/Homo-PS mixture in wet brush state. PS-*b*-PI: number-averaged molecular weight $M_n = 31.6 \times 10^3$ g/mol, $f_{PS} = 0.45$, Homo-PS: $M_n = 2.2 \times 10^3$ g/mol, $r_S = 0.15$.

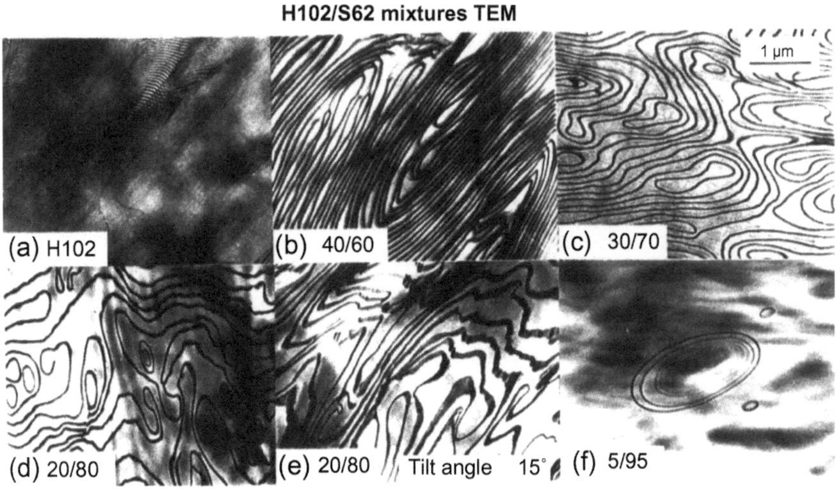

Figure 1.13 Transmission electron microscope images of PS-*b*-PI/Homo-PS mixture in dry brush state. PS-*b*-PI: $M_n = 1.0 \times 10^5$ g/mol, $f_{PS} = 0.46$, Homo-PS: $M_n = 6.2 \times 10^4$ g/mol, $r_S = 1.2$. (a) Neat PS-*b*-PI, (b) PS-PI/Homo-PS = 40/60, (c) 30/70, (d) 20/80, (e) 20/80 but tilted image, and (f) 5/95.

1.4 Dynamics of phase transition in block copolymers

Sakamoto et al. investigated the ordering processes of PS-*b*-PI, which had near-critical composition (number-averaged molecular weight $M_n = 1.5 \times 10^4$ g/mol, volume fraction of PS $f_{PS} = 0.45$, order–disorder transition temperature $T_{ODT} = 100.0$ °C) after the onset of quench from disordered state (172.0 °C) to ordered state (97.0 °C) by using time-resolved small-angle X-ray scattering (Sakamato and Hashimoto, 1998a,b). Figure 1.14 shows the change in the scattering profiles during the ordering process with time (Sakamato and Hashimoto, 1998a). The SAXS profile first changed from the initial temperature of 172 °C (dashed line) to 100.1 °C (dotted line) within 30 s, which was the shortest time covered in this experiment. After the SAXS profile reached 100.1 °C, almost no change was observed for a while as shown in the profiles at 90 and 360 s. After this period, intensity of the SAXS profile around its peak position started to increase with time and reached its constant values at ca. 3000 s. The observed change in the scattering profiles indicates that the ordering process occurs via nucleation and growth (Sakamato and Hashimoto, 1998b).

They also observed the morphology of the sample by TEM during the ordering process by freezing the sample with iced water (Sakamato and Hashimoto, 1998c). It was found that the lamellar grain with a sharp interface existed in the matrix of the disordered region at 630 s after quenching the sample from disordered (at 172.0 °C) to ordered (at 97.0 °C) states. This observation indicates the coexistence of the ordered phase (grains) having a lamellar structure with a long-range order and the disordered phase. As shown in Figure 1.15, the shapes of the lamellar grains observed

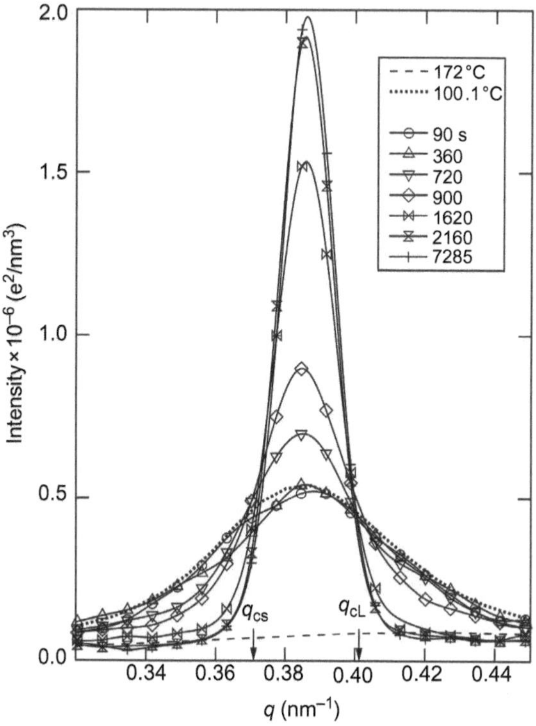

Figure 1.14 Change in the small-angle X-ray scattering profiles associating with temperature jump from disordered state (172 °C) to ordered state (97.0 °C) for poly(styrene-*block*-isoprene).
Reprinted with permission from Sakamato and Hashimoto (1998a). Copyright 1998 American Chemical Society.

were very anisotropic: the size along the lamellar normals was much larger than that parallel to the lamellar interfaces.

The shapes of the ordered grains in the matrix of the disordered phase highly depend on morphology, because the anisotropy of the interfacial free energy between the grain and matrix, which is the origin for the anisotropy of the grain shape, depends on the symmetry of the microdomain structures. Sakamoto et al. also observed the ordering processes of poly(styrene-*block*-isoprene-*block*-styrene) from disordered sphere phase to HEX and BCC phases by TEM (Sakamato and Hashimoto, 1998c). They found that both ordering processes also proceed via nucleation and growth. The SAXS analyses and TEM images indicated that the ordering process proceeds via anisotropic three-dimensional growth with homogeneous nucleation. Figure 1.16 shows the schematic pictures of grain structure of various morphologies. The LAM grains form a cigar-like prolate ellipsoidal shape. In the case of HEX grains, the ordered grains in the matrix of the disordered spheres form a lens-like oblate ellipsoidal shape, where the cylinders are packed parallel to the short axis of the oblate ellipsoid. On the other hand, the shape of the ordered grain with the spheres in a BCC lattice is almost isotropic.

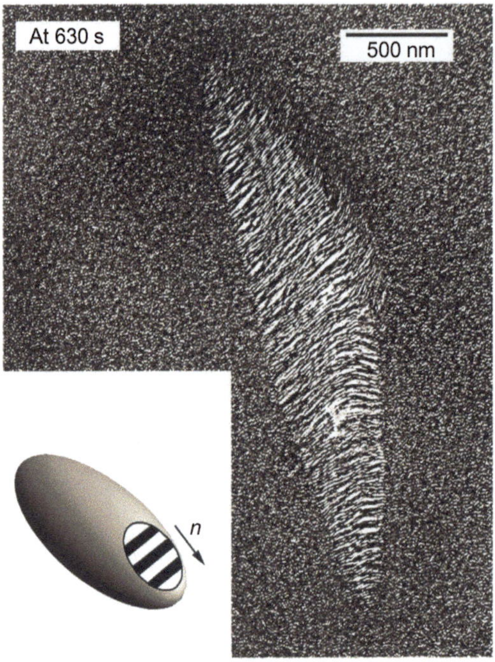

Figure 1.15 Transmission electron microscope image of a grain with lamellar structure during the ordering process of poly(styrene-*block*-isoprene).

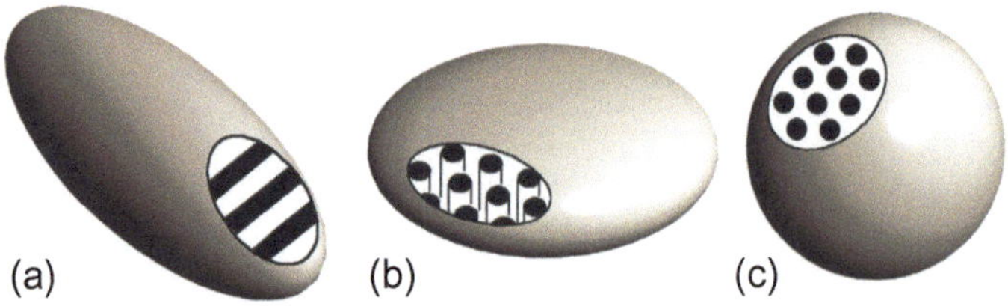

Figure 1.16 Schematic pictures of grain shape for (a) lamella (LAM), (b) cylinder (HEX), and (c) sphere (BCC) morphologies.

1.5 Structures of block copolymer in thin films

1.5.1 Free energy of block copolymer thin film

As discussed in the previous sections, the thermodynamic equilibrium morphology of microphase-separated block copolymers in bulk are determined by their primary structure (such as chemical structures of each of their blocks, N, and f) and by the interaction between the polymer chains χ. On the other hand, the morphology in thin film is strongly influenced by its boundary conditions defined mainly by the surface and interface of the film. The total free energy F of an AB diblock copolymer thin film can be expressed as a sum of the contributions from its internal structure F_{bulk} and from its surface/interface, $F_{surface}$, as follows:

$$F = F_{\text{bulk}} + F_{\text{surface}} \tag{1.5}$$

$$F_{\text{bulk}} = F_{\text{A/B}} + F_{\text{conformation}} \tag{1.6}$$

$$F_{\text{surface}} = F_{\text{A/sub}} + F_{\text{B/sub}} + F_{\text{A/air}} + F_{\text{B/air}} \tag{1.7}$$

where, $F_{\text{A/B}}$ is the interfacial energy between the A and B phases, $F_{\text{conformation}}$ is the conformational entropy of the A and B polymer chains composing the diblock copolymer, $F_{\text{A/sub}}$ and $F_{\text{B/sub}}$ are the interfacial energies of A and B phases and the substrate, and $F_{\text{A/air}}$ and $F_{\text{B/air}}$ are the interfacial energies of A and B phases and the air (free surface).

In a typical directed self-assembly process, the thickness of block copolymer thin film (t_f) deposited on the substrate is less than a few times its natural domain spacing d_0. Therefore, F_{surface} contributes significantly to the total free energy F and thus to the film's equilibrium structure. Another important factor is the effect of commensurability between d_0 and t_f. Conformational entropy of the block copolymer is smallest when the domain spacing is equal to that of the natural bulk value, or d_0. Therefore, the block copolymer film is most stable with the dimension in which domains can be piled inside the film, maintaining its natural bulk domain spacing. This effect leads to quantization of the thickness to discrete integer or half-integer values of d_0 when the domains are oriented in-plane, depending on the wetting condition as fully described in the following section. $F_{\text{conformation}}$ significantly increases when domains are confined into a film whose thickness is not commensurate with d_0 and results in transition of the morphology and/or causes spatial inhomogeneity in the film thickness. Overall, the chemical and steric effects of the film surface/interface cause complex influence on F and thus on the equilibrium morphology of the block copolymer thin film. Therefore, prediction of the morphology in thin film requires a method, such as simulation, to determine the structure with minimum F. Nevertheless, it is important to understand the physics of the typical mechanisms that govern block copolymer morphology in thin films.

Figure 1.17a and b shows top-down scanning electron micrographs (SEM) of poly(styrene-*block*-methyl methacrylate) (PS-*b*-PMMA) thin films self-assembled on the substrates, forming cylindrical and lamella microdomains, respectively. The uniform domains are perpendicularly oriented against the substrate, or oriented out-of-plane of the film, which is particularly desirable for lithographic applications. In the following section, we will focus on AB diblock copolymers and discuss the mechanism of how the thickness and the boundary conditions of block copolymer thin films influence the microdomain orientation.

1.5.2 Effect of surface energy term; $F_{surface}$

First, we discuss the influence of surface energies on the domain orientation in thin films, focusing on a lamella forming AB diblock copolymer system. As described above, orientation of the lamella domain in thin film is influenced by the interfacial energy F_{surface} between the two components of the block copolymer and the substrate and the air (here we assume that the block copolymer thin film is exposed to the air).

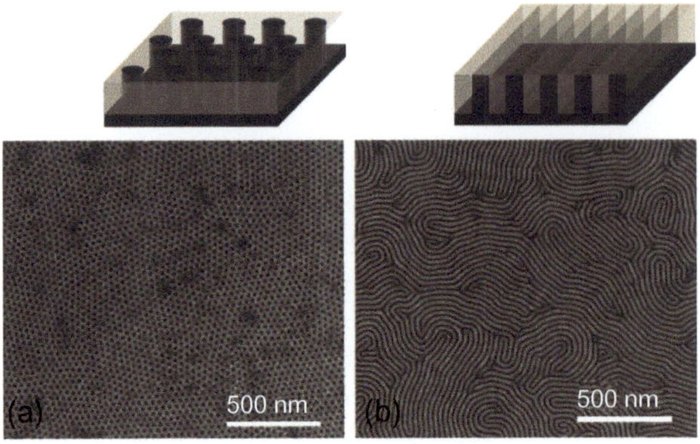

Figure 1.17 Top-view scanning electron microscope (SEM) images and illustrations of (a) cylindrical and (b) lamella microdomain structures self-assembled by poly(styrene-*block*-methyl methacrylate) (PS-*b*-PMMA) on Si substrate with out-of-plain orientation. SEM observation was carried out after PMMA removal by O_2 reactive ion etching.

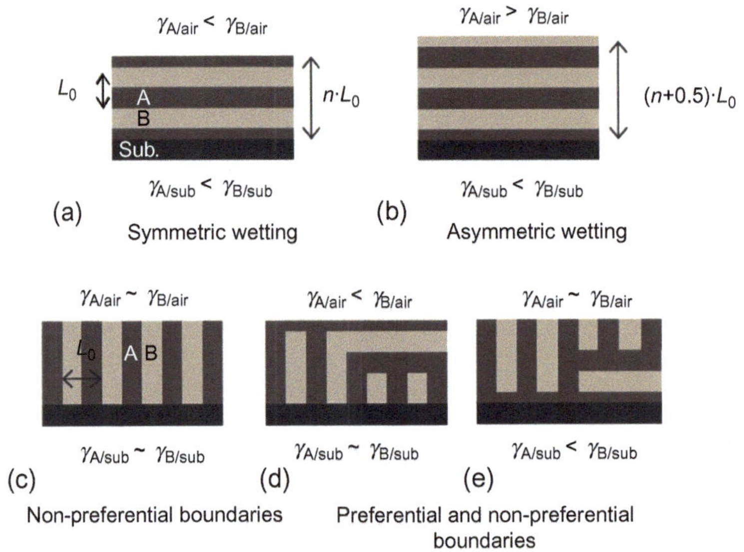

Figure 1.18 Schematic illustrations showing cross-section of AB diblock copolymer thin films with lamella structures under various interfacial affinities. Parallel orientation under (a) symmetric and (b) asymmetric wetting conditions. (c) Perpendicular orientation under non-preferential boundaries. (d) and (e) Three-dimensional structures sandwiched between preferential and non-preferential boundaries.

Schematic illustrations in Figure 1.18 show cross-sections of the lamella domains under various surface wetting conditions. The film interfaces can be classified into three categories, (i) interfacial energy of polymer A, γ_A, is higher than that of polymer B, γ_B ($\gamma_A > \gamma_B$), (ii) the opposite preference ($\gamma_B > \gamma_A$), and (iii) neutral preference ($\gamma_A \approx \gamma_B$). The interface is wetted by the domain composed with polymer having lower surface energy.

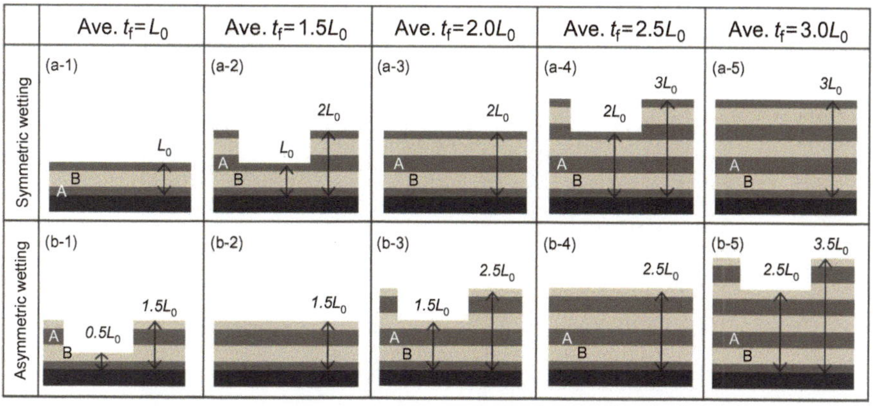

Figure 1.19 Schematic illustrations of AB diblock copolymer thin films demonstrating the effect of commensurability between lamella spacing L_0 and average film thickness t_f on the in-plain oriented lamella structures with (a) symmetric and (b) asymmetric wetting conditions.

Therefore, as illustrated in Figure 1.18a, when polymer A has lower γ against the substrate and the free air surface (i.e., $\gamma_{A/sub} < \gamma_{B/sub}$ and $\gamma_{A/air} < \gamma_{B/air}$), the substrate and the free surfaces are both wetted by the domain A. This kind of symmetric wetting causes the lamella to orient in-plane (horizontally) in the thin film. Another important aspect is the thickness t_f of the thin film. To maintain the natural bulk lamella spacing L_0 in limited thickness, film thickness should be equal to $t_f = n \cdot L_0$, where n is a non-negative integer as shown in Figure 1.19a-1, a-3, and a-5. When t_f does not fulfill this relation, lamella spacing may change from its natural value L_0 to L, which fulfills the relation $t_f = n \cdot L$, or spatial inhomogeneity in t_f may be caused as will be described in the next section. Likewise, when substrate and the free air surface possess opposite preference to polymer A and B, that is, asymmetric wetting with $\gamma_{A/sub} < \gamma_{B/sub}$ and $\gamma_{A/air} > \gamma_{B/air}$, lamella domains are oriented in-plane in the thin film (Figure 1.18b). In this case, the film thickness should fulfill $t_f = (n - 1/2) \cdot L_0$ as illustrated in Figure 1.19b-2 and b-4. When polymer A and polymer B have equal γ against the substrate and the free surface (air), or $\gamma_{A/sub} = \gamma_{B/sub}$ and $\gamma_{A/air} = \gamma_{B/air}$, the substrate and the free surface are both equally wetted by domain A and B (Figure 1.18c). In this case, the film is most stable with lamella domains orienting out-of-plane and therefore perpendicularly against the substrate. In this orientation, there is no effect of commensurability in out-of-plane direction, thus there is no limit in the film thickness. Finally, when one side of the film has preferential wetting and its opposite side has neutral wetting, as described in Figure 1.18d and e, the thin film may have T-shaped three-dimensional structures, with perpendicularly oriented lamellae against the neutral interface and parallel against the preferential interface.

Periodic nanostructures self-assembled in block copolymer thin films are often applied as a lithographic mask or template to fabricate nano-patterns. In such applications, perpendicularly oriented domains are preferable. Therefore, the substrate and the free surface should be both non-preferential against the polymers composing the diblock copolymer. The substrate can be chemically treated to alter the wetability or to create chemical pattern for directed self-assembly (see Chapter 9). However, it is often difficult to change the properties of the free surface. Figure 1.20 compares

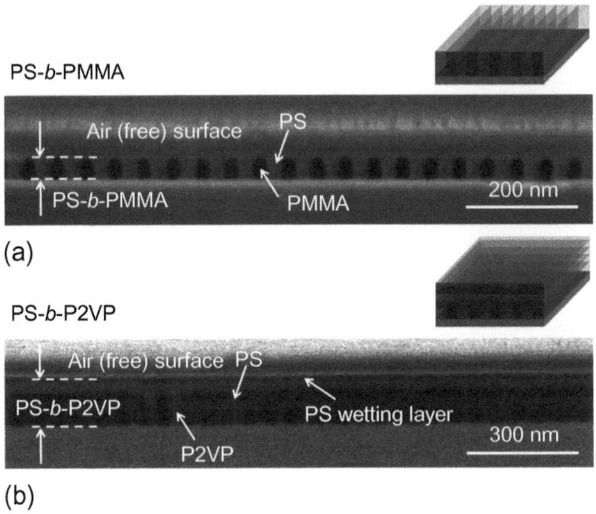

Figure 1.20 Cross-sectional scanning electron micrographs of poly(styrene-*block*-methyl methacrylate) and poly(styrene-*block*-2-vinylpyridine) (PS-*b*-P2VP) thin films, both forming lamella microdomains on the substrate with air free surface. The substrates are neutralized against PS-*b*-PMMA and PS-*b*-P2VP by grafting random copolymer of PS/PMMA and PS/P2VP, respectively.

morphologies of PS-*b*-PMMA and poly(styrene-*block*-2-vinylpyridine) (PS-*b*-P2VP), both forming lamella microdomains on the substrate with an air free surface. The substrates are neutralized against PS-*b*-PMMA and PS-*b*-P2VP by grafting random copolymer of PS/PMMA and PS/P2VP, respectively. As can be clearly seen, lamellae in PS-*b*-PMMA are oriented out-of-plane, while that of the PS-*b*-P2VP is forming T-shaped three-dimensional structures with PS domain wetting the free surface. This difference can be understood by the differences in the surface energies of PS, PMMA, and P2VP against the air at elevated temperature applied for the thermal annealing processes (~200 °C), which are $\gamma_{PS/Air}$ = ~28 (Wu, 1970; Sauer and Dee, 2002), $\gamma_{PMMA/Air}$ = ~27 (Wu, 1970), and $\gamma_{P2VP/Air}$ = 34 mN/m (Sauer and Dee, 2002). As PS and PMMA have similar γ values, the free air surface is exhibited as a neutral non-preferential interface for PS and PMMA, while it is exhibited as PS-preferential for PS and P2VP.

The film thickness for in-plane oriented domains should obey the commensurability rule for both symmetric and asymmetric wetting boundary conditions. Block copolymer thin film is commonly deposited on a substrate by spin coating from its solution and therefore, even with precise control, it is often difficult to exactly obtain the uniform film with target t_f. Figure 1.21a and b shows schematic cross-sections of an as-spun diblock copolymer thin film under symmetrically preferential wetting boundaries with initial thickness of $t_f = (1 + 0.5) \cdot L_0$, and that after microphase separation, respectively. In the as-spun film, where microphase separation is suppressed by rapid evaporation of the solvent, the film can take homogeneous thickness of $t_f = 1.5 \cdot L_0$. However, after microphase separation is completed, the film tends to be macroscopically separated into the regions with $t_f = 1.0 \cdot L_0$ and $2.0 \cdot L_0$ to fulfill the commensurability condition

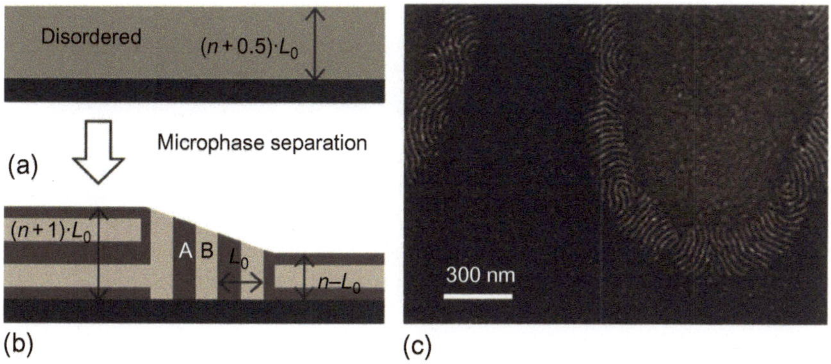

Figure 1.21 Schematic cross-sections of (a) as-spun AB diblock copolymer sandwiched between symmetrically preferential wetting boundaries with initial thickness of $t_f = (1+0.5) \cdot L_0$, and (b) after microphase separation. (c) An example of scanning electron micrograph of lamella forming block copolymer thin film with macroscopic terrace structure.

for symmetrical wetting, $t_f = n \cdot L_0$. In such a case, the film ends up with macroscopic terraces or holes with different numbers of layers as summarized in Figure 1.19a-2, a-4, b-1, b-3, and b-5 for AB block copolymer thin films with various average thickness under symmetric and asymmetric wetting conditions. Figure 1.21c shows a SEM of lamella forming block copolymer thin film. The regions with $t_f = 2.0 \cdot L_0$ and $1.0 \cdot L_0$ are coexisting, and are connected by the interfacial region with gradually decreasing thickness. As the thickness of the interfacial region does not fit the commensurability for the in-plane arrangement of the lamellae, the lamellae in the interfacial regions are oriented perpendicularly against the substrate, that is, in the orientation that has no restriction in the film thickness.

1.5.3 Effect of bulk energy term F_{bulk}

In the previous section, we discussed the influence of surface energies $F_{surface}$ on the orientation of microdomains in the thin film, assuming morphology of microdomains in a block copolymer is kept as same as that in its bulk state. In this section, we will briefly mention the effect of the bulk energy term F_{bulk} on the structure of the block copolymer thin film.

In certain cases, a decrease in interfacial energy of film surfaces $F_{surface}$ dominates the penalty of an increase in conformational entropy $F_{conformation}$, which causes changes in the microdomain morphology. One of the typical cases can be found in diblock copolymer thin film with cylindrical or spherical morphologies. As illustrated in Figure 1.22a, when free surface and substrate are both wetted by the major phase (B)—namely, by the matrix phase, spherical/cylindrical morphologies are kept as same as those in the bulk. However, when free surface of the film has smaller interfacial energy against the polymer forming the minor phase (A) than that of the major phase (B), that is, $\gamma_{A/air} < \gamma_{B/air}$, morphology of the minor phase located in adjacent to the free surface changes from their bulk to form a thin wetting layer as illustrated in

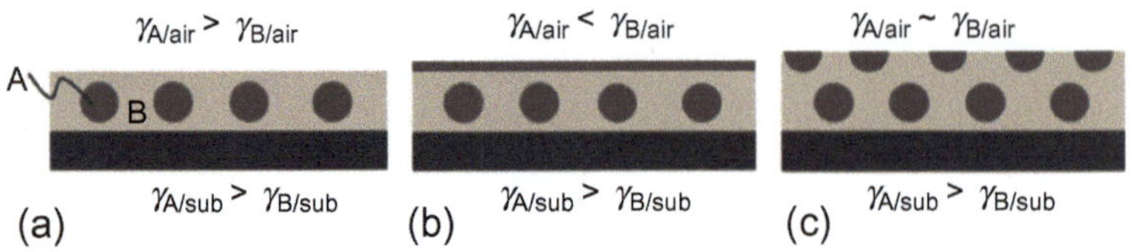

Figure 1.22 Schematic cross-sections of AB diblock copolymer thin films with spherical or in-plain oriented cylindrical domains under various wetting conditions. (a) AB diblock copolymer thin film with bulk morphology. (b) and (c) AB diblock copolymer thin film forming morphologies with non-bulk thin wetting layer and half-spheres/cylinders, respectively.

Figure 1.23 Schematic cross-sections of AB diblock copolymer forming lamella domains with asymmetric wetting with $t_f = 2.5L_0$ on (a) flat and (b) and (c) rough surfaces.

Figure 1.22b. Likewise, when substrate has a preference to the minor phase, the wetting layer is formed on the substrate. In the case when the surface is non-preferential to both of the polymers, or $\gamma_{A/air} \sim \gamma_{B/air}$, half-cylinder/half-sphere are often formed by the surface as illustrated in Figure 1.22c. The formation of these non-bulk morphologies cause an increase in conformational entropy, but the decrease of surface energies often overwhelm the penalty and decreases the total free energy F of the system.

Another example of bulk energy F_{bulk} effect can be found in lamella orientation in a thin film on a substrate with a rough surface (Sivaniah et al., 2005). Figure 1.23 illustrates cross-sections of lamella domains under asymmetric wetting boundaries with $t_f = 2.5L_0$ on flat (a) and rough surfaces (b, c). In this condition, lamellae are expected to orient in-plain in the film as shown in Figure 1.23a. However, when the substrate has a certain roughness, there is a case where the lamellae orient out-of-plain of the film. When lamellae are oriented parallel against the substrate with rough surface, F_{bulk} will increase through bending deformations of the lamellae as illustrated in Figure 1.23b. On the other hand, the principal energy penalty for perpendicularly oriented lamellae shown in Figure 1.23c arises from unfavorable contacts between one of the components and the substrate. For a sufficiently rough substrate such comparison may lead the perpendicular configuration to be more energetically favorable.

1.6 Conclusion

In this chapter, we have tried to address the basics of phase behavior of block copolymer systems. Physics governing the phase diagram and the dynamics of order–disorder transition are discussed for neat diblock copolymers, as well as for blends of diblock copolymer and homopolymer. Block copolymers are often used as thin films or in confined spaces. The significant effect of boundary conditions in such configurations, especially influences of steric confinement and surface free energy are also described.

Superior properties of block copolymers originate from their self-assembled structure with long-range order. Therefore, in-depth understanding of the physics of the self-assembly is indispensable for application. We hope this chapter will give the basis for the following chapters, in which details of block copolymers are discussed focusing on their actual use in nano-scale manufacturing.

References

Bates, F.S., Fredrickson, G.H., 1990. Block copolymer thermodynamics: theory and experiment. Annu. Rev. Phys. Chem. 41, 525. http://dx.doi.org/10.1146/annurev.pc.41.100190.002521.

Bates, F.S., Schulz, M.F., Khandpur, A.K., Forster, S., Rosedale, J.H., Almdal, K., Mortensen, K., 1994. Fluctuations, conformational asymmetry and block copolymer phase behaviour. Faraday Discuss. 98, 7. http://dx.doi.org/10.1039/FD9949800007.

Fredrickson, G.H., Helfand, E., 1987. Fluctuation effects in the theory of microphase separation in block copolymers. J. Chem. Phys. 87, 697. http://dx.doi.org/10.1063/1.453566.

Fujimura, M., Hashimoto, T., Kurahashi, K., Hashimoto, T., Kawai, H., 1981. Domain-boundary structure of styrene-isoprene block copolymer films cast from solutions. 6. Effect of temperature on spherical microdomain structure. Macromolecules 14, 1196. http://dx.doi.org/10.1021/ma50006a012.

Hamley, I.W., 1998. The Physics of Block Copolymers. Oxford University Press, Oxford. ISBN: 978-0-19-850218-0.

Hasegawa, H., Hashimoto, T., 1989. Self-assembly and morphology of block copolymer systems. Compr. Polym. Sci. Suppl. 1 (Suppl. 2), 497. http://dx.doi.org/10.1016/B978-0-08-096701-1.00253-6.

Hasegawa, H., Tanaka, H., Yamazaki, K., Hashimoto, T., 1987. Bicontinuous microdomain morphology of block copolymers. 1. Tetrapod-network structure of polystyrene-polyisoprene diblock polymers. Macromolecules 20, 1651. http://dx.doi.org/10.1021/ma00173a036.

Hashimoto, T., 1996. Thermoplastic Elastomers, second ed. Hanser, Vienna, ISBN: 978-1569903643 (Chapter15A).

Hashimoto, T., Nagatoshi, K., Todo, A., Hasegawa, H., Kawai, H., 1974. Domain-boundary structure of styrene-isoprene block copolymer films cast from toluene solutions. Macromolecules 7, 364. http://dx.doi.org/10.1021/ma60039a019.

Hashimoto, T., Todo, A., Itoi, H., Kawai, H., 1977. Domain-boundary structure of styrene-isoprene block copolymer films cast from solutions. 2. Quantitative estimation of the interfacial thickness of lamellar microphase systems. Macromolecules 10, 377. http://dx.doi.org/10.1021/ma60056a030.

Hashimoto, T., Shibayama, M., Kawai, H., 1980a. Domain-boundary structure of styrene-isoprene block copolymer films cast from solution. 4. Molecular-weight dependence of lamellar microdomains. Macromolecules 13, 1237. http://dx.doi.org/10.1021/ma60077a040.

Hashimoto, T., Fujimura, M., Kawai, H., 1980b. Domain-boundary structure of styrene-isoprene block copolymer films cast from solutions. 5. Molecular-weight dependence of spherical microdomains. Macromolecules 13, 1660. http://dx.doi.org/10.1021/ma60078a055.

Hashimoto, T., Shibayama, M., Kawai, H., 1983. Ordered structure in block polymer solutions. 4. Scaling rules on size of fluctuations with block molecular weight, concentration, and temperature in segregation and homogeneous regimes. Macromolecules 16, 1093. http://dx.doi.org/10.1021/ma00241a010.

Hashimoto, T., Ijichi, Y., Fetters, L.J., 1989. Order–disorder transition of starblock copolymers. J. Chem. Phys. 89, 2463. http://dx.doi.org/10.1063/1.455040.

Helfand, E., Wasserman, Z., 1976. Block copolymer theory. 4. Narrow interphase approximation. Macromolecules 9, 879. http://dx.doi.org/10.1021/ma60054a001.

Khandpur, A.K., Förster, S., Bates, F.S., Hamley, I.W., Ryan, A.J., Bras, W., Almdal, K., Mortensen, K., 1997. Polyisoprene-polystyrene diblock copolymer phase diagram near the order–disorder transition. Macromolecules 28, 8796. http://dx.doi.org/10.1021/ma00130a012.

Kim, M.I., Wakada, T., Akasaka, S., Nishitsuji, S., Saijo, K., Shimizu, H., Hasegawa, H., Ito, K., Takenaka, M., 2008a. Stability of the Fddd phase in diblock copolymer melts. Macromolecules 41, 7667. http://dx.doi.org/10.1021/ma801268d.

Kim, M.I., Wakada, T., Akasaka, S., Nishitsuji, S., Saijo, K., Shimizu, H., Hasegawa, H., Ito, K., Takenaka, M., 2008b. Determination of the Fddd phase boundary in polystyrene-block-polyisoprene diblock copolymer melts. Macromolecules 42, 5266. http://dx.doi.org/10.1021/ma900205s.

Koizumi, S., Hasegawa, H., Hashimoto, T., 1992. Ordered structure of block polymer/homopolymer mixtures. 4. Vesicle formation and macrophase separation. Makromol. Chem. Macromol. Symp. 62, 75. http://dx.doi.org/10.1002/masy.19920620109.

Leibler, L., 1980. Theory of microphase separation in block copolymers. Macromolecules 13, 1602. http://dx.doi.org/10.1021/ma60078a047.

Lodge, T.P., Hanley, K.J., Pudil, B., Alahapperuma, V., 2003. Phase behavior of block copolymers in a neutral solvent. Macromolecules 36, 816. http://dx.doi.org/10.1021/ma0209601.

Matsen, M.W., 2012. Effect of architecture on the phase behavior of AB-type block copolymer melts. Macromolecules 45, 2161. http://dx.doi.org/10.1021/ma202782s.

Matsen, M.W., Bates, F.S., 1996. Unifying weak- and strong-segregation block copolymer theories. Macromolecules 29, 1091. http://dx.doi.org/10.1021/ma951138i.

Matsen, M.W., Schick, M., 1994a. Stable and unstable phases of a diblock copolymer melt. Phys. Rev. Lett. 72, 2660. http://dx.doi.org/10.1103/PhysRevLett.72.2660.

Matsen, M.W., Schick, M., 1994b. Microphase separation in starblock copolymer melts. Macromolecules 27, 6761. http://dx.doi.org/10.1021/ma00101a014.

Matsen, M.W., Thompson, R.B., 1999. Equilibrium behavior of symmetric ABA triblock copolymer melts. J. Chem. Phys. 111, 7139. http://dx.doi.org/10.1063/1.480006.

Milner, S.T., Olmsted, P.D., 1997. Analytical weak-segregation theory of bicontinuous phases in diblock copolymers. J. Phys. II France 7, 249. http://dx.doi.org/10.1051/jp2:1997122.

Miyamoto, M., Sawamoto, M., Higashimura, T., 1984. Living polymerization of isobutyl vinyl ether with hydrogen iodide/iodine initiating system. Macromolecules 17, 265. http://dx.doi.org/10.1021/ma00133a001.

Mori, K., Taneka, H., Hashimoto, T., 1987. Scattering functions for disordered two-component polymer systems including block polymers. Macromolecules 20, 381. http://dx.doi.org/10.1021/ma00168a026.

Mori, K., Taneka, H., Hasegawa, H., Hashimoto, T., 1989. Small-angle X-ray scattering from block copolymers in disordered state: 2. Effect of molecular weight distribution. Polymer 30, 1389. http://dx.doi.org/10.1016/0032-3861(89)90206-1.

Ohta, T., Kawasaki, K., 1986. Equilibrium morphology of block copolymer melts. Macromolecules 19, 2621. http://dx.doi.org/10.1021/ma00164a028.

Otsu, T., Yoshida, M., Tazaki, T., 1982. A model for living radical polymerization. Makromol. Chem. Rapid Commun. 3, 133. http://dx.doi.org/10.1002/marc.1982.030030209.

Sakamato, N., Hashimoto, T., 1995. Order–disorder transition of low molecular weight polystyrene-block-polyisoprene. 1. SAXS analysis of two characteristic temperatures. Macromolecules 28, 6825. http://dx.doi.org/10.1021/ma00124a018.

Sakamato, N., Hashimoto, T., 1998a. Ordering dynamics of a symmetric polystyrene-block-polyisoprene. 1. Ordering mechanism from the disordered state. Macromolecules 31, 3292. http://dx.doi.org/10.1021/ma9717202.

Sakamato, N., Hashimoto, T., 1998b. Ordering dynamics of a symmetric polystyrene-block-polyisoprene. 2. Real-space analysis on the formation of lamellar microdomain. Macromolecules 31, 3815. http://dx.doi.org/10.1021/ma980037s.

Sakamato, N., Hashimoto, T., 1998c. Ordering dynamics of cylindrical and spherical microdomains in polystyrene-block-polyisoprene-block-polystyrene. 1. SAXS and TEM observations for the grain formation. Macromolecules 31, 8493. http://dx.doi.org/10.1021/ma980411o.

Sakurai, S., Mori, K., Okawara, A., Kimishima, K., Hashimoto, T., 1992. Evaluation of segmental interaction by small-angle X-ray scattering based on the random-phase approximation for asymmetric, polydisperse triblock copolymers. Macromolecules 25, 2679. http://dx.doi.org/10.1021/ma00036a019.

Sauer, B.B., Dee, G.T., 2002. Surface tension and melt cohesive energy density of polymer melts including high melting and high glass transition polymers. Macromolecules 35, 7024. http://dx.doi.org/10.1021/ma0202437.

Schwahn, D., Schmackers, T., Mortensen, K., 1995. Ginzburg criterion for the mean-field to three-dimensional Ising crossover in polymer blends. Phys. Rev. E 52, R1288. http://dx.doi.org/10.1103/PhysRevE.52.R1288.

Semenov, A.N., 1985. Contribution to the theory of microphase layering in block-copolymer melts. Sov. Phys. JETP 61, 733.

Shibayama, M., Hashimoto, T., Kawai, H., 1983. Ordered structure in block polymer solutions. 1. Selective solvents. Macromolecules 16, 16. http://dx.doi.org/10.1021/ma00235a005.

Sivaniah, E., Hayashi, Y., Matsubara, S., Kiyono, S., Hashimoto, T., Fukunaga, K., Kramer, E.J., Mates, T., 2005. Symmetric diblock copolymer thin films on rough substrates. Kinetics and structure formation in pure block copolymer thin films. Macromolecules 38, 1837. http://dx.doi.org/10.1021/ma0482157.

Strobl, G., 2007. The Physics of Polymers: Concepts for Understanding Their Structures and Behavior. Springer, New York. ISBN: 978-3-540-68411-4.

Szwarc, M., Levy, M., Milkovich, R., 1956. Polymerization initiated by electron transfer to monomer. A new method of formation of block polymers. J. Am. Chem. Soc. 78, 2656. http://dx.doi.org/10.1021/ja01592a101.

Takenaka, M., Wakada, T., Akasaka, S., Nishitsuji, S., Saijo, K., Shimizu, H., Kim, M.I., Hasegawa, H., 2007. Orthorhombic Fddd network in diblock copolymer melts. Macromolecules 40, 4399. http://dx.doi.org/10.1021/ma070739u.

Tanaka, H., Hasegawa, H., Hashimoto, T., 1992. Ordered structure in mixtures of a block copolymer and homopolymers. 1. Solubilization of low molecular weight homopolymers. Macromolecules 24, 240. http://dx.doi.org/10.1021/ma00001a037.

Todo, A., Uno, H., Miyoshi, K., Hashimoto, T., Kawai, H., 1977. Domain-boundary structure of styrene-isoprene block copolymer films cast from solutions. III. Preliminary results on spherical microdomains. Polym. Eng. Sci. 17, 587. http://dx.doi.org/10.1002/pen.760170816.

Tyler, C.A., Morse, D.C., 2005. Orthorhombic Fddd network in triblock and diblock copolymer melts. Phys. Rev. Lett. 94, 208302. http://dx.doi.org/10.1103/PhysRevLett.94.208302.

Wu, S., 1970. Surface and interfacial tensions of polymer melts. II. Poly(methyl methacrylate), poly(n-butyl methacrylate), and polystyrene. J. Phys. Chem. 74, 632. http://dx.doi.org/10.1021/j100698a026.

RAFT synthesis of block copolymers and their self-assembly properties

2

W.B. Farnham[*], *M.T. Sheehan*[†]
[*]E. I. duPont de Nemours and Company, Central Research and Development, Wilmington, DE, USA; [†]DuPont Electronic Polymers, Gregory, TX, USA

2.1 RAFT process description

The Commonwealth Scientific Industrial Research Organisation (CSIRO) first published its work on the development of reversible addition fragmentation chain transfer (RAFT) polymerization technology in 1998 (Chiefari et al., 1998; Moad et al., 2005). The technology offered many benefits in controlling the free radical polymerization process, including the control of size, structure, and shape of polymers; and it was capable of producing polymer dispersities <1.1 The collaboration in the field of controlled free radical polymerization between CSIRO and DuPont began more than 20 years ago and continues to be important for the further development and commercialization of RAFT. These commercial developments include applications in electroactive polymers, compatibilizers, thermoplastic elastomers, protective finishes, dispersants, polymer microgels for aqueous coatings, high-resolution patterning applications, including photoresists, and more recently in block copolymers for self-assembly (directed self-assembly (DSA)). One of the clear advantages of RAFT for continued commercial uptake is versatility in the use of a wide palette of monomers and media, enabling synthesis of polymers possessing a multitude of functional groups. Functionality that is incompatible with anionic carbon-carbon bond formation is perfectly viable. Included in this useful list are those groups that deactivate or terminate carbanions (–OH, –NH, –CH acidic groups), sensitive Si- or Sn– groups, and architecture-defining chain termini comprising carbanion-incompatible linkages. Polymers produced using RAFT technology are able to meet stringent electronics industry low metals specifications.

Several reviews and books have been published on RAFT processes (Moad et al., 2010; Keddie et al., 2012; Barner-Kowollik, 2008). This chapter is not intended to be a thorough review, but will focus on selected aspects especially relevant to electronic and lithographic applications where defectivity issues in the final application are so important.

The generally accepted mechanistic steps involved in this reversibly deactivated radical process are shown in the simplified scheme below (Figure 2.1). Diffusion-controlled radical-radical termination reactions are chief among those we seek to minimize. Other side reactions involving loss of chain-transfer end groups are possible and should be avoided for best practices. Examples include separate thermal fragmentation

Directed Self-assembly of Block Copolymers for Nano-manufacturing. http://dx.doi.org/10.1016/B978-0-08-100250-6.00002-X

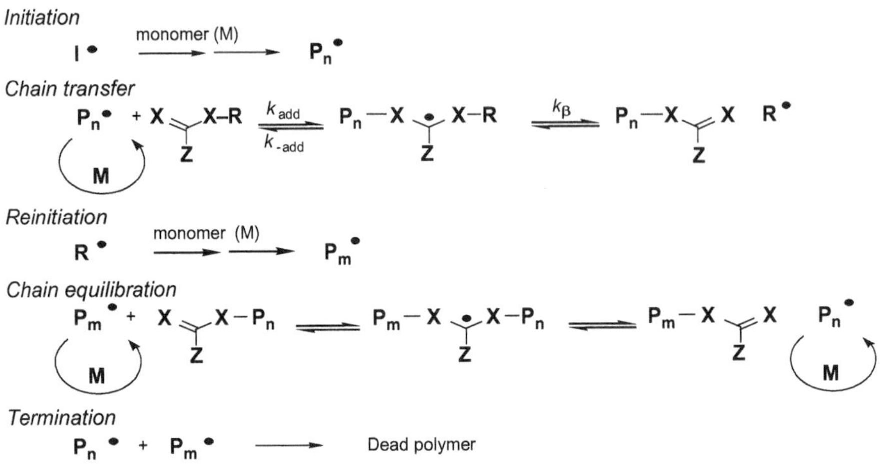

Figure 2.1 RAFT mechanism (Moad et al., 2010).

or solvolytic destruction of thiocarbonylthio groups. For selective growth from methacrylate blocks, temperature control is important as higher (>100 °C) temperatures promote both irreversible CTA fragmentation loss and monomer unzipping processes (Xu et al., 2006). Once a more stabilizing monomer (e.g., styrene) has been inserted adjacent to the thiocarbonylthio moiety, temperature constraints are reduced by comparison. Polymerization at the highest feasible monomer content, under the constraints of good mixing, adequate heat transfer, manageable viscosity, and low radical flux are general process guidelines that facilitate maximal selectivity in polymer end-group labeling. Issues involving heat transfer and high viscosity may be solved by alternate means, and others have employed RAFT processes in emulsion or mini-emulsion conditions (Liu et al., 2006; Urbani and Monteiro, 2008). In general, these multiphase processes generate additional control difficulties and seem ill-suited for the preparation of very low PD block polymers designed for complex phase behaviors. We prefer "single-phase" conditions and the use of fed initiators so as to maintain both constant initiator concentration and as high as possible ratio of [CTA]/[initiator] throughout the polymerization process. In principle, there are circumstances when fed monomer streams are needed, for example, to control disparate monomer incorporation rates or to modulate potential thermal run-aways, but such difficulties were not encountered in examples discussed in this chapter.

RAFT polymerization processes have been developed to produce several classes of materials for electronic applications, including positive- and negative-tone resists (styrenics for 248 nm photoresists; methacrylates for 193 and 193i systems), interface-controlling additives, and self-assembling diblock polymers. Important synthetic or process elements include scalable polymerization reactions, product isolation and separation schemes, and thiocarbonylthio-RAFT end-group removal methods. Low polydispersity and high compositional uniformity HCU™ were the main polymer properties sought after using the RAFT process to provide application performance improvements, for example, more uniform polymer dissolution leading to lower line edge roughness.

2.2 Polymerization process details

2.2.1 *In situ process analysis*

Raman and FTIR spectroscopy—established techniques for *in situ* reaction monitoring—have been used to good advantage in polymerization processes (Gulari et al., 1984; Dey et al., 2013). We have utilized Raman *in situ* monitoring to determine the real time trajectory of monomer content in the reactor. This allows for comparisons of reactor performance involving recipe variations (e.g., % solids, initiator feed schedules) and facilitates optimization. For the DSA application, the material target composition and block length must be reproducibly controlled within tight boundaries approaching 1%. Process monitoring and control devices are inherently necessary to achieve this. Multiple parameters have significant impacts on the RAFT process, and, among these, monomer concentration, % solids, and radical flux or initiator concentration are at the top of the list. In general, higher % solids processes are preferred, but there are limitations: reactor contents must be well-mixed and, ideally, homogeneous. Principle objectives are to produce the first block with consistent size and with intact chain-transfer agent label content as high as possible. At sufficiently high viscosity, mixing in conventional vessels becomes problematic: the Raman probe window becomes irregularly covered, and an added initiator is not uniformly distributed at an acceptable rate. In the extreme condition, reaction mass contact with reactor walls fails along with mixing and heat transfer.

Several PMMA trials targeting Mw of 30 kDa are shown in Figures 2.2 and 2.3 and illustrate some of the trends. For the 48.2% solids example, viscous flow can be managed beyond 85% conversion, but only by reducing the stirring rate so that the contents maintain adequate contact with the vessel walls for mixing and heat transfer. A small reduction in % solids allows for easier mixing and scale-up, but results in a rate penalty. Another example of insufficient mixing is illustrated in the "gray diamonds" run (Figure 2.2). In this case, which was carried out at slightly higher temperature than the others (but with 20% higher stirring rate), a torque limitation set point was reached at 93% conversion,

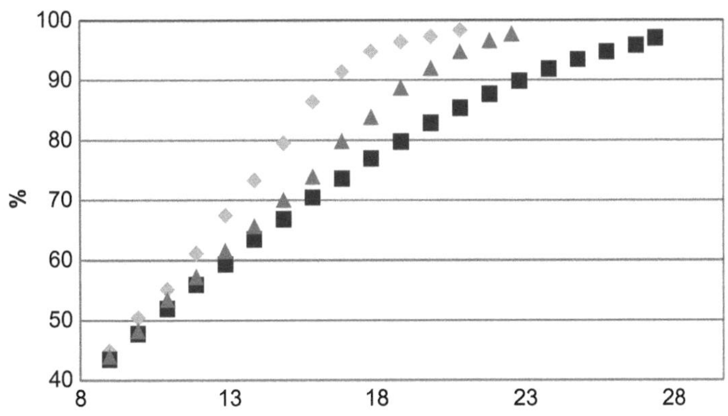

Figure 2.2 Raman MMA comparative conversion trajectories.

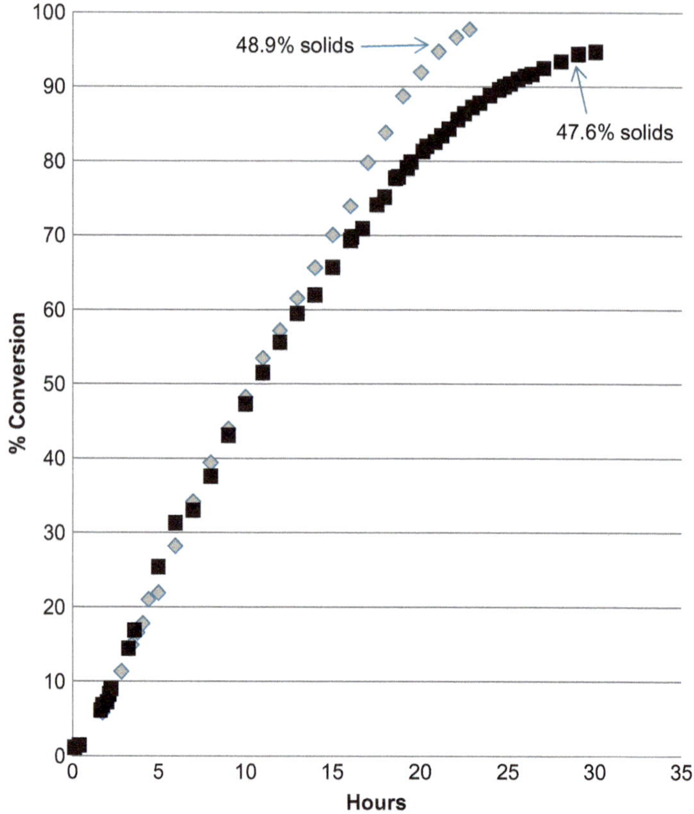

Figure 2.3 Raman MMA comparative conversion trajectories.

which disabled the stirrer. In the meantime, slow initiator feed continued and monomer was consumed at a reduced rate under poorly mixed conditions. The molecular weight distribution was adversely affected, and Mw did not reach the designed value.

2.3 RAFT end-group catalytic radical reduction

The RAFT polymers considered for DSA require an especially clean end-group reduction process to produce materials with faithfully maintained polydispersity and architecture. Recent literature discusses several apparent limitations in polymer methods (Stals et al., 2013; Vandenbergh and Junkers, 2014). Further refinements within our published catalytic radical reduction methodology have allowed for improved performance in radical reduction of thiocarbonylthiofunctionality (Farnham et al., 2010). Consideration of the RAFT forward polymerization mechanism (Figure 2.4) is useful.

Radical addition to the C=S terminal functional group is still required along with cleavage of C–S to give a polymeric radical species. However, conditions have to

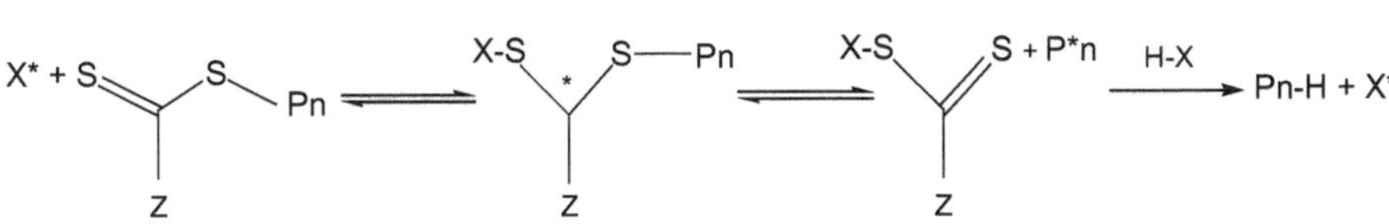

Figure 2.4 Thiocarbonythio catalytic radical reduction mechanism (Chong et al., 2007).

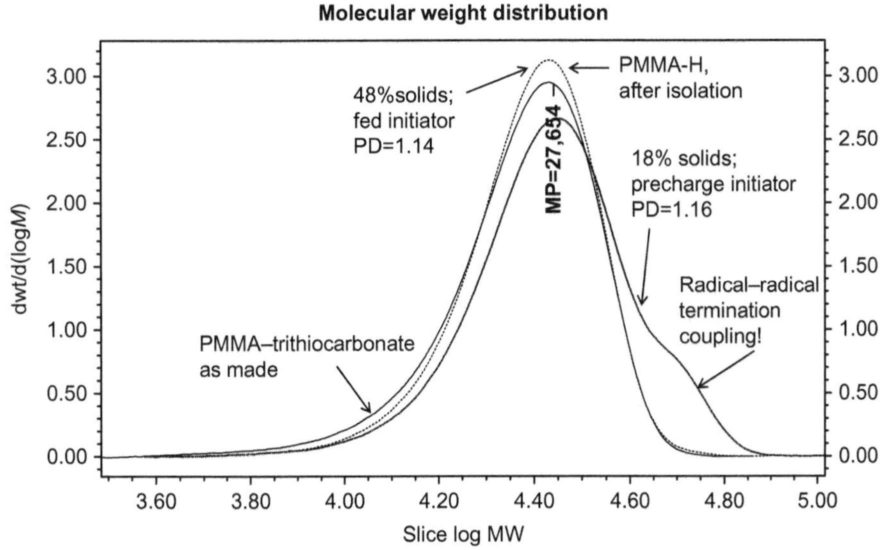

Figure 2.5 PMMA molecular weight distributions as a function of reaction conditions.

be selected with care so that bimolecular polymeric termination steps giving coupled polymer fragments, or unstable unsaturated ends, or both, are minimized. In the case of di-block or higher order RAFT polymer architectures, any coupling obviously degrades the integrity of the designed architecture. We find that the best control of this reaction is achieved by a combination of the following elements:

(1) High % solids, with reaction mixture viscosity sufficient to slow polymer diffusion, minimizing bimolecular termination without blocking other reaction paths; (2) homogeneous reaction conditions enabled by selected solvents that give high compatibility of hypophosphite salt; and (3) slow addition of selected radical initiator, allowing for catalytic quantities (3–5 mol % vs. trithiocarbonate) of initiator to be used in the process and minimizing byproducts. Low radical flux throughout the process is critical to achieving identical molecular weight distributions after reduction (Figure 2.5).

2.4 Block Copolymer *In situ* Topcoat Applications

RAFT block copolymer development was initiated for a 193 immersion application where differences in segment polarities were selected to define surface energetics at the air/solid interface while maintaining compatibility with bulk photoresist.

As 193 nm lithography was modified to incorporate water immersion technology, advances in photoresist materials were required to overcome issues at the photoresist–water interface. Application of a topcoat barrier layer reduced the amount of PAG leaching from the resist and controlled water contact angles during scanning (Sanders, 2010; Wei et al., 2006; Kim et al., 2007). We have used RAFT technology to produce block copolymers comprising a random "resist" block with composition and size based on conventional 193 dry photoresist monomers and a "low surface energy" block. Thus, random resist compositions containing uniform distributions of acid-sensitive switch monomers in the chains and trithiocarbonate end groups were converted to block structures comprising B blocks with partially fluorinated pendant chains. The relative block lengths and compositions were varied to tune solution migration behavior, surface energy, contact angles, and solubility in developer after imaging. The overall result was a diblock polymer, highly compatible with the resist composition but which stratified to the air interface upon thin film formation. This allowed both protection from water during 193 immersion processing and complete removal in the imaged areas with base developer after exposure and PEB (Sheehan et al., 2008).

An example is shown in Figure 2.6 wherein the resist component is made up of α-GBLMA, MAMA, and ECPMA, and the short second block is constructed with 2-(perfluorohexyl)ethyl methacrylate. Separately fed streams of initiator and monomer were used in the last step. No homopolymer contaminants were observed. For this application, typical resist molecular weight targets are not very high, and the desired number of fluorinated monomer units/chain is therefore low (e.g., as low as 3–5). For that reason, and further because a fluorinated homopolymer cannot be tolerated (defects), it is preferred to starve-feed the fluorinated monomer. Addition of this diblock to the corresponding photoresist composition provided significant increases in water contact angles. The advancing angle (Figure 2.7) is somewhat too high for best overall lithographic performance, so the selection of a fluorinated methacrylate monomer is made with a revised target for material design. Lithographic evaluation of the resist

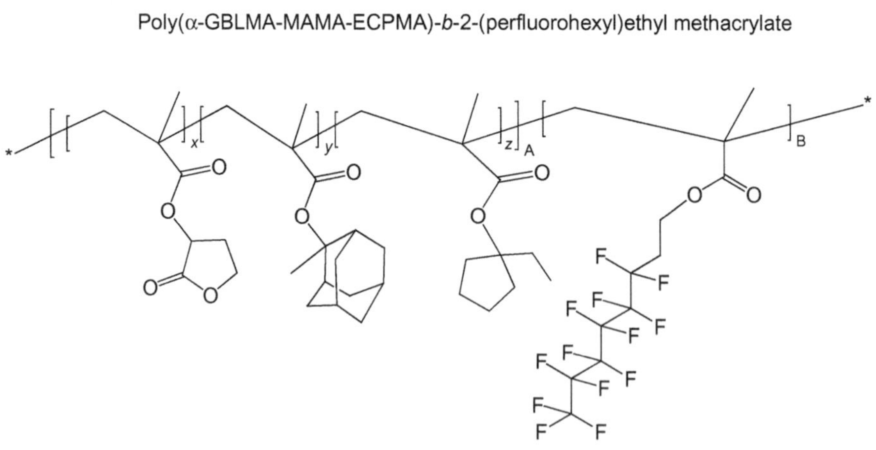

Figure 2.6 193 Immersion additive block copolymer.

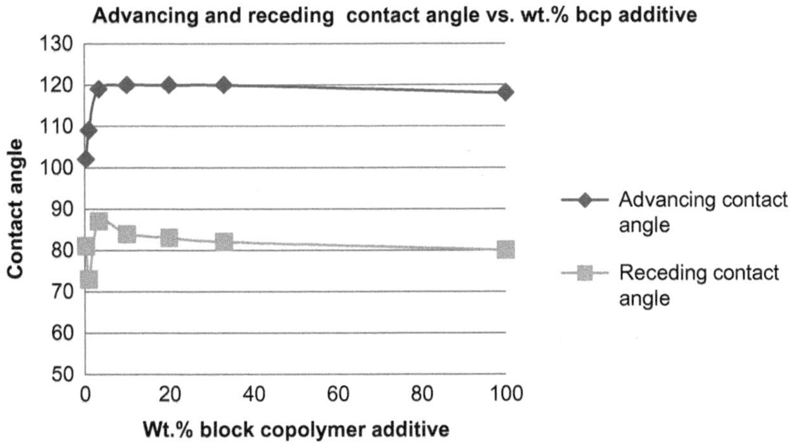

Figure 2.7 Advancing and receding contact angles.

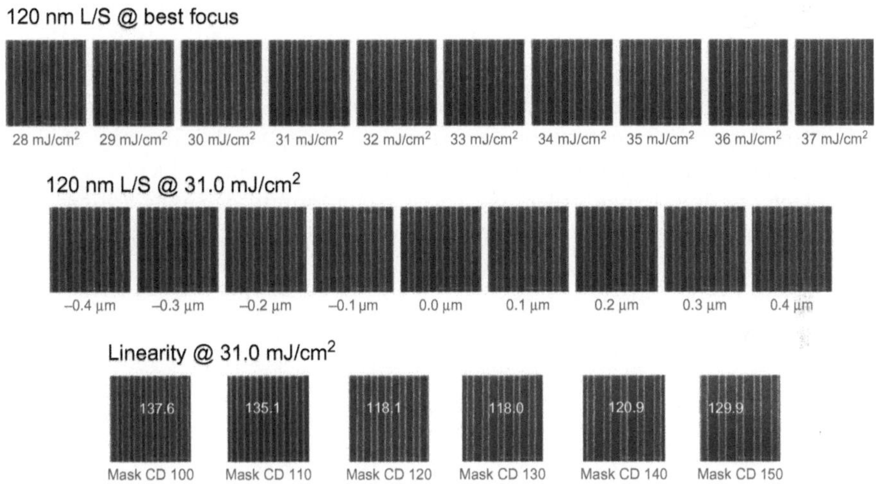

Figure 2.8 From 1 to 5 wt.% block fluoropolymer in a 193 nm photoresist (Lithography, JSR Micro) (Sheehan et al., 2008).

containing this additive showed no degradation in resist performance (Figure 2.8). Lessons learned from the above block copolymer (bcp) work were utilized in the development of initial DSA materials.

2.5 DSA Applications

DSA is proving to be an interesting and innovative method to make 2- and even 3-dimensional periodic, uniform patterns. Targeting half-pitch feature sizes ≤ 14 nm, and using a staged implementation strategy, DSA is on the lithography roadmap.

Substantial effort has taken place in assembly strategies to be employed; from a commercial standpoint, process and materials selection, suppliers, and integration efforts with end-user chip manufacturers are still in progress (Black et al., 2007; Kim et al., 2010; Van Look et al., 2014; Khaira et al., 2014). Among elements critical to acceptable performance for DSA using block copolymers (long-range order and extremely low defectivity) are the uniformity and the purity of the materials (Maher et al., 2014; Bates et al., 2014).

Known factors governing diblock copolymer morphology phase space shown in Figure 2.9 (lamellar, cylindrical, gyroid, etc.) include composition and block lengths and chi parameters. Subtle influences at interfaces are known to affect the regularity of assembly (Albert and Epps, 2010).

The first diblock system likely to see implementation in lithographic application of DSA is MMA-*b*-S. Our preliminary objectives were to qualify RAFT for satisfactory synthesis of the desired materials in our hands, implementing improvements as needed. Given success with this archetypal block system, expansion of monomer scope was anticipated to allow for higher *X* diblock systems. "Satisfactory," for the intended area of applications was not established with certainty, but general expectations were set high: reproducible compositional and molecular weight control (within 1%), excellent polydispersity control (PD <1.2, ideally <1.1), and with overall molecular weights ranging from 15 to 200 K. Target morphologies were lamellar (50/50 volume fraction) and cylindrical (e.g., 30/70 volume fraction).

For less-demanding property-response surfaces, exceptional uniformity of block size and composition is not required. For DSA, however, where objectives include low defectivity, long-range order, and so forth, uniformity demands are expected to be very high. The RAFT mechanism shown in Figure 2.1 helps to describe

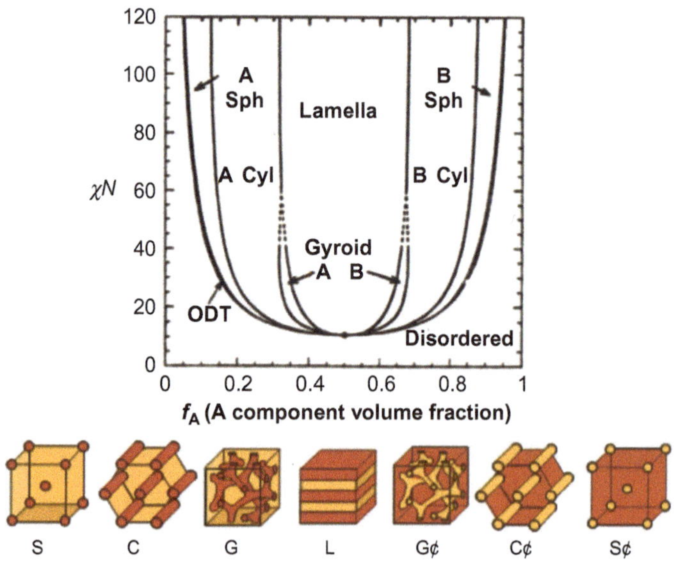

Figure 2.9 Block copolymer phase diagram.

the essential problem for reinitiation and propagation of a new block from a thiocarbonylthio-functionalized first block. Irreversible termination at any stage, and uncontrolled side reactions—mainly the formation of block B homopolymer—are to be minimized. Such competing processes were recognized early in the development of RAFT (Mayadunne et al., 2000b). Even if a fully modeled kinetic matrix for the RAFT mechanism scheme were established, monomer-dependent reaction optimization is still required for specific polymers. Several literature reports have appeared describing RAFT synthesis of MMA-*b*-S, but these lacked critical tests for product homogeneity (Moad et al., 2010; Mayadunne et al., 2000a,b; Chong et al., 1999; Bang et al., 2006).

Interaction polymer chromatography (IPC) is a critical analytical tool for assessment of product homogeneity. IPC separates macromolecules by chemical composition as well as by individual block length, so that compositional uniformity of the copolymer can be estimated (Brun and Foster, 2010). Size exclusion chromatography (SEC), which separates macromolecules by size and reveals overall molecular weight distribution, is not suited for quantitative determination of contaminant levels. Asymmetric molecular weight distributions are often indicative of lack of reaction control, but no information is provided regarding the source of "anomalies," as these could arise anywhere along the path of chain growth. Singular reliance on SEC (particularly using RI detection alone) is risky, as SEC does not distinguish between homopolymers or block copolymers (Figure 2.10). Observation of acceptable SEC traces can result from fortuitous similarities in MWDs of several components, or canceling RI signals from contaminant homopolymers with opposite RI response versus mobile phase. An example of IPC separation using a hexane/tetrahydrofuran gradient is shown in (Figure 2.11), where polystyrene and residual PMMA are the principal contaminants. Additionally, the width of the main peak in the IPC chromatogram is a measure of the compositional homogeneity of the block copolymer itself.

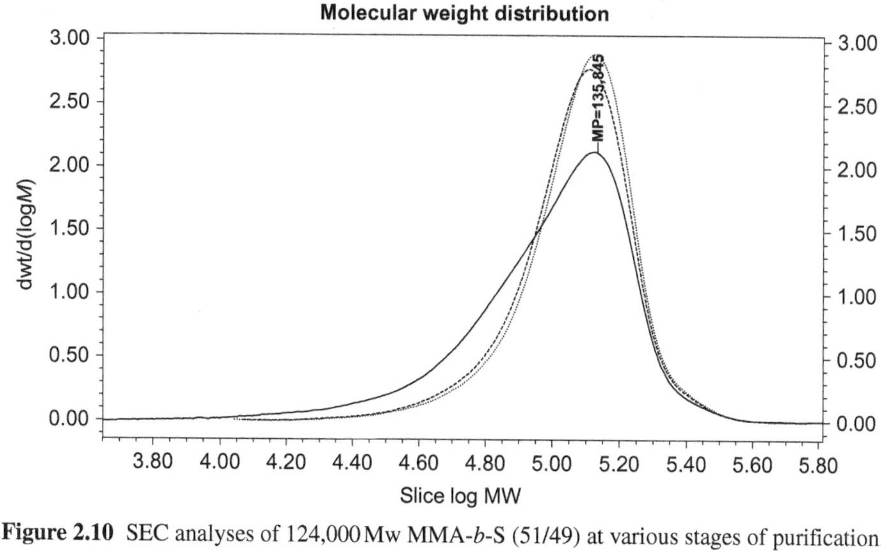

Figure 2.10 SEC analyses of 124,000 Mw MMA-*b*-S (51/49) at various stages of purification.

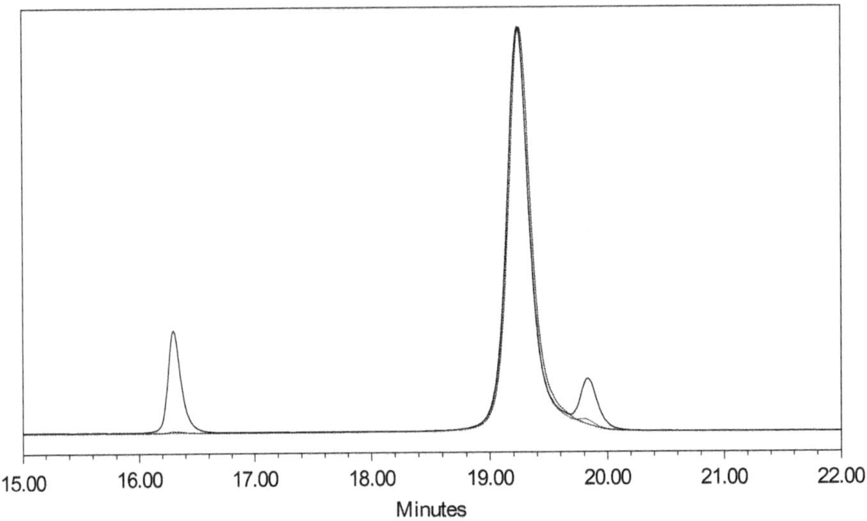

Figure 2.11 IPC analyses of 124,000 Mw MMA-*b*-S (51/49) with homopolymer contaminants, PS (16.3 min), PMMA (19.8 min). As-made, solid line; after purification, dashed line.

The radical fragmentation rate associated with the thiocarbonyl thio group in block A has to be consistent with the propagation kinetics for the monomer of block B. If this fragmentation step is too slow compared to the other processes, then selectivity for block formation will be too low. For the case of MMA and styrene, MMA must be selected as block A, at least for chain-transfer agents where Z = alkylthio groups. Low polydispersity PMMA at high monomer conversion has been discussed (Chong et al., 2003) precision and reproducibility of the scheme is upgraded by implementing *in situ* process analysis. A low-odor trithiocarbonate RAFT agent (Figure 2.12) was selected based on its excellent control of both methacrylate and styrenic monomers.

Our early efforts to prepare high molecular weight (>100 K) material quickly revealed selectivity issues in reinitiation from the poly(methyl methacrylate)-trithiocarbonate block. IPC revealed production of both homopolymers in addition to the desired diblock. In contrast to previous experience with all-methacrylate and acrylate block polymers, reinitiation and propagation with styrene monomer is best carried out at high monomer concentration but without added initiator. Styrene is

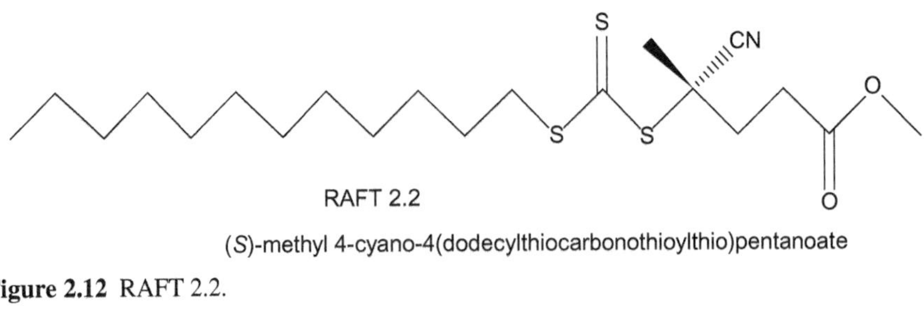

Figure 2.12 RAFT 2.2.

capable of self-initiation, and the rate of this slow radical generation process fits well enough with remaining parts of the rate matrix (Mayadunne et al., 2000a). This method provides another challenge for reaction engineering because radical flux and monomer concentration are not independent. Several reaction conditions and process variations were attempted to speed and improve the process, but resulted in unfavorable selectivities. The major shortcoming was not polydispersity alone but overall compositional uniformity. As the molecular weight target is reduced, overall selectivity is significantly improved, as expected for a system with increased concentration of the directing RAFT chain-transfer agent. A combination of polymerization process improvements and a lower molecular weight target (60 K) provided significantly improved overall selectivity, but the quantities of homopolymers coproduced with this level of thiocarbonylthio agent in the system were regarded as problematic and potentially irreproducible at scale. Reaction temperatures useful for styrene self-initiation are close to those wherein dead-end-creating trithiocarbonate cleavage from PMMA block becomes competitive. To minimize the formation of dead PMMA chains, a staged temperature ramp was utilized. This results in a lower initial styrene incorporation rate but a further reduced fraction of dead PMMA chains (Figure 2.13). Once PMMA-ttc has been consumed, the reaction temperature may be increased to gain higher polymerization rates.

SEC overlay traces developed from a trial targeting 124 K MMA-*b*-S with 50/50 composition series of three samples are shown in Figure 2.10. Except for the low molecular weight tail evident in the as-made sample, these are similar. Removal of PMMA can be assessed and quantitated by IPC, but SEC is not helpful. The SEC measurement method does not provide a quantitative assessment of compositional homogeneity, although some PDs are lower and more preferred than others. The polymer separation strategy outlined below provided material of both high molecular weight and high compositional uniformity as shown in Figures 2.10 and 2.11. In the case of MMA-*b*-S,

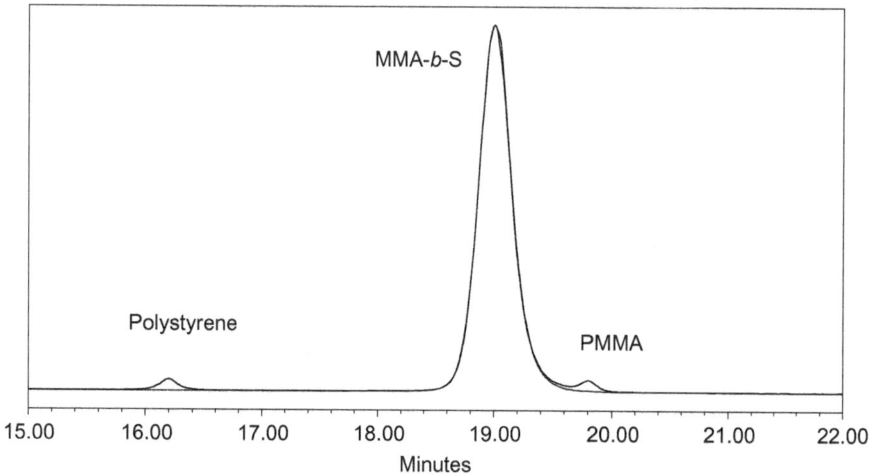

Figure 2.13 IPC analyses of 60,000 Mw MMA-*b*-S (50/50) with homopolymer contaminants, PS (16.2 min), PMMA (19.8 min). As-made, solid line; after purification, dashed line.

styrene homopolymer may be removed by the known extraction with cyclohexane—a theta solvent for polystyrene—provided the operating temperature exceeds the theta temperature. Removal of PMMA homopolymer is more difficult. In many of the block polymer cases studied, we have used "micellar aggregation," to obtain the necessary separation. Micellar aggregation is a process that features formation of block copolymer polymer micelles in a solvent medium, deliberate modulation of the solvent quality so as to favor the self-assembly of these particles into larger (often ordered) aggregates while simultaneously allowing for solubility of one of the homopolymer contaminants. This results in phase separation, providing a polymer-rich phase and a liquid phase bearing one of the homopolymer contaminants. Separation of the polymer-rich phase allows for isolation of diblock samples with a narrower composition distribution and a lower polydispersity (length distribution) than would otherwise be possible. Inherent assembly within the diblock macromolecules provides the driving force for the separation. Because there are often two homopolymer contaminants in question, and these have substantially different polarity characteristics by design, a modified process may have to be carried out using different selections of working fluids. MEK/methanol and THF/methanol systems have been used to remove PMMA from MMA-*b*-S. Tools for polymer phase separation include (1) a simple filtration membrane for sufficiently large particles that do not blind the filter, (2) a low-speed centrifuge, and (3) an open dip-tube for liquid/swollen and congealed polymer phase cases. The contaminants removed in this way correspond exactly to undesirable side reactions illustrated in the RAFT mechanism scheme: radical-radical termination and "polymer fragmentation without continued propagation." Process optimization can reduce those side reactions to surprisingly low values (<0.5%), but only within selected moderate molecular weight ranges where the RAFT process exhibits its best selectivity (Figure 2.14).

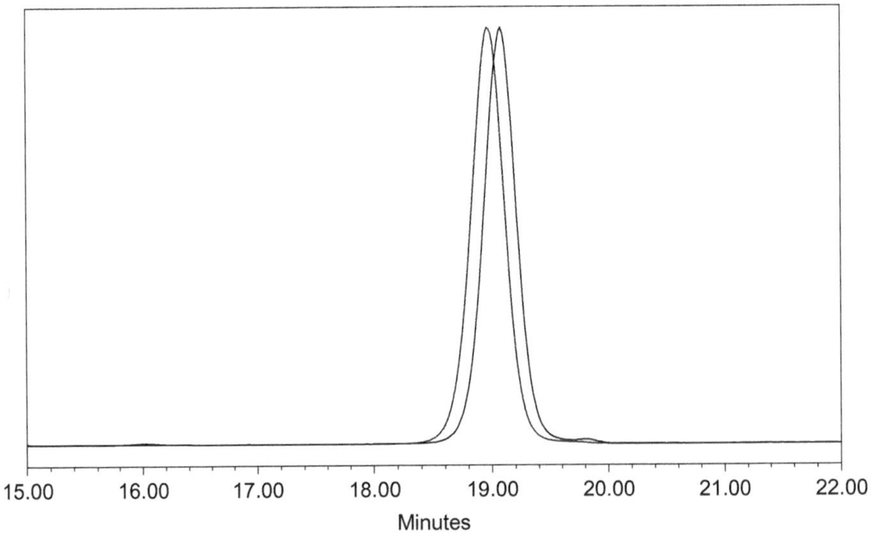

Figure 2.14 IPC analyses of 27,700 Mw MMA-*b*-S (50/50) with homopolymer contaminants, PS (16.2 min), PMMA (19.8 min). As-made, solid line; after purification, dashed line.

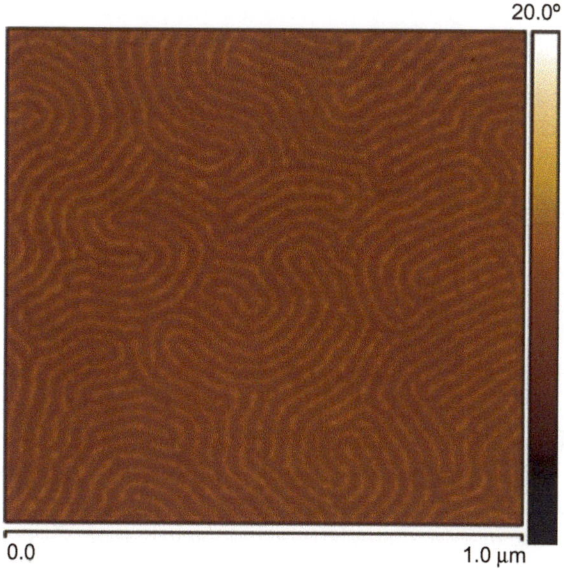

Figure 2.15 AFM phase image—undirected self-assembly—MMA-*b*-S (13.5 nm L/S).

Thin film assembly of ca. 60 K MMA-b-S on a suitable nonselective neutral layer provides vertical orientation of lamellae as demonstrated in Figure 2.15. Within a 193 nm imaged template, self-assembly of a somewhat lower Mw sample takes place in a directed fashion, with half-pitch features of 12.5 nm (Figure 2.16).

Extensions of thin film assembly strategies beyond those accessible with "simple" diblock copolymers have appeared and these can provide extraordinary, complex structures. A tunable block blend system (AB/B′C) involving additional noncovalent interactions is especially notable (Tang et al., 2010) although this is one among a large number of possibilities (Bates et al., 2012).

2.6 High *chi* block copolymers

Our strategy, and that of others, has been to seek systems to address smaller critical dimensions using a simpler selection of monomers for copolymer design. For MMA-*b*-S, a relatively low X limits the resolution of domains to 12 nm. As N is reduced to achieve smaller L_0, the system tends toward disorder. To enable sub-10 nm assembly, higher X di- or tri-block polymers are desired (Albert and Epps, 2010; Bates et al., 2014).

Segregation strengths and interfacial widths are among the self-assembly parameters that are strongly influenced by surface energetics. For the monomer selections discussed in this section, we relied on surface energy measurements as indicators of segregation strengths and likely chi magnitudes. These selected diblock polymers consist of fluorinated methacrylates (block A) and styrenics (block B). RAFT polymerization methods used are similar to those for MMA-*b*-S, except that monomer and molecular weight-dependent polymer solubility properties demand alternate solvent

Figure 2.16 SEM—directed self-assembly (12.5 nm L/S with 100 nm graphoepitaxy).

selections. In the event that homopolymer formation accompanies production of the desired diblock system, separation principles described above are analogous, but the details of media selections are completely different due to the solubility properties involved. In the case of 2-(perfluorohexyl)ethyl methacrylate, for example, best control is achieved using partially fluorinated solvent or cosolvent in the polymerization of both blocks. Without such media, phase incompatibility results in poor polydispersity control, multiple groups of polymer compositions growing at different rates, and/or complete failure to connect block A to added block B monomers.

When the phase difficulties are addressed, initiator feed conditions/reaction temperatures are properly controlled, molecular weights are in a convenient and modest range (e.g., <35 K), and selectivity in polymerization is comparable to the MMA-*b*-S examples (Figures 2.17 and 2.18). If one or more of these critical control

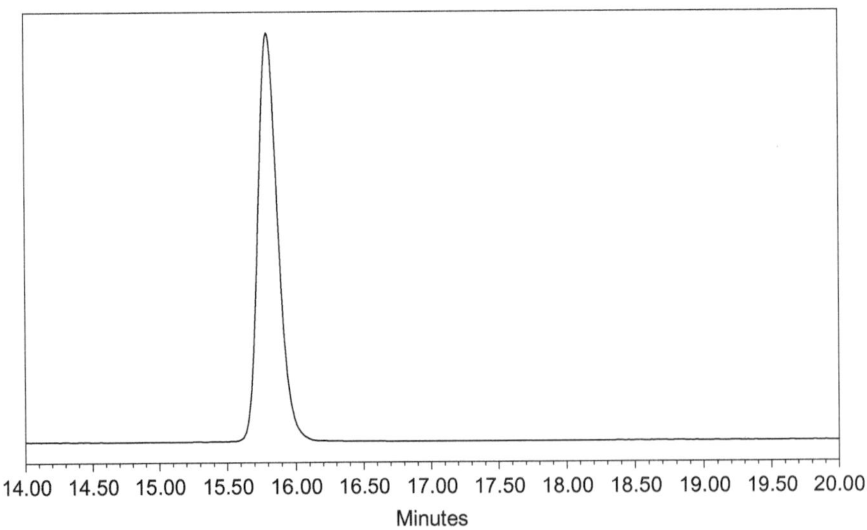

Figure 2.17 IPC analysis of OPMA-*b*-S, Mw = 19,300 Mw, PD = 1.08, demonstrating absence of homopolymers (retention time 14.5, 16.2 min).

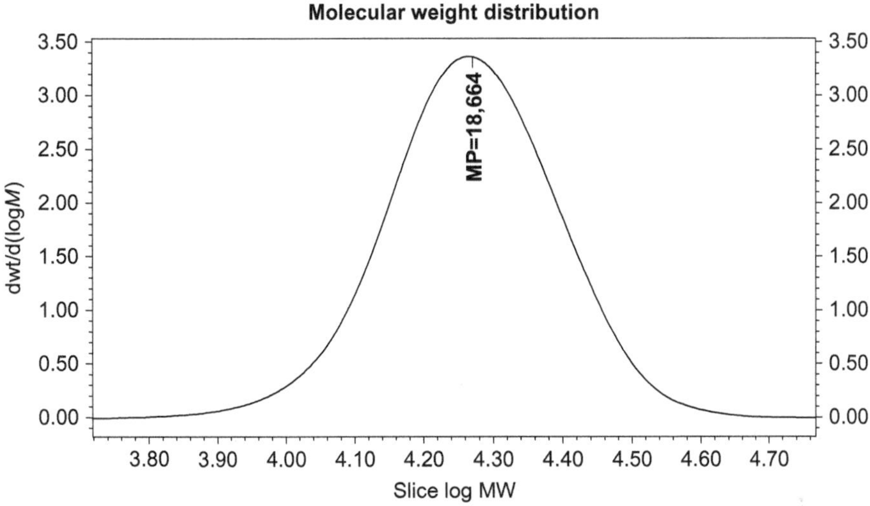

Figure 2.18 SEC of 19,300 Mw OPMA-*b*-S, PD = 1.08.

factors are not well-managed, then further separation and polymer processing is required (Figure 2.19).

The micellar aggregation process described above for the MMA-*b*-S system has been applied to 2-(perfluorohexyl)ethyl methacrylate/styrene (6,2-ZFM-*b*-S), 2-(perfluorohexyl)ethyl methacrylate/4-acetoxystyrene (6,2-ZFM-*b*-ASM), octafluoropentyl methacrylate/4-acetoxystyrene (OPMA-b-ASM), and octafluoropentyl methacrylate/styrene (OPMA-*b*-S) block systems but with different solvent pairs. These processes provided purified powder samples of 6,2-ZFM-*b*-ASM, 6,2-ZFM-*b*-S, and OPMA-*b*-S with sufficient order to obtain useful small-angle X-ray scattering

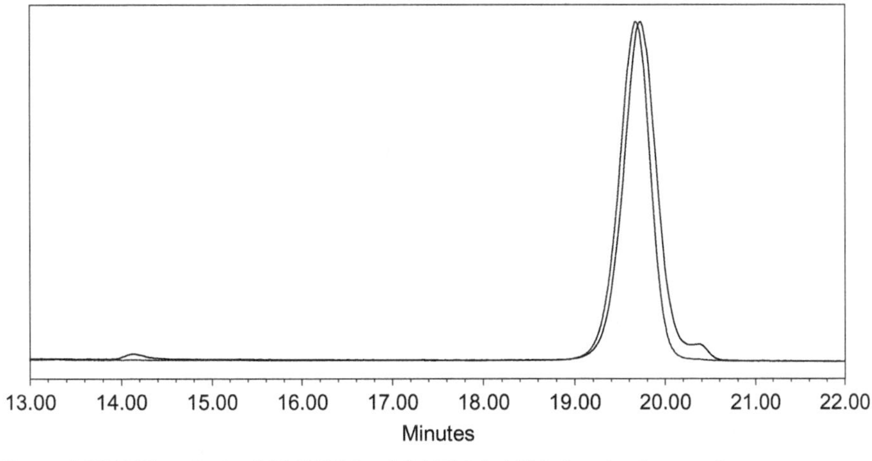

Figure 2.19 IPC analysis of 29,700 Mw, 6,2-ZFM-*b*-ASM showing homopolymer contaminants. As-made, solid line; after purification, dashed line.

(SAXS) information. A useful correlation of *d*-spacing, ranging from 14 to 24 nm, with molecular weight was identified for the 6,2-ZFM-*b*-ASM system (Sheehan et al., 2013) and this was used as a rough initial guide for preparation of the other systems. The *d*-spacing values (L_0) derived from these experiments are useful for targeting specific film thickness (Bates et al., 2014; Maher et al., 2014).

The regularity of packing within the polymer particles appears to be a function of final liquid phase composition prior to solid collection (Figure 2.20). Rate of change of the liquid phase composition is believed to be a contributing factor in the formation of ordered polymer particles, although this has not been explored in detail.

DSA application performance demands orientation control of lamellar (or, in some cases, cylindrical) phases. Preliminary thin film experiments indicate viability of the OPMA-*b*-S combination for useful self-assembly properties. Provided that a non-preferential, cross-linked, random S/OPMA/GMA composition is selected as neutral layer on silicon wafers, vertical lamellar features of promising coherence length can be generated in thin films of OPMA-*b*-S. A solvent annealing process has been employed to create the fingerprint patterns of the AFM images for this material shown in Figure 2.21. Thermodynamic stability and etch selectivity studies are in progress.

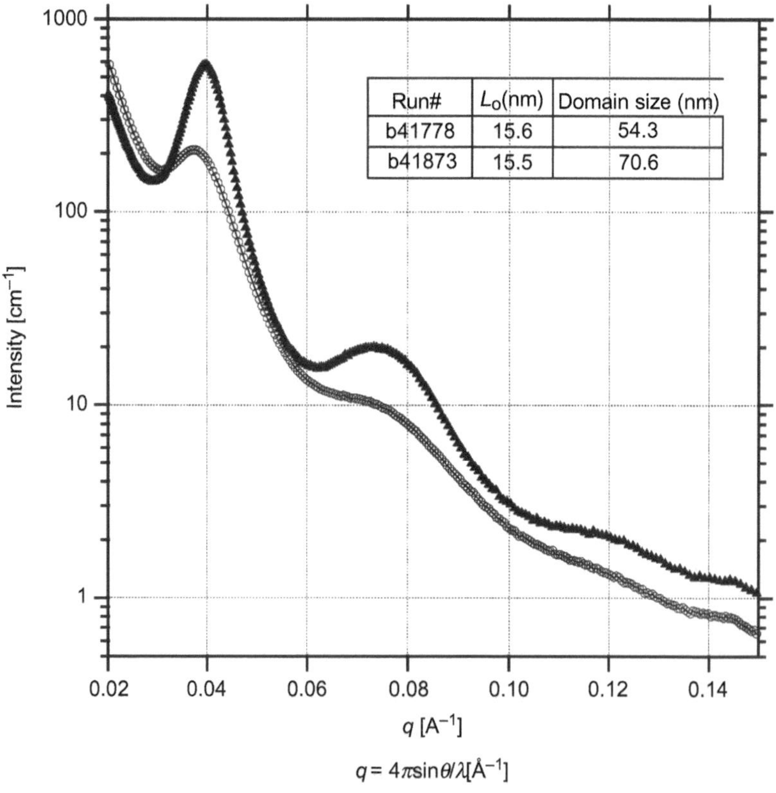

Run#	L_0(nm)	Domain size (nm)
b41778	15.6	54.3
b41873	15.5	70.6

Figure 2.20 SAXS intensity of OPMA-*b*-S powder grown under different conditions. Open circles: incomplete growth; solid triangles: >90% mass recovery of particles.

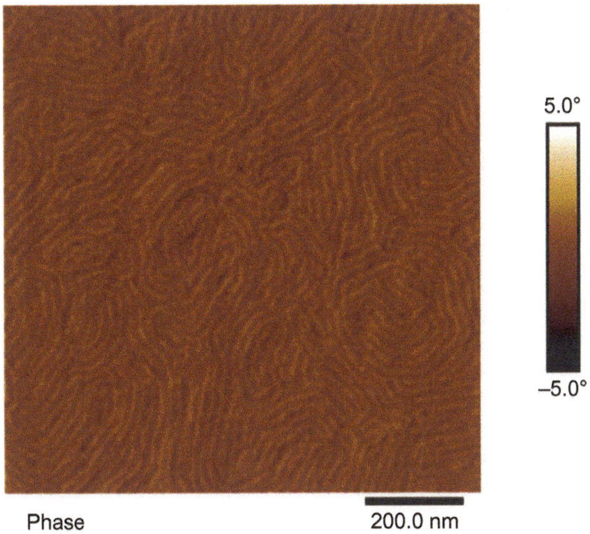

Figure 2.21 AFM phase image of OPMA-*b*-S.

2.7 Conclusions

RAFT is technology that enables the production of a variety of new block copolymers useful in DSA applications. Additional work is ongoing to optimize the assembly of block morphology in thin films with a variety of new high *chi* block copolymers, including those with more complex architectures in the context of grapho- and chemo-epitaxy control. Various target applications include not only vertically oriented lamellar structures but also contact hole shrink, contact hole multiplication, post and other 3-D structures used in microcircuit design. The full integration of low defect, reproducible structures enabled by the complete DSA multilayer design is an ongoing effort to allow commercial use of DSA in high volume microcircuit manufacturing.

Acknowledgments

The authors acknowledge the technical contributions from Yefim Brun, David Londono, Jing Li, Don Brill, Daqing Zhang, Diane Gedling, H.V. Tran, Troy Francisco, Scott Yembrick, Lois Bryman, Dominic Mancha, Charles Chambers, and Hiroshi Okazaki from DuPont. Additional technical contributions were made by Karl Berggren and Sam Nicaise from MIT and Ted Fedynyshyn from MIT Lincoln Laboratory.

References

Albert, J.N.L., Epps III, T.H., 2010. Self-assembly of block copolymer thin films. Mater. Today 13 (6), 24.

Bang, J., Kim, S.H., Drockenmuller, E., Misner, M.J., Russell, T.P., Hawker, C.J., 2006. Defect-free nanoporous thin films from ABC triblock copolymers. J. Am. Chem. Soc. 128, 7622.

Barner-Kowollik, C. (Ed.), 2008. Handbook of RAFT Polymerization. Wiley-VCH, Weinheim.

Bates, F.S., Hillmyer, M.A., Lodge, T.P., Bates, C.M., Delaney, K.T., Fredrickson, G.H., 2012. Multiblock polymers: panacea or Pandora's box? Science 336, 434.

Bates, C.M., Maher, M.J., Janes, D.W., Ellison, C.J., Willson, C.G., 2014. Block copolymer lithography. Macromolecules 47, 2–12.

Black, C.T., et al., 2007. Polymer self-assembly in semiconductor microelectronics. IBM J. Res. Dev. 51 (5), 605–632.

Brun, Y., Foster, P., 2010. Characterization of synthetic copolymers by interaction polymer chromatography: separation by microstructure. J. Sep. Sci. 33, 3501–3510.

Chiefari, J., Chong, Y.K., Ercole, F., Krstina, J., Jeffery, J., Le, T., Mayadunne Roshan, T.A., Mayadunne Roshan, T.A., Meijs, G.F., Moad, C.L., Moad, G., Rizzardo, E., Thang, S.H., 1998. Living free-radical polymerization by reversible addition-fragmentation chain transfer: the RAFT process. Macromolecules 31 (16), 5559–5562.

Chong, Y.K., Tam, P.T., Moad, G., Rizzardo, E., Thang, S.H., 1999. A more versatile route to block copolymers and other polymers of complex architecture by living radical polymerization. Macromolecules 32, 2071.

Chong, Y.K., Krstina, J., Le, T.P.T., Moad, G., Postma, A., Rizzardo, E., Thang, S.H., 2003. Thiocarbonylthio compounds [S=C(Ph)S-R] in free radical polymerization with reversible addition-fragmentation chain transfer (RAFT polymerization). Role of the free-radical leaving group (R). Macromolecules 36, 2256.

Chong, Y.K., Moad, G., Rizzardo, E., Thang, S.H., 2007. Thiocarbonylthio end group removal from RAFT synthesized polymers by radical-induced reduction. Macromolecules 40, 4446.

Dey, P., Blakey, I., Thurecht, K.J., Fredericks, P.M., 2013. Self-assembled hyperbranched polymer-gold nanoparticle hybrids: understanding the effect of polymer coverage on assembly size and SERS performance. Langmuir 29, 525.

Farnham, W., Fryd, M., Moad, G., Thang, S.H., Rizzardo, E., 2010. Removing sulfur-containing end groups from vinyl polymer without changing solvent. US Patent 7807755.

Gulari, E., McKeigue, K., Ng, K.Y.S., 1984. Raman and FTIR spectroscopy of polymerization: bulk polymerization of methylmethacrylate and styrene. Macromolecules 17, 1822.

Keddie, D.J., Moad, G., Rizzardo, E., Thang, S.H., 2012. RAFT agent design and synthesis. Macromolecules 45, 5321.

Khaira, G.S., Qin, J., Garner, G.P., Xiong, S., Wan, L., Ruiz, R., Jaeger, H.M., Nealey, P.F., 2014. Evolutionary optimization of directed self-assembly of triblock copolymers on chemically patterned substrates. ACS Macro Lett. 3 (8), 747–752.

Kim, S.S., et al., 2007. Modified polymer architecture for immersion lithography. Proc. SPIE 6519, 65191W-1.

Kim, H.-C., Park, S.-M., Hinsberg, W.D., 2010. Block copolymer based nanostructures: materials, processes, and application to electronics. Chem. Rev. 110, 146–177.

Liu, S., Hermanson, K.D., Kaler, E.W., 2006. Reversible addition-fragmentation chain transfer polymerization in microemulsion. Macromolecules 39, 4345.

Maher, M.J., Bates, C.M., Blachut, G., Sirard, S., Self, J.L., Carlson, M.C., Dean, L.M., Cushen, J.D., Durand, W.J., Hayes, C.O., Ellison, C.J., Willson, C.G., 2014. Interfacial design for block copolymer thin films. Chem. Mater. 26, 1471.

Mayadunne, R.T.A., Rizzardo, E., Chiefari, J., Krstina, J., Moad, G., Postma, A., Thang, S.H., 2000a. Living free radical polymerization with reversible addition-fragmentation chain transfer (the life of RAFT). Polym. Int. 49 (993).

Mayadunne, R.T.A., Rizzardo, E., Chiefari, J., Krstina, J., Moad, G., Postma, A., Thang, S.H., 2000b. Living polymers by the use of trithiocarbonates as reversible addition-fragmentation chain transfer (RAFT) agents: ABA triblock copolymers by radical polymerization in two steps. Macromolecules 33 (243).

Moad, G., Rizzardo, E., Thang, S.H., 2005. Living radical polymerization by RAFT process. Australian Journal of Chemistry 58 (6), 379–410.

Moad, G., Rizzardo, E., Thang, S.H., 2010. Living radical polymerization by the RAFT process—a second update. Australian Journal of Chemistry 63, 1.

Sanders, D.P., 2010. Advances in patterning materials for 193 nm immersion lithography. Chem. Rev. 110, 321.

Sheehan, M.T., Farnham, W.B., Okazaki, H., Sounik, J.R., Clark, G., 2008. RAFT technology for the production of advanced photoresist polymers. Proc. SPIE 6923, 69232E.

Sheehan, M.T., Farnham, W.B., Tran, H.V., Londono, J.D., Brun, Y., 2013. High Chi polymer development for DSA applications using RAFT technology. In: Somervell, M.H. (Ed.), Advances in Resist Materials and Processing Technology XXX, 868225/1. Proceedings of SPIE, vol. 8682. http://proceedings.spiedigitallibrary.org/proceeding.aspx?articleid=1674349. http://dx.doi.org/10.1117/12/2018255.

Stals, P.J.M., Li, Y., Burdynski, J., Nicolay, R., Nese, A., Palmans, A.R.A., Meier, E.W., Matyjasewski, K., Sheiko, S.S., 2013. How far can we push polymer architectures? J. Am. Chem. Soc. 135, 11421.

Tang, C., Hur, S., Stahl, B.C., Sivanandan, K., Dimitriou, M., Pressly, E., Fredrickson, G.H., Kramer, E.J., Hawker, C.J., 2010. Thin film morphology of block copolymer blends with tunable supramolecular interactions for lithographic applications. Macromolecules 43, 2880.

Urbani, C.N., Monteiro, M.J., 2008. RAFT-mediated polymerization in heterogeneous systems. In: Barner-Kowollik, C. (Ed.), Handbook of RAFT Polymerization. Wiley-VCH Verlag GmbH & Co, KGaA, Weinheim.

Van Look, L., Rincon Delgadillo, P., Lee, Y., Pollentier, I., Gronheid, R., Cao, Y., Lin, G., Nealey, P.F., 2014. High throughput grating qualification of directed self-assembly patterns using optical metrology. Microelectron. Eng. 123, 175–179.

Vandenbergh, J., Junkers, T., 2014. Alpha and omega: importance of the nonliving chain end in RAFT multiblock copolymerization. Macromolecules 47 (15), 5051.

Wei, Y., et al., 2006. Selection and evaluation of developer-soluble topcoat for 193 nm immersion lithography. Proc. SPIE 6153, 6153–6171.

Xu, J., He, J., Fan, D., Tang, W., Yang, Y., 2006. Thermal decomposition of dithoesters and its effect on RAFT polymerization. Macromolecules 39, 3753.

Thermal and solvent annealing of block copolymer films

3

X. Yu, Y. Han
State Key Laboratory of Polymer Physics and Chemistry, Changchun Institute of Applied Chemistry, Chinese Academy of Sciences, Changchun, P.R. China

3.1 Introduction

The self-assembly of block copolymers (BCPs) represents an efficient pathway for the fabrication of functional nanostructures based on phase separation due to the thermodynamic incompatibility between the different blocks (Kim et al., 2001; Pard et al., 2001). The ability of BCP to form a rich variety of nanoscale periodic patterns (lamellar, double gyroid, cylindrical, and spherical) with dimensions of a few nanometers and above offers the potential to fabricate high-density arrays for use in surface patterning, lithography, organic optoelectronics (Segalman et al., 2009), nanotemplating (Park et al., 1997, 2008; Pai et al., 2004; Ruiz et al., 2008; Urade et al., 2007; Hayward et al., 2005; Stoykovich and Nealey, 2006), nanoporous membranes (Thurn–Albrecht et al., 2000; Olson et al., 2007; Yang et al., 2006), and antireflection coatings (Joo et al., 2006; Li et al., 2010a, 2011; Li and Han, 2011). Many applications require the use of BCPs in thin film geometries. Compared with bulk, the boundary conditions, such as the commensurability between the natural period of the bulk microphase-separated structure of the BCP and the film thickness, the interfacial interactions between blocks and the substrate and free surface, and the surface energies of the blocks complicate the self-assembling behavior of BCP thin films (Hasegawa and Hashimoto, 1985; Coulon et al., 1989).

For many applications where regular periodic arrays are required, it will be necessary to generate long-range alignment of nanostructures in BCP films. A fine control over long-range ordering of the microdomain structures allows the fabrication of integrated microsystems and nanosystems with a high degree of complexity and functionality. However, when BCP thin films are formed from a solvent onto a substrate (e.g., prepared from spin coating or dip coating), the BCP is kinetically trapped in a nonequilibrium, disorganized state and cannot produce perfectly ordered morphologies over a large area. Obtaining the desired equilibrium morphology with eliminated defects, requires an anneal process where the polymer chains have sufficient mobility for structural reorganization. Thermal annealing and solvent vapor annealing are two efficient means to realize the optimization of the nanostructures in thin films of BCPs and have been investigated mostly. Here, we will summarize the ordered engineering of microphase separation of BCPs thin films under thermal and solvent annealing and also cover the nanostructure formation of crystalline block-containing BCPs during thermal annealing. Then, future perspectives of this research area are briefly described such as the combined annealing process of solvothermal annealing.

Directed Self-assembly of Block Copolymers for Nano-manufacturing. http://dx.doi.org/10.1016/B978-0-08-100250-6.00003-1

3.2 Thermal annealing of BCPs films

3.2.1 Fundamental consideration

BCPs are composed of two or more chemically distinct polymer chains that are covalently bound together. The interplay of the incompatibility between the different blocks and the fact that they are covalently linked on a molecular level give rise to a variety of ordered microstructures with length scales of the order of 10–100 nm. We know the propensity for BCPs to phase-separate into periodic microdomains is determined by the strength of the repulsive interaction as characterized by the product χ_{AB}N (Bates and Fredrickson, 1999; Orilall and Wiesner, 2011). N represents the degree of polymerization of the BCP and the Flory–Huggins interaction parameter, χ_{AB}, represents the driving force for microphase separation. χ_{AB} can be described by the equation below (Bates and Fredrickson, 1999):

$$\chi_{AB} = (Z / k_B T)\left[\varepsilon_{AB} - (\varepsilon_{AA} + \varepsilon_{BB})/2\right] \tag{3.1}$$

where Z is the number of nearest neighbors per monomer in the polymer, $k_B T$ is the thermal energy, and ε_{AA}, ε_{BB}, and ε_{AB} are the interaction energies per monomer between A and A, B and B, and A and B, respectively. For a given BCP system, X_{AB} gradually decreases with the external temperature. However, at elevated temperatures X_{AB} is still large enough to induce microphase separation of BCPs.

From the self-consistent field theory of a symmetric diblock copolymer, in a disordered BCP system, when the segregation strength ($X_{AB}N$) is above ≈ 10.5 (a critical value for the order–disorder transition (ODT)), the disordered system tends to phase separate into an ordered microdomain structures. This transition is termed ODT, and the transition temperature at which this phase transition occurs is referred to as the order–disorder transition temperature (T_{ODT}) (Leibler, 1980; Alexander–Katz and Fredrickson, 2007). At equilibrium, microphase-separated morphologies are formed due to the balancing of the enthalpic interfacial energy between the blocks and the entropic penalty paid due to chain stretching of the individual blocks.

3.2.2 Film thickness effect and temperature gradient

Thermal annealing is the simplest option and the most commonly used method, which consists of heating thin films in a controlled matter at a temperature above the glass transition temperatures (T_g) but below decomposition temperatures of the blocks for a time sufficient to allow approach to the equilibrium morphology. In BCP thin film, the microphase-separation morphology is strictly governed by the film thickness during the thermal annealing. When a BCP is confined to form a thin film on a substrate, the microdomain morphology is critically influenced both by the film thickness and the energetic interaction at the polymer/substrate and free surface (e.g., a gas or vacuum above the BCP film)/polymer interfaces (Wang et al., 2000; Green et al., 1991; Mayes et al., 1994; Cai, et al., 1993; Anastasiadis et al., 1990; Kim et al., 2010a). The stretching or compression of polymer chains is energetically unfavorable due to the confinement.

BCPs self-assemble into morphologies that relieve the stretching/compression of the polymer chains and reduce the entropic penalty because of the favorable enthalpic interactions at the substrate and free surface interfaces (Matsen, 1997; Albert and Epps, 2010).

Blocks of different chemical composition have different interfacial energies, which lead to preferential wetting of each interface by selective blocks of the copolymer. Generally, the block with the lower surface energy will preferentially segregate at the surface, leading to a preferred parallel orientation of microdomains. For symmetric AB-type BCPs, according to the preferred segregation of the BCP in the surfaces and interfaces, boundary conditions can be categorized as symmetric and antisymmetric. In symmetric boundary conditions, one block has the propensity to wet both the substrate and free interfaces. As a result of the symmetric boundary condition, the desired film thickness H_n is quantized as $H_n = nL_0$ (L_0 is equilibrium period, n is integer). As for asymmetric boundary conditions, where different blocks preferentially interact with one of the interface, then the commensurability condition for film thickness is $H_n = (n + 1/2)\, L_0$ (Krausch, 1995; Russell et al., 1989; Menelle et al., 1992; Anastasiadis et al., 1989; Pickett et al., 1993; Sikka et al., 1993; Smith et al., 2001, 2003; Peters et al., 2000). If the initial film thickness (h_i) meets the specified commensurability condition (e.g., $h_i = H_n$ for a symmetric boundary condition), the film is stable and the surface remains smooth. Furthermore, the ordering of the molecular chains in the film facilitated the release of elastic energy resulting in a layered structure. If h_i is incommensurate with the appropriate criteria, and deviates by an amount of δh, the reallocation of the molecular chains cause the film at the free surface to develop into discrete morphologies. Thus, these features form on the surface to minimize the total energy by quantizing the local film thickness.

In the thermal annealing process, the microphase separation feature is strictly governed by the film thickness. Limary and Green researched a series of symmetric polystyrene-*block*-poly (methyl methacrylate) (PS-*b*-PMMA) thin films having different thickness and microphase separation above T_{ODT} (Limary and Green, 1999). Different thin film morphologies have been obtained in the same film with different film thicknesses after phase separation. When the film thickness was below 3.5 nm, the films dewetted on the silicon substrate and formed bicontinuous spinodal-like patterns. For films of thickness in the range of $3.5\,\text{nm} < h < 7\,\text{nm}$, discrete holes appeared randomly throughout the film. When the film thickness increased to the range of $7\,\text{nm} < h < 35\,\text{nm}$, autophobic behavior occurred within the film. Observed surface features include bicontinuous spinodal structure, discrete holes, and stable smooth morphology. The resultant observations were a function of the initial film thickness after thermal annealing. Figure 3.1 illustrates the different observed features as a function of different film thickness.

For BCP thin films annealed above T_{ODT}, the relationship between the microdomains of the BCP and the film thickness was discovered in part from the research of Limary and Green's. However, the microphase separation of BCP films is a complex phenomenon and there are many parameters that influence the process. Smith et al. proposed a combinatorial experiment to improve experimental efficiency and throughput by varying several different key parameters simultaneously. The authors ingeniously prepared a series of symmetric PS-*b*-PMMA films with varying molecular

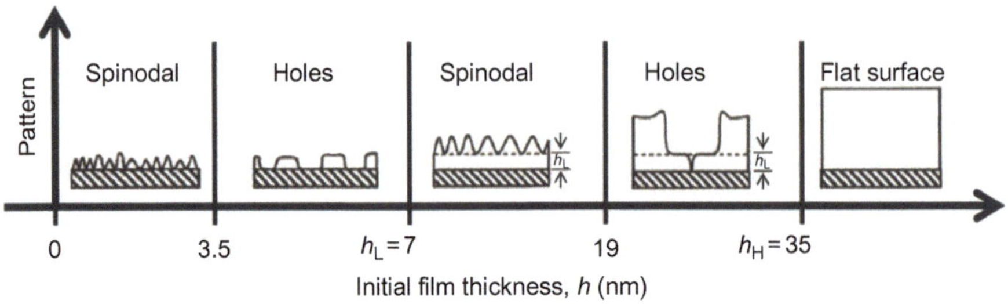

Figure 3.1 Surface feature as a function of initial film thicknesses in PS-*b*-PMMA diblock copolymer. These films were annealed at 170 °C, a temperature above the T_{ODT}. The topography of the films prior to annealing is flat (Limary and Green, 1999).

mass (M) and film thicknesses on a single substrate. By taking this approach, they strategically avoided the difficult and time-consuming process of fabricating a large number of different BCP films with different thicknesses on individual substrates (Smith et al., 2007). Additionally, their experiments thermo-annealed the samples using a temperature gradient orthogonal to the thickness-gradient direction (Figure 3.2). The morphologies of microdomains at different points along the substrate have been obtained. As a result of this clever experimental design, a series of structural

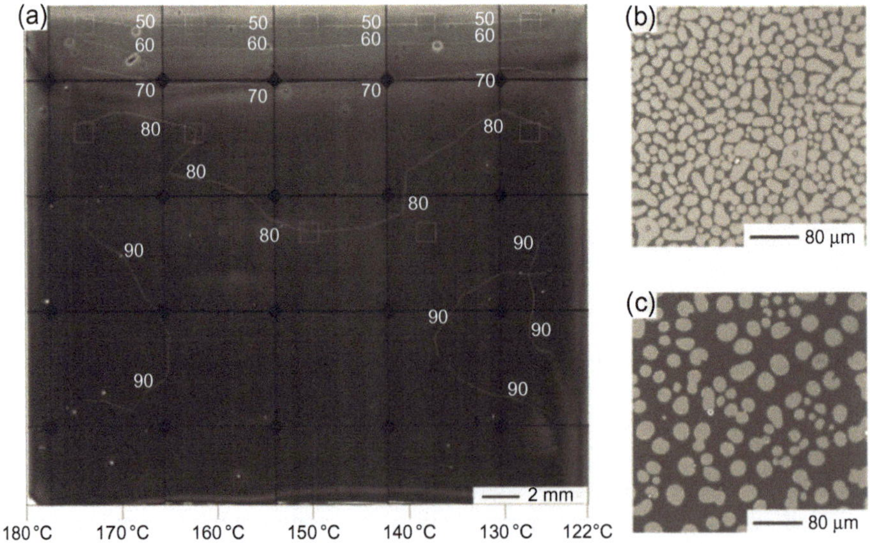

Figure 3.2 (a) Images of PS-*b*-PMMA film morphology with a temperature (T) gradient orthogonal to the thickness gradient. (b) and (c) Representative micrographs showing the morphologies observed at higher magnification after an 84.5 h annealing time. Micrographs are acquired for the regions (b) film thickness of 53 nm, T = 175 °C and (c) film thickness of 80 nm and T = 163 °C (Smith et al., 2007).

property relationships between microphase-separation morphologies as a function of film thickness, annealing time, annealing temperature, and the molecular mass of the BCP were established.

As evident by Smith et al.'s work, temperature gradient annealing has become a potential technique in creating highly ordered microdomains over a large area. Hashimoto's group developed a temperature-gradient heating apparatus to orient the microdomains of the BCP thick films (Bodycomb et al., 1999; Hashimoto et al., 1999). Samples were prepared in a "zone heating device," which consists of a pair of central heating blocks surrounded by two pairs of cooling blocks. The heating blocks were set above the T_{ODT} of the BCP, and the cooling blocks were set to a temperature well below the T_{ODT} of the BCP. This arrangement produced a sharp temperature gradient along the *z*-axis. Ordered lamellae grew from the disordered phase when the BCP cools within the gradient. Meanwhile, the lamellar planes aligned normal to the direction of the moving temperature gradient. Utilizing this process rendered orientation of lamellar microdomains in a PS-block-poly(isoprene) (PS-*b*-PI) BCP. Register and coworkers further explored a novel variation of this method involving an oscillating temperature gradient (Angelescu et al., 2007). They demonstrated that the ordering and reduction in defect density were achieved when the solidification front was advanced in an oscillatory fashion.

The studies described above substantiate that significant improvement in long-range order was achieved when zone annealing the samples above the T_{ODT}. Subsequently, Berry et al. reported the ordering and alignment of BCP films with zone annealed below the T_{ODT}—referred to as "cold zone annealing" (CZA) (Berry et al., 2007). CZA (defined by a sample temperature range, $T_g < T_{Hot} < T_{ODT}$) yielded cylinder-forming BCP films (PS-PMMA) with preferential alignment. Additionally, the CZA improved the coarsening of the BCP because long orientational correlation lengths were obtained after short annealing times. The authors also observed a significant decrease in defect density of the cylindrical microdomains in PS-*b*-PMMA thin films when decreasing the velocity at a given temperature and temperature gradient. Recently, Hashimoto et al. applied a moving temperature gradient field to BCPs creating hexagonally packed cylindrical (hex-cyl) microdomains. In this work, unique columnar grains macroscopically extended along the gradient axis (Mita et al., 2008). They found the moving temperature gradient field became the dominant effect over those of the glass surface-induced field (Mita et al., 2007).

3.2.3 Crystallization behavior induced by thermal annealing

BCPs containing a crystalline block component have been investigated widely because of their interesting hierarchical structure driven by couples and the competition between microphase separation of the disparate blocks and crystallization (Rangarajan et al., 1995a,b; Ryan et al., 1995; Hamley et al., 1996a,b, 1998; Quiram et al., 1997). The investigation of crystalline BCP thin films focuses on (i) whether crystallization disrupts the previously existing melt microphase-separated structure and (ii) the orientation of crystalline. In most cases, coupling and competition between crystallization and microphase separation were researched under thermal annealing conditions.

In the cases for crystalline-amorphous diblock copolymers, different and unique morphologies were observed depending on the T_{ODT}, the crystallization temperature T_c of the crystalline block, and the glass-transition temperature T_g of the amorphous block (Nandan et al., 2006). A hard nanoconfinement can be created if the T_g of amorphous blocks is higher than the T_m of crystalline (Zhu et al., 1999). The morphology development becomes highly complex when the amorphous block is liquid-like or rubbery during the crystallization process. Studies on crystalline-amorphous diblock copolymers ($T_g < T_c$) carried out on weakly segregated systems (PCL-*b*-PB, PEO-*b*-PB, PI-*b*-PEO) showed that breakout crystallization was favored in weakly segregated systems (Nojima et al., 1992; Lee et al., 2002; Floudas et al., 2001). With the increasing of the segregation strength of diblock copolymers, crystallization can be confined, templated, or can breakout depending on melts structure and thermal treatment process. The "classification map" compiled by Loo et al. and Xu et al. uncovered three modes of crystallization in the phase diagram of these BCP systems (Loo et al., 2002; Xu et al., 2002). By carefully tuning molecular length and temperature, possible control of a wide range of phase behaviors from pure "breakout" were made possible (Hobbs and Register, 2006).

In one example, Fu et al. investigated the effect of crystallization on pre-established relief structures as a function of crystallization temperature on poly(L-lactic acid)-block-polystyrene (PLLA-*b*-PS) (Fu et al., 2004). Crystallization was severely confined in the mesophase for $T_c < T_g^{PS}$. For $T_g^{PS} < T_c < T_m$ and $T_c \rightarrow T_m$, crystallization increased the lamellar spacing and severely broke the relief structures. Crystal lamellae were found perpendicular to the free surface. At very high temperature of 180 °C, PLLA forms metastable crystals during prolonged periods of annealing. Due to a decrease in segregation strength, some of copolymer chains were driven, by the line tension at edges, through the lamellae to form terrace structures.

Owing to the advances in synthetic technology, various functional BCPs containing rigid-rod block were synthesized recently (de Boer et al., 2001; Liu et al., 2002; Olsen and Segalman, 2005; Tao et al., 2007). The phase behavior of BCPs with rigid-rod block was fundamentally different from that of coil BCPs due to the difference in chain rigidity of the stiff rod-like conjugated blocks and flexible coil-like blocks (Kim et al., 2010b). For the most favored all-conjugated BCPs of the rod-rod type, the rod-like nature of most conjugated polymers complicates the self-assembly of corresponding BCPs through a competition between crystalline and liquid crystalline interactions and nanophase separation (Scherf et al., 2008, 2009).

Jenekhe and coworkers synthesized all-conjugated diblock copolymer of poly (3-butylthiophene)-*b*-poly(3-octylthiophene) (P3BT-*b*-P3OT) and investigated the phase behavior and solid-state morphology with wide-angle X-ray scattering and small-angle X-ray scattering (Wu et al., 2009). After annealing at 280 °C under high vacuum for 12 h, the all-conjugated diblock copolymer thin films were found to microphase separate from the melt phase into two distinct crystalline domains that correspond to P3BT and P3OT blocks with a lamellar structure.

Han and coworkers investigated the microphase separation and crystallization in thin films of all-conjugated diblock copolymers poly(2,5-dihexyloxy-*p*-phenylene)-block-(3-hexylthiophene) (PPP-*b*-P3HT) under thermal annealing (Yu et al., 2012;

Yang et al., 2014). Through the modulation of the competition characteristics between the microphase separation and crystallization, confined crystallization in BCPs with a higher block ratio of PPP and breakout crystallization in BCPs with a higher block ratio of P3HT were obtained. They also reported that during thermal annealing process, the "face-on" chain conformations of PPP moiety changed into an "edge-on." Therefore, the PPP and P3HT blocks formed a crystal structure both uniformly consisting of "edge-on" orientation.

3.3 Solvent annealing of BCPs films

In the thermal annealing process, the temperature window between the T_g of the blocks and the degradation temperature of the involved components might only be very small. Meanwhile, for many high molar mass BCP systems, the T_{ODT} increases greatly. Thus, thermal treatments alone are not effective because long annealing times are often required. Additionally, many copolymers contain blocks that are susceptible to thermal degradation and so high temperatures and longer thermal annealing times are undesirable. Solvent–vapor annealing has been proven in experiments to be a prominent alternative approach to producing long-range ordered BCP thin films (Kim et al., 2004; Peng et al., 2004; Xuan et al., 2004; Fukunaga et al., 2002). In the solvent–vapor annealing process, neutral or selective free surface can be created, and mobility is easily induced without the danger of degradation. In addition to inducing orientation and ordering, the self-assembled morphology may be reversibly tuned as well (Peng et al., 2004; Paik et al., 2010).

3.3.1 Fundamental consideration

The BCP assembly behavior in thin films becomes even more complicated when compared to thermal annealing (Albert and Epps, 2010). The presence of a solvent in BCP films performs several important functions. First, the solvent swells the BCP film, effectively reducing the T_g of the blocks and markedly enhancing the chain mobility (Zettl et al., 2010; Di et al., 2009; Cavicchi et al., 2005). Second, the solvent adsorbed in the film affects the interaction parameter between blocks (χ_{eff}), varying the effective volume fraction of the blocks (Xuan et al., 2004; Knoll et al., 2002; Li et al., 2010b; Jung and Ross, 2009). Third, the solvent mediates surface energies and tends to change the preferential wetting of free surface by different blocks of the copolymer (Xuan et al., 2004; Cavicchi et al., 2005; Lin et al., 2002). Fourth, solvent interactions with the substrate can screen undesirable polymer interactions with the substrate or cause film dewetting (Kim et al., 2004; Cavicchi et al., 2005). Finally, the swelling and the confinement of the film results in the deflection of the effective L_0 from bulk, and alters the previously established film thickness commensurability conditions (Chen et al., 2004; Knoll et al., 2007). Another consideration that cannot be overlooked is the evaporation rate of the solvent and the resultant order and orientation of the BCP thin film. In this process, consideration must be given to the solvent evaporation kinetics balanced against the diffusion of the polymer chains.

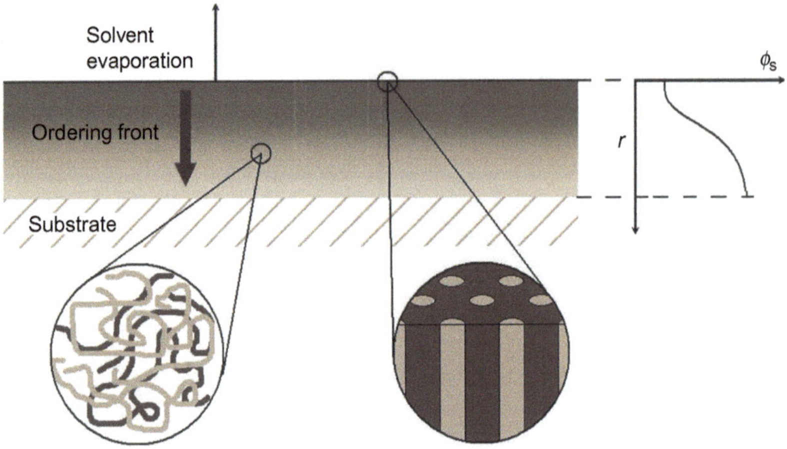

Figure 3.3 Schematic diagram of the solvent evaporation in a thin block copolymer film (Kim et al., 2004).
Copyright 2004 Wiley-VCH. Reproduced with permission.

In the solvent–vapor annealing, as-prepared BCP thin films are held in a controlled atmosphere containing solvent vapors at temperatures typically well below the bulk T_g of both blocks for a certain period of time. On subsequent evaporation of the solvent, polymer chains form well-organized nanostructures. Turturrot and coworkers observed cylindrical and lamellar microdomains in thin films perpendicular to the surface due to the solvent evaporation (Turturrot et al., 1995). Kim and Libera investigated this process with more detail using a similar triblock copolymer (Kim and Libera, 1998a,b). A reasonable mechanism of orientation was proposed: in the subsequent solvent evaporation process, a gradient of the solvent concentration was formed with solvent concentration being lowest at the surface and highest in the interior of the film. Consequently, the ordering of the BCP with microphase separation occurred initially at the film surface as a response to the gradient in solvent concentration. As solvent was continuously removed, an ordering front propagated into the film that served as a template to which the ordered structure was formed. Figure 3.3 provides a depiction of this process (Kim et al., 2004).

3.3.2 Factors affecting the annealing process

In the solvent–vapor annealing process, the final thin film morphology is sensitive to a number of parameters, including solvent selectivity (Fukunaga et al., 2000; Elbs et al., 2002), vapor pressure, the annealing time (Rodwogin et al., 2010), and solvent evaporation rate (Phillip et al., 2010; Gong et al., 2006). One aspect that governs the effectiveness of solvent–vapor annealing is the selectivity of each block for the solvent. The miscibility between a polymer and a solvent can be controlled by the polymer–solvent interaction parameter, $\chi_{P\text{-}S}$ (P = polymer and S = solvent) (Brandrup et al., 1999). For nonpolar systems, $\chi_{P\text{-}S}$ can be estimated by the expression

$$\chi_{\text{P-S}} = \frac{V_{\text{S}}\left(\delta_{\text{S}} - \delta_{\text{P}}\right)^2}{RT} + 0.34 \tag{3.2}$$

where V_S is the molar volume of the solvent, R is the gas constant, T is the temperature, and δ_S and δ_P are the solubility parameters of the solvent and polymer, respectively. For polar systems, $\chi_{\text{P-S}}$ can be described by the equation (Chen et al., 2004):

$$\chi_{\text{P-S}} = \frac{V_{\text{S}}\left[\left(\delta_{\text{dS}} - \delta_{\text{dP}}\right)^2 - \left(\delta_{\text{pS}} - \delta_{\text{pP}}\right)^2\right]}{RT} \tag{3.3}$$

where δ_d is the dispersion solubility parameter and δ_p is the polar solubility parameter. Using the Flory–Huggins criterion, the complete solvent–polymer miscibility can be realized when $\chi_{\text{P-S}} < 0.5$. The affinity between solvent and polymer is stronger when this value is smaller. For a given copolymer system, solvent usually can be classified as neutral or selective, according to whether it is a good solvent for both blocks, or a good solvent for one but a poor, or nonsolvent, for the other (Hanley et al., 2000; Huang et al., 1998; Pedersen et al., 2000). In general, a neutral solvent distributes itself nearly equally between microdomains and can screen the unfavorable contact of blocks. If the solvent is highly selective to one block, then the nonfavorable interactions can increase.

By manipulating the solvent selectivity, the degree of microphase separation shifts. In one example, Xuan et al. monitored the time development of the microdomain structure in symmetric diblock copolymer PS-*b*-PMMA ultrathin films (less than the 1/2 L_0) annealed with PMMA-selective solvent (chloroform or acetone) and PS-selective solvent (toluene or carbon disulfide), as shown in Figure 3.4 (Xuan et al., 2004). Films annealed with PMMA-selective solvents produced well-ordered hexagonally packed domains. As for the films annealed with PS-selective solvents, flat film with small protrusions were obtained. After preparation, the BCP film on a substrate preferentially attracted the PMMA block, the PMMA block dominated the substrate, and the PS block jockeyed for position at the film-free surface interface. Under PMMA-selective solvents vapor treatment, a strong attractive interaction existed between the PMMA block and the solvent and thus, PMMA was pulled toward the film surface to

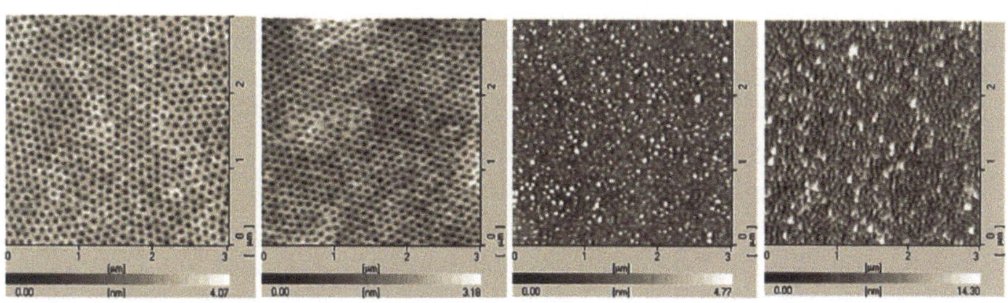

(a) Chloroform-annealed (b) Acetone-annealed (c) Toluene-annealed (d) CS_2-annealed

Figure 3.4 AFM images of symmetrical PS-*b*-PMMA thin film morphology after annealing in different solvent vapors (Xuan et al., 2004).

maximize the PMMA-solvent contacts. Here, the solvent preference for PMMA was the key point for overcoming the preference aggregation of the lower surface tension PS block to the free surface to fabricate well-ordered microstructures.

By choosing solvent with different selectivity, reversible switching of microdomains was reported by Peng et al. (2004). Specifically, their work investigated the reversible switching of microdomains on annealing PS-*b*-PMMA thin films in solvent vapors with different selectivity. Films annealed with acetone vapor at high pressure for 5 h produced a well-ordered hexagonally packed nanoporous structure after exposing the treated film to carbon disulfide (CS_2) vapor (PS selective) for 80 min. By switching the film boundary condition to a PS-selective solvent vapor, their morphologies and surface properties were reversibly switchable. The switching of the film morphology was influenced by the different behaviors of the PS and PMMA chains in their selective solvent vapors. When annealed in acetone vapor for different times, both PS and PMMA blocks become present at the depressions or striped film surface. After annealing in PS-selective solvent, the PMMA blocks migrate away from the solvent interface and approach a collapsed conformation to avoid contact with solvent. Meanwhile, the PS blocks are swollen and diffuse to the solvent interface to form a shield around the PMMA aggregates. These two effects lead to the opposite movement of the two blocks, and eventually result in an ordered hexagonally packed nanoprotuberant feature in the film. Additionally, the morphology reversibly switched back to the original state after annealing the film again in acetone vapor.

Because the nature of the solvent vapor is of prime importance in the final morphology, the use of solvent mixtures offers a promising alternative for tuning solvent selectivity and provides an approach for achieving the desired BCP morphology. With mixed solvents, the selectivity and solvent vapor pressure can be gradually and systematically tuned. The solvent mixture approach has been successfully used in systems such as poly(styrene-*b*-ethylene oxide) (PS-*b*-PEO)/benzene/water and poly (styrene-*b*-dimethylsiloxane) (PS-*b*-PDMS)/toluene/heptanes. Figure 3.5 gives the highly ordered and defect-free perpendicular cylinder morphologies of PS-*b*-PEO annealed with benzene and water (preferential for PS and PEO, respectively) (Kim et al., 2004).

The solvent–vapor annealing process is usually carried out in a chamber that contains a reservoir of liquid solvent(s). In such a reservoir system, it is difficult to independently control the partial vapor pressures of the solvents. In contrast, continuous flow systems can provide flow, mixing flowing species simultaneously, and this enables control over the vapor concentration and has been used to investigate solvent vapor pressure and deswelling rate effects on microdomain morphology and orientation (Cavicchi and Russell, 2007; Gotrik et al., 2015). Albert et al. designed a new type of microfluidic device to perform solvent–vapor annealing with different solvent vapor compositions in a single experiment (Albert et al., 2011). Such microfluidic devices are composed of six annealing chambers and one isolated chamber for reference. They were able to mix two gas streams in two different solvent "injection ports" to produces discrete gradients in solvent vapor composition and/or concentration. The device enables real-time monitoring of film thickness to connect how film swelling during the annealing process alters film morphology. In the work of Albert et al., the self-assembly

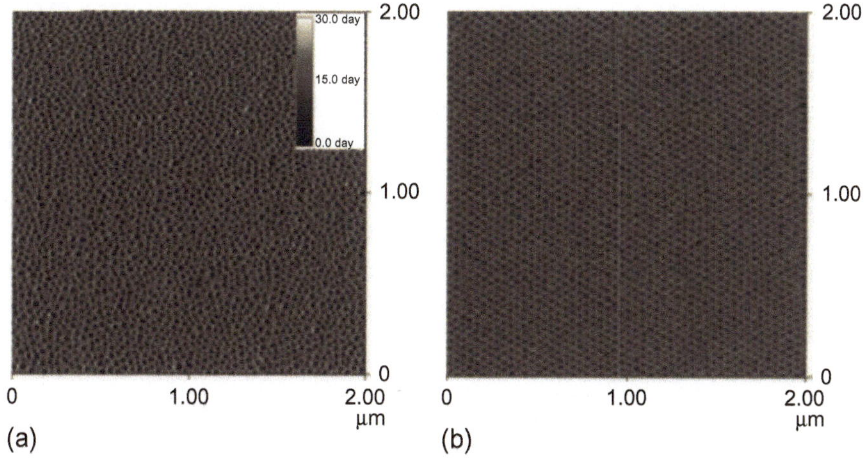

Figure 3.5 As-cast (a) and solvent–vapor annealed (b) thin films of poly(styrene)-*b*-poly(ethylene oxide) block polymers (Kim et al., 2004).

behavior of the cylinder forming poly(styrene)-*b*-poly(isoprene)-*b*-poly(styrene) BCP was investigated. Solvents with different selectivity *n*-hexane (poly(isoprene) selective) and tetrahydrofuran (a neutral solvent for both blocks) were used. The mole fraction of *n*-hexane in the chambers can be varied from >0.99 to 0.69, and the solvent concentration in the film can be changed from 60% to 40% (v/v.) Varying the solvent composition during annealing enabled morphological transformations from spheres to perpendicular cylinders or parallel cylinders. Additionally, it could also result in dewetting the BCP films.

During the solvent–vapor annealing process, another commonly highlighted aspect of the process is the time dependence of annealing on the final thin film morphology. During annealing, two dynamic processes take place: (i) diffusion of solvent within the thin film and (ii) polymer chain diffusion. Increasing the solvent–vapor annealing time will allow for greater solvent uptake (until the solvent in the film is at chemical equilibrium with the solvent in the free surface) and for more time for the polymer chains to rearrange themselves in their desired, lowest energy thermodynamic state (Jung and Ross, 2009). It is important to point out that metastable features are usually formed with shorter annealing times. In some systems, many well-ordered metastable microphase separation features haven been observed, and the microdomains evolve continuously with increased annealing time (Kim et al., 2004).

Meanwhile, the vapor pressure of the solvent also dictates the swelled film thickness providing control over the relative concentrations of polymer and solvent in the film (Sinturel et al., 2013; Hur et al., 2015). The solvent vapor pressure can be manipulated by temperature, mass flow controllers, and other means such as applying backpressure (i.e., controlling the total pressure of the system). The vapor pressure of the solvent used for annealing must be precisely controlled to realize the fabrication of the different self-assembled structures. In one example, when annealed, two PS-*b*-PMMA thin films of the same composition with acetone vapor of different vapor pressure and two series of nanostructures transformation processes have been observed by varying the annealing time. With a high vapor pressure, the BCP film morphology experienced the evolution of disordered state to nanoscale depression morphology and at last to a striped morphology. Under a low vapor pressure, different microdomains formed, including highly ordered spherical domains, a novel so-called flower-like pattern composed of six PS spheres wherein each PS sphere belongs to three "flowers," and, at last, striped morphology. See Figure 3.6 to see this particular morphological pattern (Peng et al., 2005).

In some systems, lengthening annealing time does not result in the long-range ordered structures, and the final morphology will depend on the solvent evaporation step. The solvent evaporation rate is one of the key factors that determines the film morphologies (Phillip et al., 2010; Pedersen et al., 2000). Fast solvent evaporation generally results in the trap of nonequilibrium structures, whereas slower evaporation usually approaches the thermodynamic equilibrium morphology. Kim and Libera investigated the effect of the evaporation rate on the self-assembled microdomains of PS/PB/PS triblock copolymer thin film with a ~100 nm thickness

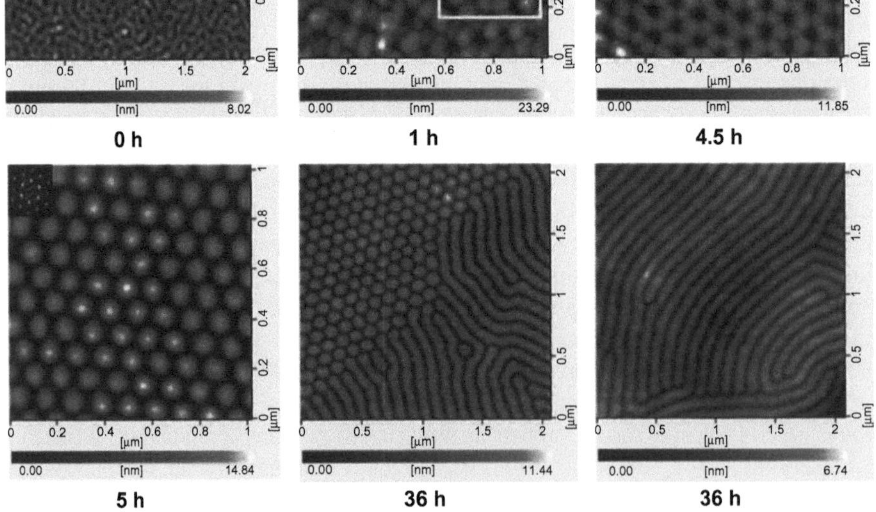

Figure 3.6 Feature evolution of PS-*b*-PMMA microphase separation in solvent–vapor annealing with various annealing times (Peng et al., 2005).

(Kim and Libera, 1998a). Their results demonstrated that vertically aligned cylindrical PS microdomains were obtained when an intermediate evaporation rate (~ 5 nl/s) was used. Microphase-separated structure without long-range order and a duplex microstructure of PS cylinders aligned vertically or in-plane, were observed when fast (~ 200 nl/s) and slow (~ 1.5 nl/s) evaporation rates were used, respectively. Fukunaga et al. investigated the effect of solvent removal rate in solvent annealed lamellar poly(styrene-*b*-2-vinylpryidine-*b*-*tert*-butyl methacrylate) (PS-*b*-P2VP-*b*-P*t*BMA) films and found that parallel orientations and perpendicular orientations occurred for slow and fast solvent removal rates, respectively (Fukunaga et al., 2000). The investigation of the effect of different solvent removal rates in the solvent annealing process is a research direction requiring further investigation and understanding.

3.3.3 Combination of solvent annealing and thermal annealing

In the solvent and thermal annealing process, several hours are typically required for the fabrication of highly ordered periodic structures with low defect density over large areas. From a manufacturing consideration, implementing an annealing process with short processing times is preferred. To achieve such a feat, some researchers are examining a solvothermal annealing process (i.e., the combination of the solvent annealing and thermal annealing). In a solvothermal annealing approach, described by Zhang et al., the BCP film was placed on a microwave-safe glass support (extending out of the solvent-filled volume) in a sealed vial with the presence of selected solvent in it (Zhang et al., 2010). The BCP film was simultaneously exposed to solvent vapor and irradiated with microwave energy. This process significantly reduced the time needed to generate highly organized cylinder morphologies in thin films of poly(styrene)-*b*-poly(2-vinylpyridine) (PS-*b*-P2VP), and it also created PS-*b*-PMMA nanostructures with lower defect densities and larger domain sizes.

Another report describing a solvothermal annealing process was given by Ross et al. where they introduced a two-stage annealing process for BCP films (Gotrik and Ross, 2013). In this approach, the BCP film was first exposed to a saturated solvent vapor at ambient temperature (23 °C) to swell the films, and then a thermal "quenching" process was applied by heating the films to drive out the solvent. By heating the BCP film directly without raising the temperature of the entire chamber, changes in the ambient vapor pressure of the solvent was avoided. This two-stage annealing process helped decouple the effects of solvent and heating on the final BCP film morphology. The comparison between thermal, solvent, and solvothermal annealing of the bilayer cylinder samples are given in Figure 3.7. We can see from the images, with a 5 min solvothermal anneal (270 s solvent incorporation + 30 s thermal quench), well-organized two perpendicular layers of cylinders structure were obtained, which was orders of magnitude faster than a thermal anneal or a solvent–vapor annealing alone. Overall, it is evident that the solvothermal anneal process provides a rapid and simple self-assembly strategy making it an attractive strategy for annealing BCP films.

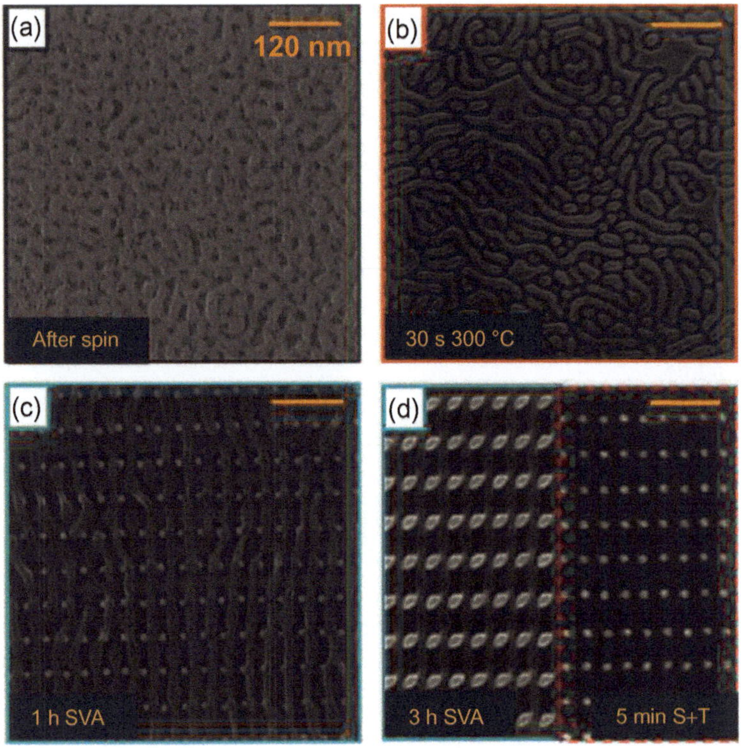

Figure 3.7 Micrographs of films resulting from different annealing processes applied to a cylinder, forming 45 kg/mol PS-PDMS templated by a rectangular array of PS-coated posts. (a) Micelles after spin coating from propylene glycol methyl ether acetate; (b) short cylinders after a high temperature thermal anneal; (c) partial separation into two layers of cylinders after a 1 h solvent vapor anneal; and (d) a comparison between a 3 h solvent vapor anneal (left) and a 5 min solvothermal anneal (right), both showing a templated bilayer of cylinders (Gotrik and Ross, 2013). Copyright 2014 American Chemical Society. Reproduced with permission.

3.4 Summary and outlook

Today, preparing micro/nano-patterned surfaces by microphase separation of BCP is one of the most promising techniques for the fabrication of nanoscale arrays, not only because it is a simple process but also because it is lower cost than traditional methods. It is essential to control both the orientation and lateral ordering of the nanoscopic domains to fully realize the potential use of these structures. As we have described, thermal annealing and solvent annealing are the two powerful and most commonly used methods that allow for the generation of highly organized nanostructures. The annealing processes impart mobility to the copolymers and enable the removal of defects. The obtained highly ordered nanoscopic arrays hold tremendous promise in a variety of advanced nanotechnologies. In recent years, a number of approaches have been developed to enhance the lateral ordering of BCP microdomains by applying external stimuli such as electric fields, shear alignment, graphoepitaxy, and chemically patterned substrate. From a manufacturing viewpoint, more basic studies will be required in the future to understand and optimize the processing condition for self-assembly of various BCPs during annealing. Meanwhile, it is necessary to establish a comprehensive understanding regarding BCP thin film self-assembly, which can be applied to manipulate the phase behavior of new functional materials.

References

Albert, J.N.L., Epps III, T.H., 2010. Self-assembly of block copolymer thin films. Mater. Today 13, 24–33.

Albert, J.N.L., Bogart, T.D., Lewis, R.L., Beers, K.L., Fasolka, M.J., Hutchison, J.B., Vogt, B.D., Epps, T.H., 2011. Gradient solvent vapor annealing of block copolymer thin films using a microfluidic mixing device. Nano Lett. 11, 1351–1357.

Alexander–Katz, A., Fredrickson, G.H., 2007. Diblock copolymer thin films: a field–theoretic simulation study. Macromolecules 40, 4075–4087.

Anastasiadis, S.H., Russell, T.P., Katija, S.K., Majkrzak, C.F., 1989. Neutron reflectivity studies of the surface-induced ordering of diblock copolymer films. Phys. Rev. Lett. 62, 1852–1855.

Anastasiadis, S.H., Russell, T.P., Satija, S.K., Majkrzak, C.F., 1990. The morphology of symmetric diblock copolymers as revealed by neutron reflectivity. J. Chem. Phys. 92, 5677–5691.

Angelescu, D.E., Waller, J.H., Adamson, D.H., Register, R.A., Chaikin, P.M., 2007. Enhanced order of block copolymer cylinders in single-layer films using a sweeping solidification front. Adv. Mater. 19, 2687–2690.

Bates, F.S., Fredrickson, G.H., 1999. Block copolymers-designer soft materials. Phys. Today 52, 32–38.

Berry, B.C., Bosse, A.W., Douglas, J.F., Jones, R.L., Karim, A., 2007. Orientational order in block copolymer films zone annealed below the order-disorder transition temperature. Nano Lett. 7, 2789–2794.

Bodycomb, J., Funaki, Y., Kimishima, K., Hashimoto, T., 1999. Single-grain lamellar micro-domain from a diblock copolymer. Macromolecules 32, 2075–2077.

Brandrup, J., Immergut, E.H., Grulke, E.A., Abe, A., Bloch, D.R., 1999. Polymer Handbook, fourth ed. Wiley, New York.

Cai, Z.-H., Huang, K., Montano, P.A., Russell, T.P., Bai, J.M., Zajac, G.W., 1993. Experimental study of the surface structure of diblock copolymer films using microscopy and x-ray scattering. J. Chem. Phys. 98, 2376–2386.

Cavicchi, K.A., Russell, T.P., 2007. Solvent annealed thin films of asymmetric polyisoprene-polylactide diblock copolymers. Macromolecules 40, 1181–1186.

Cavicchi, K.A., Berthiaume, K.J., Russell, T.P., 2005. Solvent annealing thin films of poly(isoprene–b–lactide). Polymer 46, 11635–11639.

Chen, Y.Z., Huang, H.Y., Hu, Z.J., He, T.B., 2004. Lateral nanopatterns in thin diblock copolymer films induced by selective solvents. Langmuir 20, 3805–3808.

Coulon, G., Deline, V.R., Russell, T.P., Green, P.F., 1989. Surface–induced orientation of symmetric, diblock copolymers: a secondary ion mass–spectrometry study. Macromolecules 22, 2581–2589.

de Boer, B., Stalmach, U., van Hutten, P.F., Melzer, C., Krasnikov, V.V., Hadziioannou, G., 2001. Supramolecular self-assembly and opto-electronic properties of semiconducting block copolymers. Polymer 42, 9097–9109.

Di, Z., Posselt, D., Smilgies, D.M., Papadakis, C.M., 2009. Structural rearrangements in a lamellar diblock copolymer thin film during treatment with saturated solvent vapor. Macromolecules 43, 418–427.

Elbs, H., Drummer, C., Abetz, V., Krausch, G., 2002. Thin film morphologies of ABC triblock copolymers prepared from solution. Macromolecules 35, 5570–5577.

Floudas, G., Vazaiou, B., Schipper, F., Ulrich, R., Wiesner, U., Iatrou, H., Hadjichristidis, N., 2001. Poly(ethylene oxide-*b*-isoprene) diblock copolymer phase diagram. Macromolecules 34, 2947–2957.

Fu, J., Cong, Y., Li, J., Luan, B., Pan, C.Y., Yang, Y.M., Li, B.Y., Han, Y.C., 2004. Hole nucleation and growth induced by crystallization and microphase separation in thin semicrystalline diblock copolymer films. Macromolecules 37, 6918–6925.

Fukunaga, K., Elbs, H., Magerle, R., Krausch, G., 2000. Large–scale alignment of ABC block copolymer microdomains via solvent vapor treatment. Macromolecules 33, 947–953.

Fukunaga, K., Hashimoto, T., Elbs, H., Krausch, G., 2002. Self-assembly of a lamellar ABC triblock copolymer thin film. Macromolecules 35, 4406–4413.

Gong, Y.M., Huang, H.Y., Hu, Z.J., Chen, Y.Z., Chen, D.J., Wang, Z.B., He, T.B., 2006. Inverted to normal phase transition in solution–cast polystyrene–poly(methyl methacrylate) block copolymer thin films. Macromolecules 39, 3369–3376.

Gotrik, K.W., Ross, C.A., 2013. Solvothermal annealing of block copolymer thin films. Nano Lett. 13, 5117–5122.

Gotrik, K.W., Hannon, A.F., Son, J.G., Keller, B., Alexander-Katz, A., Ross, C.A., 2015. Morphology control in block copolymer films using mixed solvent vapors. ACS Nano 6, 8052–8059.

Green, P.F., Christensen, T.M., Russell, T.P., 1991. Ordering at diblock copolymer surfaces. Macromolecules 24, 252–255.

Hamley, I.W., Fairclough, J.P.A., Terrill, N.J., Ryan, A.J., Lipic, P.M., Bates, F.S., Towns–Andrews, E., 1996a. Crystallization in oriented semicrystalline diblock copolymers. Macromolecules 29, 8835–8843.

Hamley, I.W., Fairclough, J.P.A., Ryan, A.J., Bates, F.S., Towns–Andrews, E., 1996b. Crystallization of nanoscale–confined diblock copolymer chains. Polymer 37, 4425–4429.

Hamley, I.W., Fairclough, J.P.A., Bates, F.S., Ryan, A.J., 1998. Crystallization thermodynamics and kinetics in semicrystalline diblock copolymers. Polymer 39, 1429–1437.

Hanley, K.J., Lodge, T.P., Huang, C.-I., 2000. Phase behavior of a block copolymer in solvents of varying selectivity. Macromolecules 33, 5918–5931.

Hasegawa, H., Hashimoto, T., 1985. Morphology of block polymers near a free surface. Macromolecules 18, 589–590.

Hashimoto, T., Bodycomb, J., Funaki, Y., Kimishima, K., 1999. The effect of temperature gradient on the microdomain orientation of diblock copolymers undergoing an order-disorder transition. Macromolecules 32, 952–954.

Hayward, R.C., Chmelka, B.F., Kramer, E.J., 2005. Crosslinked poly(styrene)–block–poly (2–vinylpyridine) thin films as swellable templates for mesostructured silica and titania. Adv. Mater. 17, 2591–2595.

Hobbs, J.K., Register, R.A., 2006. Imaging block copolymer crystallization in real time with the atomic force microscope. Macromolecules 39, 703–710.

Huang, C.-I., Chapman, B.R., Lodge, T.P., Balsara, N.P., 1998. Quantifying the "neutrality" of good solvents for block copolymers: poly(styrene-b-isoprene) in toluene, benzene, and THF. Macromolecules 31, 9384–9386.

Hur, S.-M., Khaira, G.S., Ramírez-Hernández, A., Müller, M., Nealey, P.F., de Pablo, J.J., 2015. Simulation of defect reduction in block copolymer thin films by solvent annealing. ACS Macro Lett. 4, 11–15–11–15.

Joo, W., Park, M.S., Kim, J.K., 2006. Block copolymer film with sponge-like nanoporous strucutre for antireflection coating. Langmuir 22, 7960–7963.

Jung, Y.S., Ross, C.A., 2009. Solvent–vapor-induced tunability of self-assembled block copolymer patterns. Adv. Mater. 21, 2540–2545.

Kim, G., Libera, M., 1998a. Morphological development in solvent–cast polystyrene–polybutadiene–polystyrene (SBS) triblock copolymer thin films. Macromolecules 31, 2569–2577.

Kim, G., Libera, M., 1998b. Kinetic constraints on the development of surface microstructure in SBS thin films. Macromolecules 31, 2670–2672.

Kim, H.C., Jia, X.Q., Stafford, C.M., Kim, D.H., McCarthy, T.J., Tuominen, M., Hawker, C.J., Russell, T.P., 2001. A Route to nanoscopic SiO2 posts via block copolymer templates. Adv. Mater. 13, 795–797.

Kim, S.H., Misner, M.J., Xu, T., Kimura, M., Russell, T.P., 2004. Highly oriented and ordered arrays from block copolymers via solvent evaporation. Adv. Mater. 16, 226–231.

Kim, H.C., Park, S.M., Hinsberg, W.D., 2010a. Block copolymer based nanostructures: materials, processes, and applications to electronics. Chem. Rev. 110, 146–177.

Kim, J.K., Yang, S.Y., Lee, Y., Kim, Y., 2010b. Functional nanomaterials based on block copolymer self-assembly. Prog. Polym. Sci. 35, 1325–1349.

Knoll, A., Horvat, A., Lyakhova, K.S., Krausch, G., Sevink, G.J.A., Zvelindovsky, A.V., Magerle, R., 2002. Phase behavior in thin films of cylinder-forming block copolymers. Phys. Rev. Lett. 89, 035501.

Knoll, A., Tsarkova, L., Krausch, G., 2007. Nanoscaling of microdomain spacings in thin films of cylinder-forming block copolymers. Nano Lett. 7, 843–846.

Krausch, G., 1995. Surface induced self-assembly in thin polymer films. Mater. Sci. Eng. R14, 1–94.

Lee, W., Chen, H.L., Lin, T.L., 2002. Correlation between crystallization kinetics and microdomain morphology in block copolymer blends exhibiting confined crystallization. J. Polym. Sci. Polym. Phys. 40, 519–529.

Leibler, L., 1980. Theory of microphase separation in block copolymers. Macromolecules 13, 1602–1617.

Li, X., Han, Y.C., 2011. Tunable wavelength antireflective film by non–solvent–induced phase separation of amphiphilic block copolymer micelle solution. J. Mater. Chem. 21, 18024–18033.

Li, X., Gao, J.P., Xue, L.J., Han, Y.C., 2010a. Porous polymer film with gradient–refractive–index structure for broad–band and omnidirectional antireflection coating. Adv. Funct. Mater. 20, 259–265.

Li, Y.H., Huang, H.Y., He, T.B., 2010b. The effect of the preferential affinity of the solvent on the microstructure of solution–cast block copolymer thin films. J. Phys. Chem. B 114, 1264–1270.

Li, X., Xue, L.J., Han, Y.C., 2011. Broadband antireflection of block copolymer/homopolymer blends films with gradient–refractive–index structures. J. Mater. Chem. 21, 5817–5826.

Limary, R., Green, P.F., 1999. Hierarchical pattern formation in thin film diblock copolymers above the order-disorder transition temperature. Macromolecules 32, 8167–8172.

Lin, Z.Q., Kim, D.H., Wu, X.D., Boosahda, L., Stone, D., LaRose, L., Russell, T.P., 2002. A rapid route to arrays of nanostructures in thin films. Adv. Mater. 14, 1373–1376.

Liu, J., Sheina, E., Kowalewski, T., McCullough, R.D., 2002. Tuning the electrical conductivity and self-assembly of regioregular polythiophene by block copolymerization: nanowire morphologies in new di- and triblock copolymers. Angew. Chem. Int. Ed. 41, 329–332.

Loo, Y.L., Register, R.A., Ryan, A.J., 2002. Modes of crystallization in block copolymer microdomains: breakout, templated, and confined. Macromolecules 35, 2365–2374.

Matsen, M.W., 1997. Thin films of block copolymer. J. Chem. Phys. 106, 7781–7791.

Mayes, A.M., Russell, T.P., Bassereau, P., Baker, S.M., Smith, G.I., 1994. Evolution of order in thin block copolymer films. Macromolecules 27, 749–755.

Menelle, A., Russell, T.P., Anastasiadis, S.H., Katija, S.K., Majkrzak, C.F., 1992. Ordering of thin diblock copolymer films. Phys. Rev. Lett. 68, 67–70.

Mita, K., Tanaka, H., Saijo, K., Takenaka, M., Hashimoto, T., 2007. Cylindrical domains of block copolymers developed via ordering under moving temperature gradient. Macromolecules 40, 5923–5933.

Mita, K., Tanaka, H., Saijo, K., Takenaka, M., Hashimoto, T., 2008. Ordering of cylindrical domains of block copolymers under moving temperature gradient: separation of T–induced ordering from surface–induced ordering. Macromolecules 41, 6787–6792.

Nandan, B., Hsu, J.Y., Chen, H.L., 2006. Crystallization behavior of crystalline–amorphous diblock copolymers consisting of a rubbery amorphous block. J. Macro. Sci. C Polym. Rev. 46, 143–172.

Nojima, S., Kato, K., Yamamoto, S., Ashida, T., 1992. Crystallization of block copolymers. 1. Small-angle X-ray scattering study of a epsilon–caprolactone–butadiene diblock copolymer. Macromolecules 25, 2237–2242.

Olsen, B.D., Segalman, R.A., 2005. Structure and thermodynamics of weakly segregated rod–coil block copolymers. Macromolecules 38, 10127–10137.

Olson, D.A., Chen, L., Hillmyer, M.A., 2007. Templating nanoporous polymers with ordered block copolymers. Chem. Mater. 20, 869–890.

Orilall, M.C., Wiesner, U., 2011. Block copolymer based composition and morphology control in nanostructured hybrid materials for energy conversion and storage: solar cells, batteries, and fuel cells. Chem. Soc. Rev. 40, 520–535.

Pai, R.A., Humayun, R., Schulberg, M.T., Sengupta, A., Sun, J.N., Watkins, J.J., 2004. Mesoporous silicates prepared using preorganized templates in supercritical fluids. Science 303, 507–510.

Paik, M.Y., Bosworth, J.K., Smilges, D.M., Schwartz, E.L., Andre, X., Ober, C.K., 2010. Reversible morphology control in block copolymer films via solvent vapor processing: an in situ GISAXS study. Macromolecules 43, 4253–4260.

Pard, M., Chaikin, P.M., Register, R.A., Adamson, D.H., 2001. Large area dense nanoscale patterning of arbitrary surfaces. Appl. Phys. Lett. 79, 257.

Park, M., Harrison, C., Chaikin, P.M., Register, R.A., Adamson, D.H., 1997. Block copolymer lithography: periodic arrays of $\sim 10^{11}$ holes in 1 square centimeter. Science 276, 1401–1404.

Park, S., Wang, J.Y., Kim, B., Xu, J., Russell, T.P., 2008. A simple route to highly oriented and ordered nanoporous block copolymer templates. ACS Nano 2, 766–772.

Pedersen, J.S., Hamley, I.W., Ryu, C.Y., Lodge, T.P., 2000. Contrast variation small–angle neutron scattering study of the structure of block copolymer micelles in a slightly selective solvent at semidilute concentrations. Macromolecules 33, 542–550.

Peng, J., Xuan, Y., Wang, H.F., Yang, Y.M., Li, B.Y., Han, Y.C., 2004. Solvent-induced microphase separation in diblock copolymer thin films with reversibly switchable morphology. J. Chem. Phys. 120, 11163–11170.

Peng, J., Wei, Y.H., Wang, H.F., Li, B.Y., Han, Y.C., 2005. Solvent induced sphere development in symmetric diblock copolymer thin films. Macromol. Rapid Commun. 26, 738–743.

Peters, R.D., Yang, X.M., Kim, T.K., Sohn, B.H., Nealey, P.F., 2000. Using self-assembled monolayers exposed to X-rays to control the wetting behavior of thin films of diblock copolymers. Langmuir 16, 4625–4631.

Phillip, W.A., O'Neill, B., Rodwogin, M., Hillmyer, M.A., Cussler, E.L., 2010. Self-assembled block copolymer thin films as water filtration membranes. Appl. Mater. Interfaces 2, 847–853.

Pickett, G.T., Witten, T.A., Nagel, S.R., 1993. Equilibrium surface orientation of lamellae. Macromolecules 26, 3194–3199.

Quiram, D.J., Register, R.A., Marchand, G.R., 1997. Crystallization of asymmetric diblock copolymers from microphase-separated melts. Macromolecules 30, 4551–4558.

Rangarajan, P., Register, R.A., Fetters, L.J., Bras, W., Naylor, S., Ryan, A.J., 1995a. Crystallization of a weakly segregated polyolefin diblock copolymer. Macromolecules 28, 4932–4938.

Rangarajan, P., Register, R.A., Adamson, D.H., Fetters, L.J., Bras, W., Naylor, S., Ryan, A.J., 1995b. Dynamics of structure formation in crystallizable block copolymers. Macromolecules 28, 1422–1428.

Rodwogin, M.D., Spanjers, C.S., Leighton, C., Hillmyer, M.A., 2010. Polylactide–poly(dimethylsiloxane)–polylactide triblock copolymers as multifunctional materials for nanolithographic applications. ACS Nano 4, 725–732.

Ruiz, R., Kang, H., Detcheverry, F.A., Dobisz, E., Kercher, D.S., Albrecht, T.R., de Pablo, J.J., Nealey, P.F., 2008. Density multiplication and improved lithography by directed block copolymer assembly. Science 321, 936–939.

Russell, T.P., Coulon, G., Deline, V.R., Miller, D.C., 1989. Characteristics of the surface-induced orientation of symmetric diblock PS/PMMA copolymers. Macromolecules 22, 4600–4606.

Ryan, A.J., Hamley, I.W., Bras, W., Bates, F.S., 1995. Structure development in semicrystalline diblock copolymers crystallizing from the ordered melt. Macromolecules 28, 3860–3868.

Scherf, U., Gutacker, A., Koenen, N., 2008. All-conjugated block copolymers. Acc. Chem. Res. 41, 1086–1097.

Scherf, U., Adamczyk, S., Gutacker, A., Koenen, N., 2009. All-conjugated, rod–rod block copolymers-generation and self-assembly properties. Macromol. Rapid Commun. 30, 1059–1065.

Segalman, R.A., McCulloch, B., Kirmayer, S., Urban, J.J., 2009. Block copolymers for organic optoelectronics. Macromolecules 42, 9205–9216.

Sikka, M., Singh, N., Karim, A., Bates, F.S., Satija, S.K., Majkrzak, C.F., 1993. Entropy-driven surface segregation in block copolymer melts. Phys. Rev. Lett. 70, 307–310.

Sinturel, C., Vayer, M., Morris, M., Hillmyer, M.A., 2013. Solvent vapor annealing of block polymer thin films. Macromolecules 46, 5399–5415.

Smith, A.P., Douglas, J.F., Meredith, J.C., Amis, E.J., Karim, A., 2001. High–throughput characterization of pattern formation in symmetric diblock copolymer films. J. Polym. Sci. Part B: Polym. Phys. 39, 2141–2158.

Smith, A.P., Sehgal, A., Douglas, J.F., Karim, A., Amis, E.J., 2003. Combinatorial mapping of surface energy effects on diblock copolymer thin film ordering. Macromol. Rapid Commun. 24, 131–135.

Smith, A.P., Douglas, J.F., Amis, E.J., Karim, A., 2007. Effect of temperature on the morphology and kinetics of surface pattern formation in thin block copolymer films. Langmuir 23, 12380–12387.

Stoykovich, M.P., Nealey, P.F., 2006. Moore's law: the future of Si microelectronics. Mater. Today 9, 20–25.

Tao, Y., Olsen, B.D., Ganesan, V., Segalman, R.A., 2007. Domain size control in self-assembling rod–coil block copolymer and homopolymer blends. Macromolecules 40, 3320–3327.

Thurn–Albrecht, T., Steiner, R., DeRouchey, J., Stafford, C.M., Huang, E., Bal, M., Tuominen, M., Hawker, C.J., Russell, T.P., 2000. Nanoscopic templates from oriented block copolymer films. Adv. Mater. 12, 787–791.

Turturrot, A., Gattiglia, E., Vacca, P., 1995. Free surface morphology of block copolymers: I. Styrene–butadiene diblock copolymers. Polymer 36, 3987–3996.

Urade, V.N., Wei, T.C., Tate, M.P., Kowalski, J.D., Hillhouse, H.W., 2007. Nanofabrication of double–gyroid thin films. Chem. Mater. 19, 768–777.

Wang, Q., Nath, S.K., Graham, M.D., Nealey, P.F., de Pablo, J.J., 2000. Symmetric diblock copolymer thin films confined between homogeneous and patterned surfaces: simulations and theory. J. Chem. Phys. 112, 9996–10010.

Wu, P.T., Ren, G.Q., Li, C.X., Mezzenga, R., Jenekhe, S.A., 2009. Crystalline diblock conjugated copolymers: synthesis, self-assembly, and microphase separation of poly(3-butylthiophene)-b-poly(3-octylthiophene). Macromolecules 42, 2317–2320.

Xu, J.T., Turner, S.C., Fairclough, J.P.A., Mai, S.M., Ryan, A.J., Chaibundit, C., 2002. Isothermal crystallization kinetics and melting behavior of poly(oxyethylene)-b-poly(oxybutylene)/poly(oxybutylene) blends. Macromolecules 35, 6937–6945.

Xuan, Y., Peng, J., Cui, L., Wang, H.F., Li, B.Y., Han, Y.C., 2004. Morphology development of ultrathin symmetric diblock copolymer film via solvent vapor treatment. Macromolecules 37, 7301–7307.

Yang, S.Y., Ryu, I., Kim, H.Y., Kim, J.K., Jang, S.K., Russell, T.P., 2006. Nanoporous membranes with ultrahigh selectivity and flux for the filtration of viruses. Adv. Mater. 18, 709–712.

Yang, H., Wang, L., Yu, X.H., Zhang, J.D., Geng, Y.H., Han, Y.C., 2014. Nanowire shish–kebab structures and molecular orientation control of all-conjugated diblock copolymer. Chem. Asian J. 9, 2239–2248.

Yu, X.H., Yang, H., Wu, S.P., Geng, Y.H., Han, Y.C., 2012. Microphase separation and crystallization of all-conjugated phenylene–thiophene diblock copolymers. Macromolecules 45, 266–274.

Zettl, U., Knoll, A., Tsarkova, L., 2010. Effect of confinement on the mesoscale and macroscopic swelling of thin block copolymer films. Langmuir 26, 6610–6617.

Zhang, X., Harris, K.D., Wu, N.L.Y., Murphy, J.N., Buriak, J.M., 2010. Fast assembly of ordered block copolymer nanostructures through microwave annealing. ACS Nano 4, 7021–7029.

Zhu, L., Chen, Y., Zhang, A., Calhoun, B.H., Chun, M., Quirk, R.P., Cheng, S.Z.D., Hsiao, B.S., Yeh, F., Hashimoto, T., 1999. Phase structures and morphologies determined by competitions among self-organization, crystallization, and vitrification in a disordered poly(ethylene oxide)-b-polystyrene diblock copolymer. Phys. Rev. B 60, 10022–10031.

Field-theoretic simulations and self-consistent field theory for studying block copolymer directed self-assembly

4

V.V. Ginzburg, J.D. Weinhold*[†], *P.D. Hustad*[‡], *P. Trefonas*[‡], *B. Kim*[§], *N. Laachi*[§], *G.H. Fredrickson*[§]

[*]The Dow Chemical Company, Midland, MI, USA; [†]The Dow Chemical Company, Freeport, TX, USA; [‡]Dow Electronic Materials, Marlborough, MA, USA; [§]University of California, Santa Barbara, CA, USA

4.1 Introduction

In recent years, block copolymer directed self-assembly (DSA) has become a promising new approach to printing sub-40nm features (Luo and Epps, 2013; Albert and Epps, 2010; Sanders et al., 2010). In DSA, nanoscale patterns are obtained as a result of thermodynamic microphase separation between two or more chemically distinct blocks covalently bonded into block copolymers (for more detailed description of block copolymer fundamentals and the concept of microphase separation, see Bates and Fredrickson, 1999). By manipulating the chemical nature of each of the individual blocks and molecular weight of the block copolymer, one can vary the symmetry of the equilibrium structure (lamellar, hexagonal, body-centered-cubic, gyroid, etc.) and its equilibrium period (often referred to as "pitch"). In general, the phase behavior of simple diblock copolymers is understood fairly well (Leibler, 1980; Semenov, 1985; Matsen and Schick, 1994b; Matsen and Bates, 1996a,b; Drolet and Fredrickson, 1999; Fredrickson et al., 2002), although some quantitative disagreements between theory and experiment still remain. The situation is, in general, more complicated for block copolymer mixtures, blends of copolymers with homopolymers, triblocks, and tetrablocks, and so forth. The number of possible degrees of freedom and combinations increases very rapidly, and it becomes, on a practical basis, impossible to enumerate all possible morphologies for all possible formulations (Bates et al., 2012). Even so, it would seem that the same theoretical approaches that worked for the case of diblocks are equally applicable to blends, triblocks, and so forth.

The predictive challenge becomes more complex for the case of thin films of block copolymers, where the morphology is influenced by the interplay between film thickness, equilibrium pitch, and energies of interaction between top (air), bottom (substrate), and possibly side surfaces with various polymer species (e.g., polystyrene (PS) and poly(methylmethacrylate) (PMMA) blocks). Thus, in the case of undirected self-assembly, while the desired outcome would be to have an array of vertical

Directed Self-assembly of Block Copolymers for Nano-manufacturing. http://dx.doi.org/10.1016/B978-0-08-100250-6.00004-3

lamellae or hexagonal pattern of vertical cylinders, in many cases, the block copolymer prefers a horizontal or mixed morphology (Kikuchi and Binder, 1994; Matsen, 1997; Geisinger et al., 1999a,b; Knoll et al., 2002; Lyakhova et al., 2004; Yang et al., 2006). Perpendicular morphologies can be favored if the top and bottom surfaces are both nonpreferential ("neutral") in case of lamella-forming diblocks, or if they are slightly majority-block preferential in case of cylinder-forming diblocks. Thus, for the case of PS-*b*-PMMA diblocks, Russell and coworkers (Mansky et al., 1997a,b; Ham et al., 2008) and Nealey and coworkers (Ruiz et al., 2008; Han et al., 2009; Suh et al., 2010) demonstrated thin films with the desired perpendicular morphology by carefully preparing the "neutral" bottom surface using a PS-*r*-PMMA random copolymer brush with appropriate PS to PMMA weight ratio. Even under such conditions, achieving long-range order of lamellae or cylinders is nearly impossible, as the energy of dislocation or disclination defects is fairly low. In addition, for applications in semiconductor patterning, the position of the block copolymer domains must be highly controlled to enable device fabrication. Hence, there is a need for an external directing force to guide the self assembly, so-called "directed self-assembly" or DSA. Approaches to DSA can be generalized into two categories, chemoepitaxy and graphoepitaxy. In chemoepitaxy, selected areas on a relatively flat substrate are made preferential to one of the blocks; graphoepitaxy usually involves forming a topographic prepattern and then using a combination of confinement forces and surface preferentiality to cause the block copolymers to "multiply" or "rectify" it (Sanders et al., 2010; Cheng et al., 2010; Ginzburg et al., 2015; Millward et al., 2014; Segalman et al., 2001; Sundrani and Sibener, 2002; Jung and Ross, 2007). Either way, the interactions between polymers and substrate become spatially nonuniform, and predicting the film morphology as a function of polymer formulation becomes very challenging. Several examples of such predictions using modern field-based computer simulation tools are described below.

4.2 Overview of field-theory-based simulations of block copolymer DSA

Before discussing specific applications of SCFT to DSA, it is necessary to briefly review the concept of field-theoretic simulations (FTS)—SCFT, of course, being one of the simplest and most widely used FTS techniques. There are numerous references describing the development and role of field-theoretic approaches in polymer science—see, for example, Matsen and Schick (1994a), Schmid (2011), Shi (2004), Fredrickson et al. (2002), Fredrickson (2006, 2007), Bosse et al. (2007), and Hur et al. (2009). All such approaches describe polymers in a coarse-grained fashion, so that their "building blocks" are not atoms but "statistical segments," each segment having length b and volume v_0. The polymer chains are usually considered flexible and modeled using an Edwards Hamiltonian (Edwards, 1965). Various species (e.g., PS, PMMA, etc.) can be labeled by Latin letters A, B, C, and so forth. For an AB-diblock, the volume fraction of the A-block is usually denoted f, and the total number of statistical segments, N. The interaction between segments A and B is characterized by a Flory–Huggins (Flory, 1941; Huggins, 1941) parameter χ_{AB}. (Note that f and the product $\chi_{AB}N$ fully determine the equilibrium bulk morphology of a diblock, at least in the mean-field approximation.) The radius of gyration of a chain is given by, $R_g = b(N/6)^{1/2}$.

Below, we use the derivation of Matsen (1997) to arrive at the SCFT equations for an AB-diblock in a confined geometry; while other authors use a slightly different notation, the results are generally the same. Consider a melt of AB diblocks containing n chains. Each polymer chain is parameterized with a variable, s, that increases continuously along its length; at the A end of the chain, $s = 0$, at the junction, $s = f$, and at the B end of the chain, $s = 1$. Then, a function $\mathbf{r}_\alpha(s)$ describes the configuration of the αth chain in space. Next, one can define segment density operators,

$$\hat{\varphi}_A(\mathbf{r}) = v_0 N \sum_{\alpha=1}^{n} \int_0^f \mathrm{d}s\, \delta(\mathbf{r} - \mathbf{r}_\alpha(s)) \tag{4.1a}$$

$$\hat{\varphi}_B(\mathbf{r}) = v_0 N \sum_{\alpha=1}^{n} \int_f^1 \mathrm{d}s\, \delta(\mathbf{r} - \mathbf{r}_\alpha(s)) \tag{4.1b}$$

It is typically assumed in these models that a block copolymer melt is nearly incompressible, implying a constraint that can be written as,

$$\hat{\varphi}_A(\mathbf{r}) + \hat{\varphi}_B(\mathbf{r}) = \Phi_0(\mathbf{r}) \tag{4.2}$$

In Equation (4.2), $\Phi_0(\mathbf{r})$ is the so-called "cavity function"; $\Phi_0(\mathbf{r}) = 1$ inside the film, $\Phi_0(\mathbf{r}) = 0$ in air or resist or substrate, and $0 < \Phi_0(\mathbf{r}) < 1$ in narrow interfacial or surface regions. A convenient choice for $\Phi_0(\mathbf{r})$ can be written as,

$$\Phi_0(\mathbf{r}) = \frac{1}{2}\left[1 + \tanh\left(S\frac{d(\mathbf{r})}{\Delta}\right)\right]$$

where $S = 1$ outside the polymer region and -1 inside; $d(\mathbf{r})$ is the distance from point $\mathbf{r}$ to the nearest surface, and Δ is the width of the transition region (usually on the order of 0.2–0.3R_g). Through such cavity functions or "masks," it is possible to conduct simulations of complex-shaped polymer domains relevant for DSA studies in geometrically simpler simulation cells such as a cube or parallelepiped.

The partition function of the system can be written as,

$$\begin{aligned} Z = &\int \left(D\mathbf{r}_\alpha P[\mathbf{r}_\alpha;0,1]\right)\delta\left[\hat{\varphi}_A + \hat{\varphi}_B - \Phi_0\right] \\ &\times \exp\left\{-v_0^{-1}\int \mathrm{d}\mathbf{r}\left[\chi_{AB}\hat{\varphi}_A(\mathbf{r})\hat{\varphi}_B(\mathbf{r}) - h_A(\mathbf{r})\hat{\varphi}_A(\mathbf{r}) - h_B(\mathbf{r})\hat{\varphi}_B(\mathbf{r})\right]\right\} \end{aligned} \tag{4.3}$$

The functional integral is over all possible configurations of all chains; the probability distribution for flexible Gaussian chains is,

$$P[\mathbf{r}_\alpha; s1, s2] \propto \exp\left\{-\frac{3}{2Nb^2}\int_{s_1}^{s_2} \mathrm{d}s \left|\frac{\mathrm{d}}{\mathrm{d}s}\mathbf{r}_\alpha(s)\right|^2\right\} \tag{4.4}$$

The surface fields h_A and h_B are zero everywhere except the surface layers; they can be written as,

$$h_{A(B)}(\mathbf{r}) = \begin{bmatrix} N\chi_{A(B),P}, d(\mathbf{r}) = 0 \\ 0, d(\mathbf{r}) \neq 0 \end{bmatrix}$$

(here, label P designates the species corresponding to the surface; if there are several different surfaces, or if the surface is nonuniform, there will be more than one type of P).

The next step is to insert integrals $\int (D\Phi_A)\delta[\Phi_A - \hat{\varphi}_A]$ and $\int (D\Phi_B)\delta[\Phi_B - \hat{\varphi}_B]$, both of which are unity, into Equation (4.3), and replace the delta functionals with integral representations, leading to,

$$Z \propto \int (D\Phi_A DW_A D\Phi_B DW_B D\Xi)\exp\left\{-\frac{F(\Phi_A, W_A, \Phi_B, W_B, \Xi)}{k_B T}\right\} \quad (4.5)$$

The Hamiltonian F is described by,

$$\begin{aligned} \frac{F}{nk_B T} &= -\ln Q[W_A, W_B] \\ &+ \frac{1}{V}\int_V \mathrm{d}\mathbf{r}\left[\chi_{AB} N\Phi_A(\mathbf{r})\Phi_B(\mathbf{r}) - h_A(\mathbf{r})\Phi_A(\mathbf{r}) - h_B(\mathbf{r})\Phi_B(\mathbf{r})\right] \\ &+ \frac{1}{V}\int_V \mathrm{d}\mathbf{r}\left[-W_A\Phi_A(\mathbf{r}) - W_B\Phi_B(\mathbf{r}) - \Xi(\mathbf{r})(\Phi_0(\mathbf{r}) - \Phi_A(\mathbf{r}) - \Phi_B(\mathbf{r}))\right] \end{aligned} \quad (4.6)$$

Equation (4.6) describes free energy per chain as functional of the densities and conjugate fields. The additional field $\Xi(\mathbf{r})$ is the Lagrange multiplier inserted to ensure incompressibility. An alternative approach utilized by Fredrickson and coworkers (Bosse et al., 2007) uses "mixed" fields W_+ and W_- which allows a description without the use of a Lagrange multiplier—the two approaches are essentially equivalent. The single-chain partition function Q is given by,

$$Q[W_A, W_B] = \int D\mathbf{r}_\alpha P[\mathbf{r}_\alpha; 0,1]\exp\left\{-\int_0^f \mathrm{d}s\, W_A(\mathbf{r}_\alpha(s)) - \int_f^1 \mathrm{d}s\, W_B(\mathbf{r}_\alpha(s))\right\} \quad (4.7)$$

As usual in thermodynamics, the free energy of the system is expressed as $G = -k_B T \ln Z$, with Z given by Equation (4.5). However, computing the functional integral (4.5) is generally very difficult, in part because the W and Ξ fields are purely imaginary. One way of calculating such integral is through a simulation method known as the complex Langevin technique (Man et al., 2014; Fredrickson et al., 2002; Audus et al., 2013; Delaney and Fredrickson, 2013; Lennon et al., 2008a,b; Lee et al., 2008; Fredrickson, 2002, 2007; Duchs et al., 2003; Alexander-Katz et al., 2003a,b). A more common approach, however, is to approximate the integral by the extremum of the integrand, and thus the free energy would be given by $F(\phi_A, w_A, \phi_B, w_B, \xi)$, where ϕ_A, w_A, ϕ_B, w_B, and ξ are the functions for which F attains its minimum. This "mean-field" approximation is what is known as self-consistent field theory (SCFT), and today it is one of the most widely used numerical methods to study block copolymer morphology in DSA. Other methods include dissipative particle dynamics (Sato et al., 2013; Kodera et al., 2013; Morita, 2013, 2014; Nakano et al., 2013; Kim, 2014), dynamic density functional theory (Sevink et al., 2001), and Monte Carlo simulations (Yoshida et al., 2013; Liu et al., 2012a,b, 2013; Ramirez-Hernandez et al., 2010, 2012; Peters et al., 2012; Nagpal et al., 2011a,b, 2012; Ji et al., 2011; Detcheverry et al., 2008, 2009a,b, 2010a–c; Daoulas et al., 2008; Wang et al., 2000a–c, 2001, 2003). Detailed review of the other methods is beyond the scope of this chapter, but they are certainly widely used and, in some cases, provide useful insights on dynamics as well as equilibrium structure. In general, however, SCFT has the advantage of speed, the ability to easily compute the free energy (albeit approximate) of any field configuration, and the compatibility with "string methods" to compute free energy barriers for escaping defect states.

Going back to the SCFT methodology, the last step is to obtain self-consistency equations for densities and fields. They are given by,

$$w_{\mathrm{A}}(\mathbf{r}) = \chi_{\mathrm{AB}} N \varphi_{\mathrm{B}}(\mathbf{r}) - h_{\mathrm{A}}(\mathbf{r}) + \xi(\mathbf{r}) \tag{4.8a}$$

$$w_{\mathrm{B}}(\mathbf{r}) = \chi_{\mathrm{AB}} N \phi_{\mathrm{A}}(\mathbf{r}) - h_{\mathrm{B}}(\mathbf{r}) + \xi(\mathbf{r}) \tag{4.8b}$$

$$\varphi_{\mathrm{A}}(\mathbf{r}) + \varphi_{\mathrm{B}}(\mathbf{r}) = \Phi_0(\mathbf{r}) \tag{4.8c}$$

$$\varphi_{\mathrm{A}}(\mathbf{r}) = -\frac{V}{Q}\frac{\partial Q}{\partial W_A} = \frac{V}{Q}\int_0^f \mathrm{d}s\left[q^+(\mathbf{r},s)q(\mathbf{r},s)\right] \tag{4.8d}$$

$$\varphi_{\mathrm{B}}(\mathbf{r}) = -\frac{V}{Q}\frac{\partial Q}{\partial W_{\mathrm{B}}} = \frac{V}{Q}\int_f^1 \mathrm{d}s\left[q^+(\mathbf{r},s)q(\mathbf{r},s)\right] \tag{4.8e}$$

In Equations (4.8d) and (4.8e), we introduced forward and backward propagators, $q(\mathbf{r},s)$ and $q^+(\mathbf{r},s)$. They are computed using modified diffusion equations,

$$\frac{\partial q(\mathbf{r},s)}{\partial s} = \left[\nabla^2 - w_{t(s)}(\mathbf{r})\right]q(\mathbf{r},s) \tag{4.9a}$$

$$\frac{\partial q^+(\mathbf{r},s)}{\partial s} = -\left[\nabla^2 - w_{t(s)}(\mathbf{r})\right]q^+(\mathbf{r},s) \tag{4.9b}$$

The index $t(s)$ is equal to A if $0<s<f$, and B otherwise. The initial conditions are simply $q(\mathbf{r},s=0)=1$ and $q^+(\mathbf{r},s=1)=1$ indicating that either end of the diblock is free and could be anywhere with equal probability. All length scales have been set in units of the polymer radius of gyration, R_g. The partition function, Q, is expressed in terms of the forward propagator as,

$$Q = \int_V \mathrm{d}\mathbf{r}\left[q(\mathbf{r},1)\right] \tag{4.10}$$

Thus, given the confining geometry (function $\Phi_0(\mathbf{r})$) and information about the interactions between various surfaces with blocks A and B (functions $h_A(\mathbf{r})$ and $h_B(\mathbf{r})$), one can solve Equations (4.8a)–(4.10) to obtain morphology maps, or species density fields, determining the locations of A-rich and B-rich regions. It is important to recognize that solutions of the SCFT equations are local minima of the free energy functional F. Thus, the morphologies obtained can be either stable equilibrium states (the global minimum of F) or metastable states (which are often defect states). Different solutions can be accessed by starting SCFT simulations from a range of initial field configurations. In DSA practice, one of the blocks (e.g., the A-block) would be removable by etching, and another block (B) would constitute the matrix. Though etching processes would likely shift the interface boundaries compared to the pre-etched polymer morphology, equilibrium SCFT structures should still provide a good estimate for how the system would look after etch.

At this stage, we conclude the description of the method and turn to specific examples of its application. Note that solving Equations (4.8a)–(4.10) is nontrivial and is usually done in an iterative fashion using either a simple Picard algorithm or some more complex but sometimes faster semi-implicit approaches (Ceniceros and Fredrickson, 2004; Sides and Fredrickson, 2003; Rasmussen and Kalosakas, 2002). Depending on the hardware used, degree of parallelization, and complexity of the problem, calculations can take from several seconds to multiple days. The UCSB group has developed a flexible pseudo-spectral SCFT code base that can run on single GPUs, multiple cores on a CPU, or across multiple CPUs. The GPU implementation is particularly well suited to large 3D SCFT simulations of DSA processes (Delaney and Fredrickson, 2013).

4.3 Chemoepitaxy modeling

One approach for line-space patterning, termed chemoepitaxy, utilizes a sparse chemical pattern that aligns a block copolymer morphology (see Figure 4.1; Millward et al., 2014). An exemplary chemoepitaxy process for line-space formation involves a lamellar morphology of PS-*b*-PMMA guided by lithographically patterned thin stripes of PS mat spaced apart with a so-called "neutral" brush (Liu et al., 2011). The neutral brush is typically a PS-*r*-PMMA random copolymer. For optimal results, the PS stripes are spaced at an integer multiple of the pitch of the block copolymer, L_0.

In the past decade, there have been a number of papers and reviews (Liu et al., 2013; Chang et al., 2013; Edwards et al., 2007; Millward et al., 2014) discussing both advantages and shortcomings of chemoepitaxy. The chemical pattern parameters and those of the subsequent DSA are shown in Figure 4.2. The brush is usually near-neutral (PS-*r*-PMMA random copolymer whose optimal composition depends on several factors as will be discussed below). As discussed by multiple authors (Doerk et al., 2012; Stoykovich et al., 2010; Daoulas et al., 2008; Park et al., 2007; Edwards et al., 2004, 2007; Kim et al., 2006), the ability of the chemical pattern to form a defect-free array of vertical lamellae depends crucially on the width of the guiding stripe, w, the chemistry of the brush, film thickness, and the line multiplication factor, $n = L_x/L_0$. The "neutral" brush can be further optimized so that the f_{PS} is approximately the compositional area weighted average of the surface between the stripes, $f_{PS} \sim 1 - [n/(2n - 1)]$, while the lines are strongly PS-preferential. Because of the complexity of the system, it is very

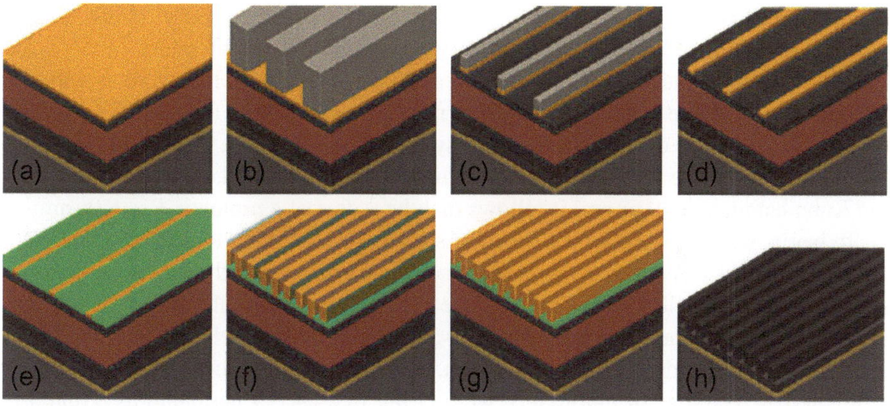

Figure 4.1 Schematics of line-space chemoepitaxy process flow. (a) PS pinning mat (yellow) grafted to wafer with SiON (dark gray)/carbon (maroon) stack, unbound polymer rinsed. (b) 193i litho patterning of resist (light gray) above pinning mat. (c) Resist trim and PS mat etch. (d) Residual resist strip. (e) Neutral underlayer (NUL) coated, grafted, and rinsed. (f) PS PMMA coated and annealed (PMMA domains translucent). (g) PMMA selective dry develop etch. (h) Pattern transfer etches through NUL, SiON, Carbon and into substrate with clean of mask materials.
Reprinted from Millward et al. (2014). Copyright (2014) Society of Photo Optical Instrumentation Engineers.

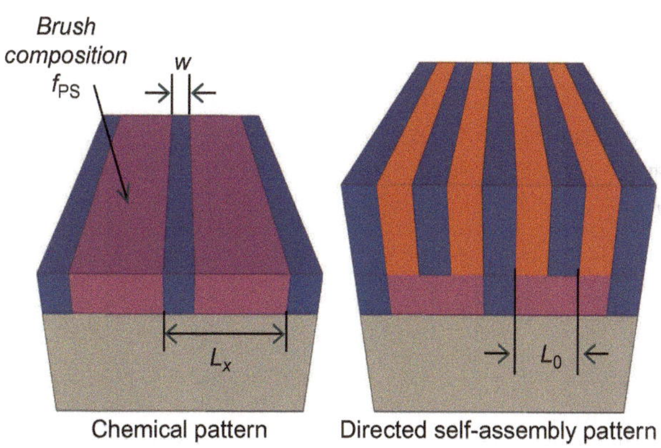

Figure 4.2 Parameters of chemoepitaxial DSA.
Reprinted from Ginzburg et al. (2013a). Copyright (2013) Society for Photopolymer Science and Technology.

difficult to map out a universal "phase diagram." Previous modeling efforts included strong segregation theory (Pereira and Williams, 1998), Monte Carlo (Liu et al., 2013), single-chain-in-mean-field (Daoulas et al., 2008), and coarse-grained molecular dynamics (Lawson et al., 2013). Earlier, SCFT had been extensively applied to the related problem of lamellar morphologies in nonpatterned thin films (Pickett and Balazs, 1997; Matsen, 1997; Geisinger et al., 1999a,b). Here, we present our initial efforts to study chemoepitaxy DSA using SCFT (Ginzburg et al., 2013).

We considered a PS–PMMA symmetric (f_A=0.5) monodisperse diblock with M_n=50 kg/mol. Based on the estimates of $\chi_{PS\text{–}PMMA}$ given in the literature (Zhao et al., 2008), we estimate the segregation strength as $\chi N \approx 19$. (Here and below, we use simply χ in place of χ_{AB}.) The equilibrium lamellar period is estimated as $L_0 \approx$ 26–27 nm (based on the reported value of L_0=25 nm for the symmetric PS–PMMA with M_n=44 kg/mol) (Liu et al., 2013). We set the reference volume, v_{ref}=105 cm^3/mol, and the corresponding Kuhn length is a=0.68 nm; it follows that the radius of gyration is R_g = 6.06 nm.

To begin, we simulate the bulk lamellar structure and attempt to determine the bulk lamellar period, L_0. This initial step is necessary to correctly select the mesh size in future simulations of thin films (the box size needs to be an exact multiple of the lamellar period L_0 in order to promote perpendicular alignment of the lamellae). In this analysis, we fix the mesh size $dx = 0.125R_g \approx 0.76$ nm, and perform one-dimensional simulations with varying box size. In Figure 4.3, we plot the free energy per chain as a function of normalized lamellar period $\lambda = L/L_0$, with L_0 chosen so that the minimum of the curve is located at $\lambda=1$. In this way, we obtain L_0=36 lattice units=$4.5R_g \approx 26.5$ nm. The SCFT free energy (blue line) is very close to the free energy based on the strong segregation theory (red line), as calculated by Semenov (1985):

$$g = g_0 + \kappa\left[\lambda^2 + 2\lambda^{-1} - 3\right] \tag{4.11}$$

where g_0 is the free energy at equilibrium, and κ is given by:

$$\kappa = \frac{1}{2}\left[\frac{\pi^2}{24}\chi N\right]^{1/3} \tag{4.12}$$

Here, χN=19, so $\kappa \approx 1.0$.

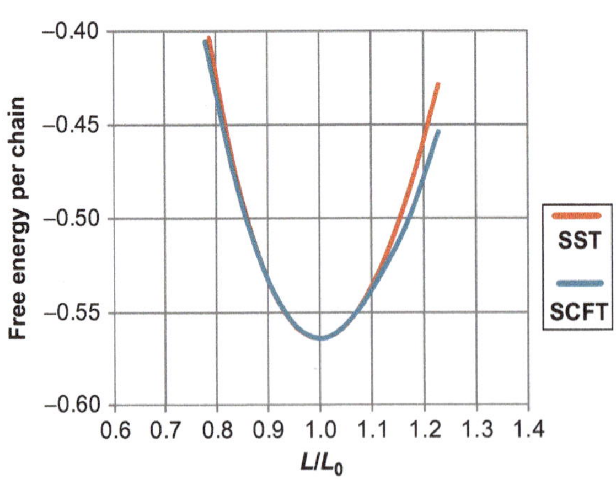

Figure 4.3 Free energy per chain as function of lamella stretching or compression. The blue line is SCFT, and the red line is the strong segregation theory estimate.
Reprinted from Ginzburg et al. (2013a). Copyright (2013) Society for Photopolymer Science and Technology.

We next consider chemoepitaxy in thin films with the height $H=0.55L_0\approx 16\,\text{nm}$. The top (air) surface is assumed to be neutral, that is, $\chi_{Aa}N-\chi_{Sa}N=0$. The brush outside the stripe is also neutral, $\chi_{AR}N-\chi_{SR}N=0$, and the stripe is PS-like, $\chi_{Ag}N=\chi_{AS}N=19.0$; $\chi_{Sg}N=0$. The box size in the X-dimension is varied, taking integer multiples of L_0: 36 lattice units (l.u.) ($1L_0$), 72 l.u. ($2L_0$), 108 l.u. ($3L_0$), 144 l.u. ($4L_0$), and 180 l.u. ($5L_0$); those correspond to line multiplication ratios, n, from 1 to 5.

We now consider the influence of the stripe width on the morphology. The simulations are performed as follows. For a selected multiplication, $n=L_x/L_0$, we vary the PS-stripe width in a wide range. The initial conditions in this simulation are uniform with small random noise. For stripe width $w\approx 0.5L_0$, the resulting morphology always turns out to be perpendicular lamellae, with PS-lamella in the center aligned with the PS-stripe in the center. We save the chemical potential fields and pressure configuration after achieving a chosen convergence criterion (average density deviation from unity, $\left|\varphi_A(\mathbf{r})+\varphi_B(\mathbf{r})-\Phi_0(\mathbf{r})\right|<5\times10^{-4}$), and denote such configurations the "positive" morphology. Separately, we perform a similar calculation but with the central guide stripe being PMMA-like; again, results are saved after reaching the same convergence criterion. This is denoted the "negative" morphology. Next, we run two separate SCFT simulations for each stripe width, w, using "negative" and "positive" morphologies as initial conditions. Those are also run until the same convergence criterion is reached. Afterward, we compare free energies of the two states and plot the difference between the two as function of w. All simulations were conducted in 2D (xz plane), assuming homogeneity in the y direction to reduce the computational effort.

The use of positive versus negative initial morphologies is a valuable tool for evaluating the effect of pitch multiplication. Starting with a positive morphology, one would expect very little change as the simulation proceeds. With a low level of multiplication and a stripe width similar to the domain width, a simulation starting with a negative morphology is expected to rearrange the domains such that a PS domain overlays the stripe. However, with higher degrees of multiplication, the directing force of the stripe acts on a smaller fraction of the total volume. Therefore, one can expect that negative morphologies at high multiplication will no longer undergo a large-scale shift, but instead, PS domains near the stripe will deform locally to cover at least part of the stripe.

The morphologies (PS-density map, shown in XZ-plane) are plotted in Figure 4.4 (top—"positive" morphology; bottom—"negative" morphology) for the case of 3× multiplication; similar patterns were also obtained for other values of n. The behavior seems to be fairly independent of the pattern pitch and primarily dictated by the width of the PS-stripe.

We can compute the free energy difference between the "positive" and "negative" morphologies as a function of the stripe width and the pattern pitch (or line multiplication). In Figure 4.5, the free energy difference (per chain), f (not to be confused with the block-A volume fraction labeled the same way earlier in the text), is plotted as a function of the stripe width; where $f<0$, the "positive" arrangement (with PS domain in the center) is preferential, while $f>0$ indicates preference for the "negative" arrangement (with PMMA domain in the center). In general, the positive solution is favored for all $w<L_0$, while the negative solution is likely favored for larger stripe width (we have not investigated that region in the current study). The guide width

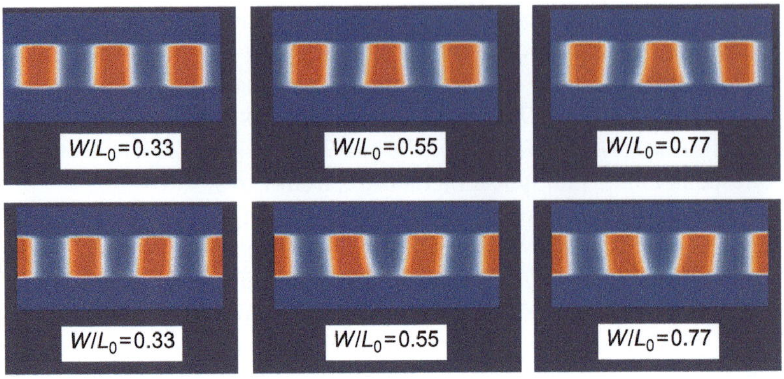

Figure 4.4 PS-block density maps for 3× multiplication. (Top) "positive" and (bottom) "negative" initial conditions.
Reprinted from Ginzburg et al. (2013b). Copyright (2013) Society for Photopolymer Science and Technology.

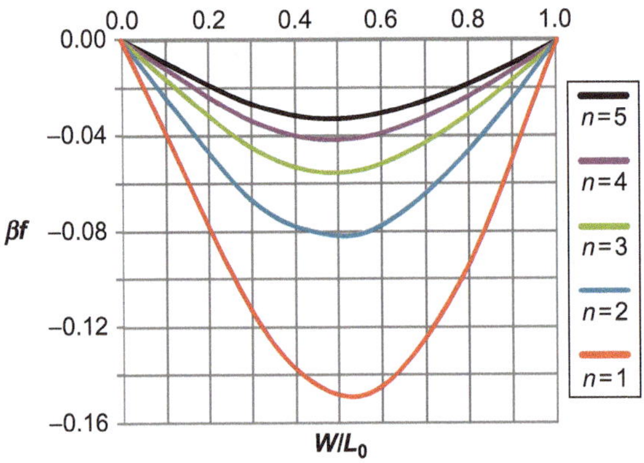

Figure 4.5 Free energy (per chain, in units of kT) difference between "positive" and "negative" morphologies as function of the guide width for various line multiplication factors.
Reprinted from Ginzburg et al. (2013a). Copyright (2013) Society for Photopolymer Science and Technology.

corresponding to the maximum stability of the "positive" morphology is located at $w_0 = (0.55 \pm 0.05)L_0$, in excellent agreement with the results of Nealey and coworkers (Liu et al., 2013). We also can see that as the pattern pitch L_x increases from 1× to 5× multiplication, the depth of the minimum at w_0 decreases as the inverse of the multiplication parameter n (Figure 4.6, diamonds). The curve in Figure 4.6 is a fit through the data using the following equation: $U = \frac{0.159}{n}$. Thus, if the 1:1 chemoepitaxy creates a free energy gain of ~0.159 kT/chain for the "positive" morphology compared to an alternative, the 3:1 pattern favors the "positive" morphology at about 0.055 kT/chain, and the 4:1 pattern only at 0.04 kT/chain. Thus, one expects that the defect probability would increase dramatically as the multiplication factor increases.

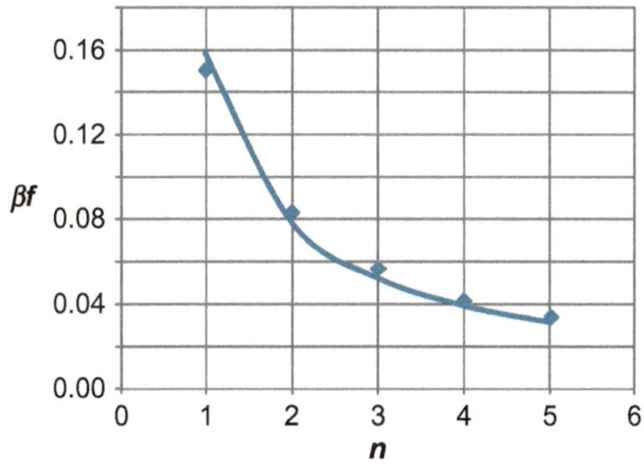

Figure 4.6 Free energy (per chain, in units of *kT*) difference between "positive" and "negative" morphologies as function of the guide width for various line multiplication factors. Reprinted from Ginzburg et al. (2013a). Copyright (2013) Society for Photopolymer Science and Technology.

We can estimate the upper bound on the "registration defect density" in the following way. Let us assume that the registration error is localized on a single line and that the adjacent lines are not affected (this assumption is not unreasonable if one looks at morphologies in Figure 4.4). In that case, one can compute the free energy cost of a registration defect *per unit length* of each line. To estimate the actual density of such defects, we still need to know the linear dimension of each defect. The simplest possible assumption would be to suggest that it should be on the order of L_0 so as to have enough room to incorporate two new interfaces. Thus, the total free energy of a defect would be estimated as the free energy difference per chain multiplied by the number of chains in a volume element HL_0L_0. It is assumed that the neighboring elements do not interact, so that the probability of any element to flip from "positive" to "negative" is determined by the Boltzmann statistics. Given these assumptions, we can estimate the defect density (in cm^{-2}), as shown in Figure 4.7. This analysis predicts an increase

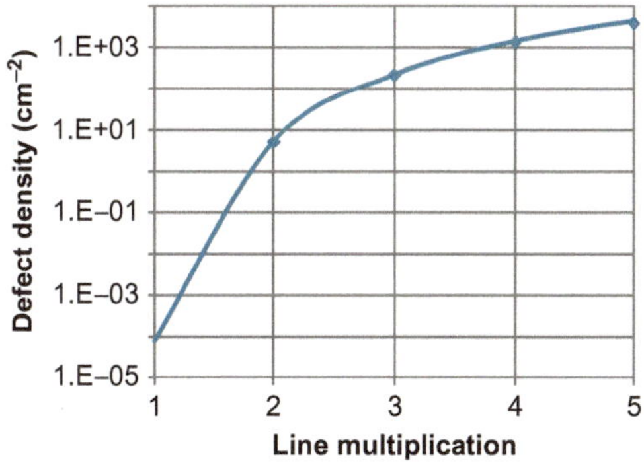

Figure 4.7 Estimated defect density (in cm^{-2}) as function of the line multiplication parameter *n*. See text for calculation details.
Reprinted from Ginzburg et al. (2013a). Copyright (2013) Society for Photopolymer Science and Technology.

in defect density as the line multiplication factor n increases from 1 to 5, from a low defect density of just 1×10^{-4} defects/cm^2 for 1× multiplication to a somewhat high defect density for 5× multiplication of about 5000 defects/cm^2. Actual experimental numbers for defect densities are likely lower as we did not include the energetic costs of grain boundaries or other interfaces in this estimate (and we note in passing that other studies (Nagpal et al., 2012) reported defect densities on the order of 10^{-120} cm^{-2} and even lower, so the issue is far from settled). The trend, however, is very real: going to 1× to 2× multiplication would increase the defect density by a factor approaching 10^5; from 2× to 3×, by another factor of 40; from 3× to 4×, by yet another factor of 5. It will become very interesting to compare these predictions to experimentally derived defect densities when experimental values become available from meticulous full-wafer studies. (We note that our calculated values are in reasonable agreement—within one order of magnitude—with preliminary measurements of Bencher et al. (2011) who estimated the density of DSA defects for 3× chemoepitaxy of 12 nm half-pitch PS–PMMA at 25 defects/cm^2 or less.)

The above calculations were performed in two dimensions, so the treatment of defects was only approximate and qualitative. More recently, Izumi et al. (2014) extended SCFT of similar lamellar PS-*b*-PMMA on striped surfaces in three dimensions and investigated the energies of specific defects (dislocations and disclinations). For the case of $\chi N=25$, multiplication factor $n=4$, film thickness $H=4R_g$, and stripe width $w_0=0.5L_0$, the free energy of a dislocation pair was estimated at ~50–100 k_BT, depending on the pattern pitch (P/L_0), as shown in Figure 4.8. The ITRS target for defect density, $n_d\sim0.01$ defects/cm^2 implies that the defect free energy should be greater than $30k_BT$ for a defect of core size ~30 nm. Therefore, SCFT simulations suggest that the density of dislocation pairs should be sufficiently small even if the pitch of the pattern becomes slightly incommensurate with that of the block copolymer, provided, again, that the stripe width is reasonably close to the optimum of $0.5L_0$. We note that the χN and H values in the Izumi study were larger than those used by Ginzburg and coworkers. Since the defect formation energy increases with both parameters (linearly

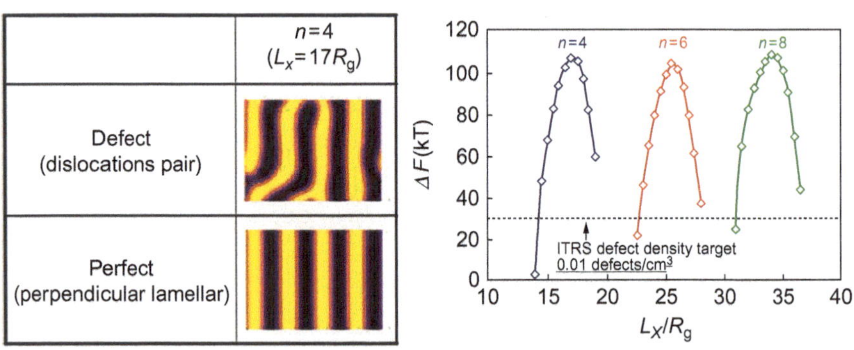

Figure 4.8 PMMA density map (top view) for the $n=4$, $L_x=17R_g$ (left) and the free energy of the dislocation pair defect as function of the pattern pitch (right).

with H for defects extending through the height of the film), there is apparently room for designing systems that can meet or exceed the ITRS defect target. Of course, there could be other possible defect types than elementary isolated dislocation pairs, so the estimate here should be viewed as a lower bound.

We next turn our attention to graphoepitaxy.

4.4 Graphoepitaxy modeling

In graphoepitaxy, block copolymers are used to "multiply" or "rectify" prepatterns in a photoresist or hard mask. As discussed by Cheng et al. (2010) some of the widely used examples include "contact hole shrink" and "line-space pattern in a rectangular trench," and we will primarily review those applications and corresponding SCFT modeling. Other interesting examples of graphoepitaxy include the use of posts to template PS-*b*-PDMS patterns (Cheng et al., 2001, 2002, 2003, 2004, 2006a–c; Chuang et al., 2006, 2008; Bita et al., 2008; Ross and Cheng, 2008; Ross et al., 2008; Mickiewicz et al., 2010; Hannon et al., 2013), but those would be outside the scope of this chapter.

4.4.1 Cylinders in a rectangular trench

The purpose of this application is similar to the previous case—namely, one wants to print a repeating line-space pattern; instead of perpendicular lamellae, however, one now uses parallel (horizontal) cylinders as shown in Figure 4.9 (Millward et al., 2014; Mishra et al., 2012; Sundrani and Sibener, 2002; Jung and Ross, 2007). Potential advantages of this method are as follows. First of all, there is no longer a need to carefully balance the surface affinity of the two blocks; therefore, one can use polymer pairs other than PS and PMMA to tailor the new patterns. A second advantage—closely related to the first—is that one can increase the incompatibility between the two blocks, χ, by combining blocks with more dissimilar polarities (e.g., PS and PDMS). Accordingly, it would allow one to reduce the total molecular weight while keeping the total segregation, χN, sufficiently large for quality microphase separation and reasonably low line-edge roughness. But a reduction in molecular weight also implies a smaller domain period. Thus, one can print patterns with smaller pitch (<20 nm), not accessible to lower-χ PS–PMMA systems. According to Mishra et al. (2012), cylinder-forming PS–P2VP block copolymers could produce monolayers with cylinder spacing close to 15 nm, and for higher-χ block copolymers, the limit can probably be pushed a little more (maybe down to 12–14 nm pitch). The goal, then, is to understand the influence of various factors (trench width, trench depth, film thickness, surface polarity, etc.) on the ability of the polymers to form ideal defect-free patterns of horizontal cylinders lying parallel to trench walls. Recently, Fredrickson and coworkers (Kim et al., 2013, 2014a,b) studied this system in depth, and we will review their approach and main conclusions below.

The system considered is a rectangular trench filled with a thin layer of AB diblock (A is the minority block, $f_A \sim 0.23$–0.30 so that the equilibrium morphology in the bulk is cylindrical hexagonal). The top (air) surface can be neutral (if the diblock is PS–PMMA) or A-preferential (if the diblock is PS–PDMS) or B-preferential

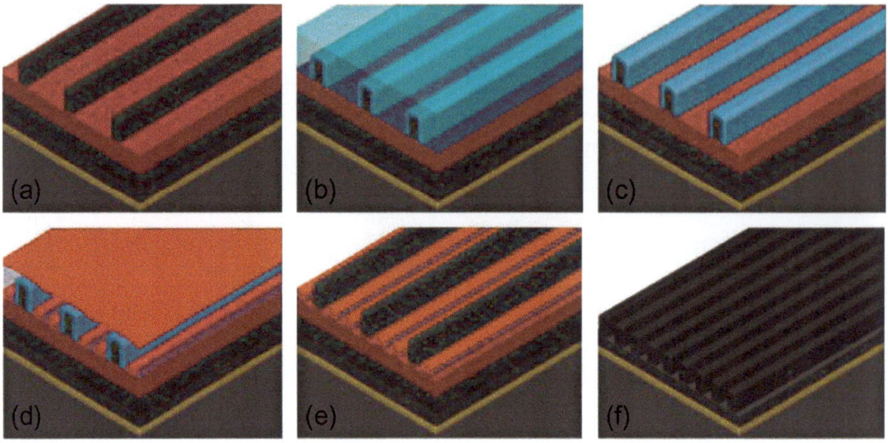

Figure 4.9 Schematics of cylinder graphoepitaxy process flow. (a) Guidelines for templates formed through 193i lithography and dry etch trim and pattern transfer processes; maroon colored film below guidelines is carbon. (b) Brush coated and grafted to guidelines; un-grafted material shown as translucent. (c) Un-grafted brush polymer rinsed away with PGMEA. (d) Si-BCP coated and annealed; organic domain shown as translucent, silicon domain is red. (e) Dry develop etch of Si-BCP and brush. (f) Pattern transfer through carbon and into substrate with clean of mask materials.
Reprinted from Millward et al. (2014). Copyright (2014) Society of Photo Optical Instrumentation Engineers.

(in case of PS–P2VP). The walls and the substrate could be neutral, A-preferential, or B-preferential, depending on the polymer, walls, and substrate chemistry and presence or absence of grafted brush layers.

The first question to be addressed is what would be the optimal film height and trench width to result in the most stable configuration of horizontal cylinders parallel to the walls? For the case of neutral top and bottom surfaces, and A-preferential walls (PS–PMMA diblock in a polar resist and with a neutral PS-*r*-PMMA brush at the bottom), one can use SCFT to compute the free energy as function of the trench width and film height. The calculations were performed for the case of $\chi N = 25, f = 0.25$. It was found that the optimal film height corresponded to ~$3.75R_g \approx 0.93L_0$; the optimal trench width depended on the desired number of cylinders. The lowest free energy states corresponded to cases where the trench width, w, was an integer multiple of the pitch, $L_0 = 4R_g$. Thus, the formation of a given number of cylindrical periods is limited to a well-defined window of commensurability. Inside each domain of width approximately $4R_g$, the excess free energy has a minimum at the center of the window whereby the elastic component of the excess free energy is zero while the residual free energy is due to surface tension. Away from the optimal width, with the application of either compressive or tensile strains, elastic stresses build up in the polymer and ultimately lead to the creation or removal of a cylindrical period. Changing the boundary conditions, of course, would modify the locations of the minima, possibly due to altered surface wetting layers, but the width of the commensurability window would still remain approximately the same.

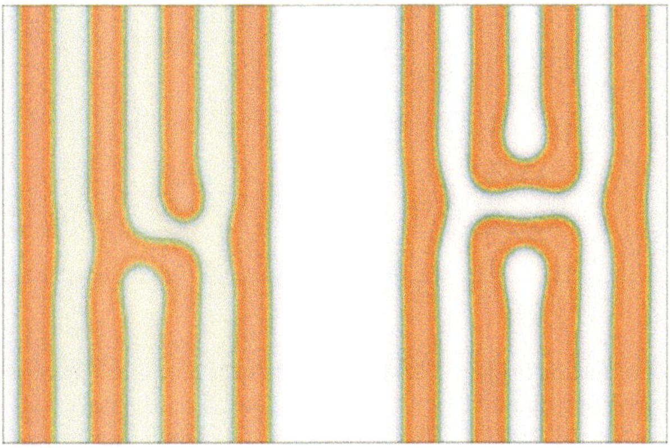

Figure 4.10 Density plot of the minor component in block copolymer of four periods of cylindrical DSA. (Left) dislocation (DL) and (right) disclination (DC).
Reprinted from Kim et al. (2014a). Copyright © (2014) Wiley Periodicals, Inc., A Wiley Company.

Once the boundary conditions, trench width, and film height are fixed, the next question is what is the probability of defects such as dislocations or disclinations (Figure 4.10)? To simulate that, one needs to prepare an initial field configuration corresponding to a defect in question, and then equilibrate that initial configuration with an SCFT run. The free energy of the resulting morphology ("dislocation" or "disclination") is then compared to that of the ideal one. Kim et al. (2014a) computed the free energies of several defects under different boundary conditions and as a function of the "elastic strain" ε (defined as $\left[\frac{w}{w_0}-1\right]$, where w_0 is the optimal trench width for the specified number of cylinders per trench under specific boundary conditions). Thus, for example, in the case of A-preferential walls and bottom, B-preferential top surface (mimicking PS-*b*-P2VP diblock with A=P2VP, B=PS), and assuming $\chi N=30$, the free energies of dislocations and disclinations were shown to range between 2 and $25k_BT$ (dislocations) and 5 and $40k_BT$ (disclinations) as a function of ε and f (Figure 4.11). As discussed before, however, to achieve desired low defectivity, one needs to have defect-free energies $>30k_BT$, so the defect density is likely to be high even if the trench width is perfectly commensurate with the block copolymer pitch ($\varepsilon=0$). The defect-free energy increases as a function of both f and χN, so one can probably improve the defectivity by moving to higher segregation strengths ($\chi N>35$), as shown in Figure 4.12. The simulations also showed that higher defect-free energies (better defectivity) were obtained with boundary conditions where the top surface was A-preferential, and the rest were B-preferential (something similar to PS-*b*-PDMS with A=PDMS, B=PS). Even so, the simulations suggested that the defectivity in line-space patterns formed by horizontal cylinders would almost always be larger than if the lines and spaces were created by vertical lamellae with film thickness comparable to L_0 (Takahashi et al., 2012).

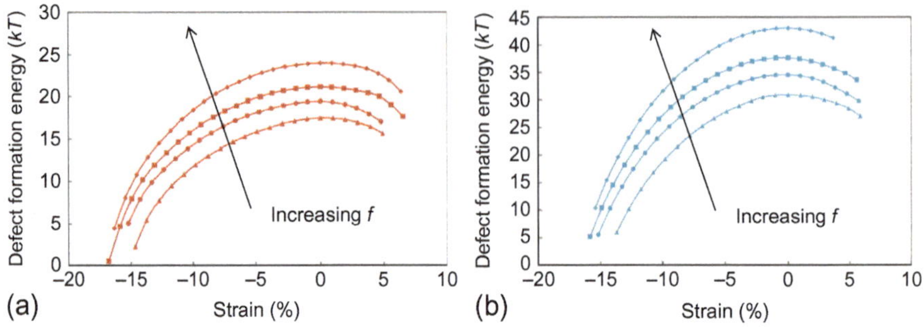

Figure 4.11 Plots of the dislocation (left) and the disclination (right) formation energies for confined cylinder-forming block copolymers with various minority block fractions as function of the dimensionless strain. Triangles, circles, squares and diamonds correspond to fractions, $f_A = 0.23$, 0.24, 0.25, and 0.27, respectively. Here, $\chi N = 30$, $L_z = 5.6R_g$, the sidewalls and bottom substrates are A-attractive ($\chi_{wA} = -32$) and the top surface is B-attractive ($\chi_{wB} = -32$). Reprinted from Kim et al. (2014a). Copyright © (2014) Wiley Periodicals, Inc., A Wiley Company.

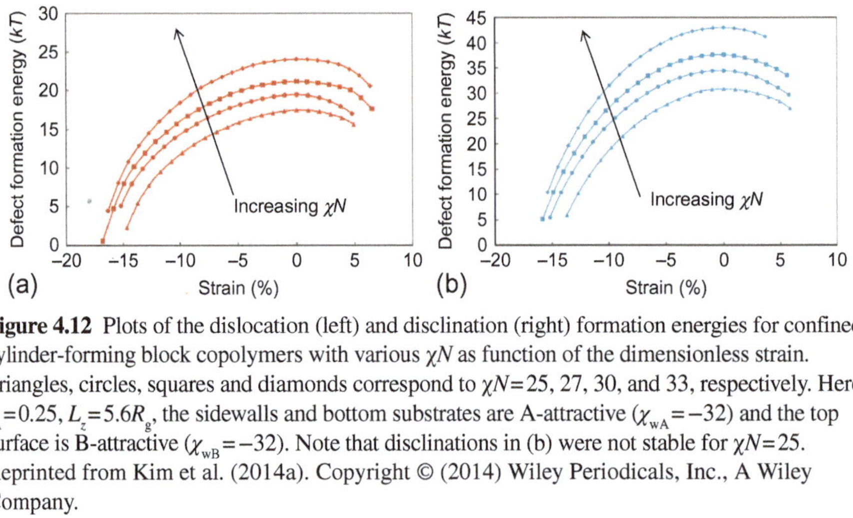

Figure 4.12 Plots of the dislocation (left) and disclination (right) formation energies for confined cylinder-forming block copolymers with various χN as function of the dimensionless strain. Triangles, circles, squares and diamonds correspond to $\chi N = 25$, 27, 30, and 33, respectively. Here, $f_A = 0.25$, $L_z = 5.6R_g$, the sidewalls and bottom substrates are A-attractive ($\chi_{wA} = -32$) and the top surface is B-attractive ($\chi_{wB} = -32$). Note that disclinations in (b) were not stable for $\chi N = 25$. Reprinted from Kim et al. (2014a). Copyright © (2014) Wiley Periodicals, Inc., A Wiley Company.

The above analysis assumed that the trench cross-section was perfectly rectangular. In reality, however, there could be two additional complications. First, there could be some intrinsic surface roughness due to the imperfections of the UV-lithography used to print the graphoepitaxial pattern. This issue has been addressed in a recent study by Kim et al. (2014b), although detailed discussion is outside the scope of this review. Second, the use of a brush to modify the surface could also cause a change in the optimal trench width and defect-free energy. This topic has not yet been fully understood, though initial studies are underway.

In summary, early SCFT results indicate that using DSA for line-space patterning and multiplication remains a challenge that requires very accurate formulation preparation and block copolymer synthesis. Much also depends on the surface preparation and the choice of brush or other surface modification method. Further studies will help to better understand ways to reduce defectivity and maintain/improve line-edge roughness and line-width roughness.

4.4.2 Contact hole shrink

The contact hole shrink (alternatively known as vertical interconnect access or VIA lithography, Figure 4.13) DSA has been discussed by various authors (Cheng et al., 2010; Sanders et al., 2010; Laachi et al., 2013, 2014). A VIA is the component of an integrated circuit that connects different conducting layers and allows the current to pass through the layers. A major goal of contact hole shrink is to produce high-resolution cylindrical holes with reduced critical dimensions (CD) relative to a cylindrical prepattern created with top-down lithography. The DSA of block copolymers under cylindrical confinement thus offers a route to pattern such features at the desired length scale. In a typical graphoepitaxial approach, a melt of asymmetric diblock copolymers segregates into microdomains whereby the minority block ideally forms a cylindrical domain in the center of the guiding hole. Following a selective etch process, the inner cylindrical domain is removed, leaving behind a reduced-size feature for further processing. In addition to hole shrinking, great interest lies in the use of DSA for rectification, that is, the ability to produce robust cylinders that are immune to variations and nonuniformities in the guiding hole dimensions. The use of

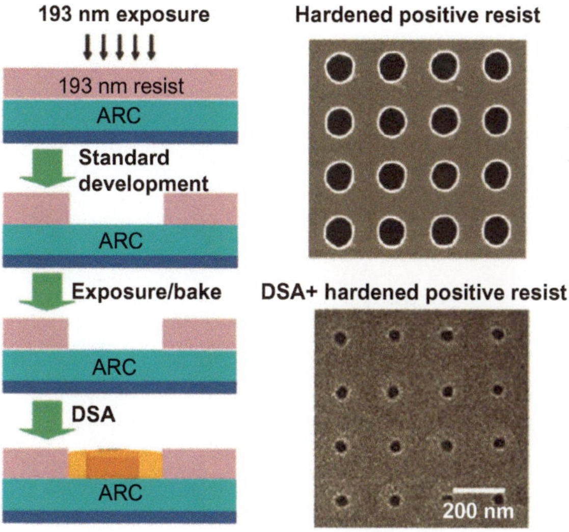

Figure 4.13 Contact hole shrink graphoepitaxy. (Left) DSA prepatterns made from a positive tone development scheme featuring a subsequent flood exposure and bake step to switch the polarity of the photoresist. (Right, top) 100 nm diameter VIA prepatterns in a photoresist after UV flood exposure and bake. (Right, bottom) resulting 45 nm VIAs formed by DSA of cylinder-forming PS-*b*-PMMA block copolymer in the prepattern.

computational modeling to predict the formation of the central cylinder is discussed in several recent publications (Laachi et al., 2013, 2014; Reilly et al., 2013; Muramatsu et al., 2014; Nakano et al., 2013)

Let us consider the case of a cylinder-forming PS-*b*-PMMA diblock, with PMMA being the minority, cylinder-forming block. The PMMA volume fraction was set at $f = 0.3$, and the segregation strength varied over the range, $\chi N = 15, 20, 25$. The diameter of the prepattern was varied from 50 to 70 nm and the depth was maintained at 100 nm. Note that chains with $\chi N = 15$, 20, and 25 have a radius of gyration of $R_g = 5.6$, 6.5, and 7.2 nm, respectively. Likewise, in units of R_g, the 50–70 nm range translates to 9–12R_g for $\chi N = 15$, 8–11R_g for $\chi N = 20$, and 7–10R_g for $\chi N = 25$. Two different types of surface chemistry were considered: PMMA-preferential and neutral. The morphology maps for both cases are shown in Figures 4.14 and 4.15. One can observe that for the case of PMMA-selective walls and substrate, there is a reasonably broad window where the central cylinder is formed and reaches almost all the way to the floor; however, other morphologies can also be observed as the ratio of prepattern diameter to the copolymer pitch is increased. For the case of neutral surfaces, the morphology map is more complex, with mixed cylinder and ring structures. Note that for the case of PMMA-preferential surfaces, the central cylinder does not fully reach to the floor; the morphology is often referred to as a "bowl" and, depending on the type of etch conditions used, could be undesirable in the application. The "bowl" can be eliminated if one uses neutral surfaces; however, the process window is then narrower and the free energy of defects such as a "broken cylinder" can be lower. Blending block copolymer with a PMMA homopolymer can serve to increase the process window in the PMMA-attractive case, although the defect-free energy may either increase or decrease compared to the pure diblock case. Finally, it was clearly demonstrated that the diameter of

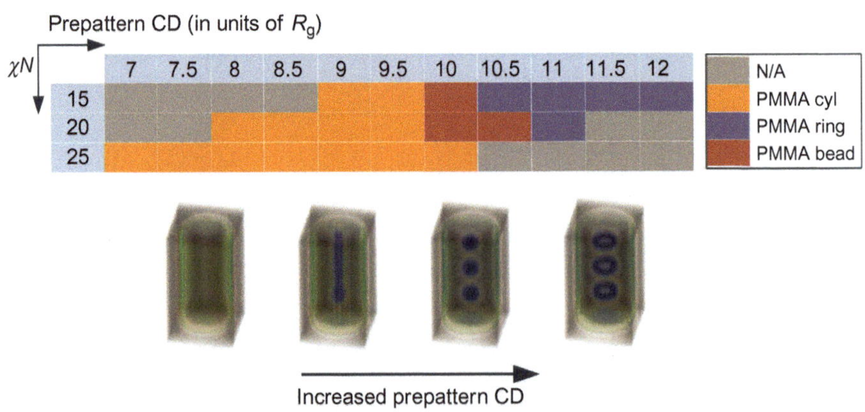

Figure 4.14 (Top) basic morphologies arising in PMMA-attractive cylindrical prepatterns of various CD and for various segregation strengths χN. (Bottom) three-dimensional illustrations of the PMMA density profiles, shown here in dark blue, corresponding to various morphologies with increasing prepattern CD.

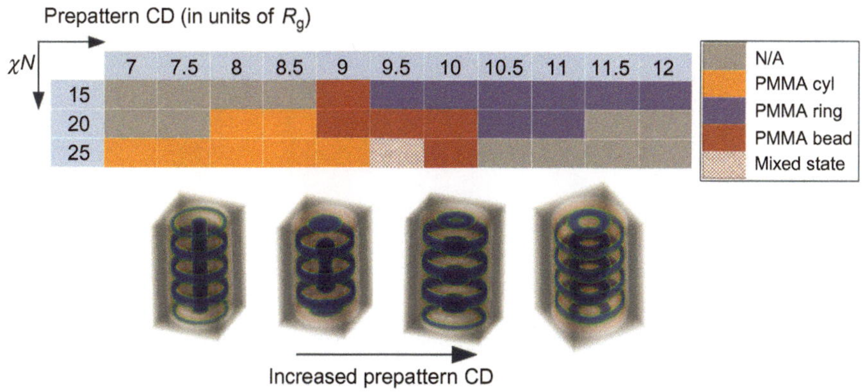

Figure 4.15 Same as Figure 4.14, but for neutral prepattern walls and floor. Reprinted from Laachi et al. (2014). Copyright (2014) Society of Photo Optical Instrumentation Engineers.

the inner cylinder was directly proportional to the diameter of the prepattern (Laachi et al., 2014), so there was no "rectification" (defined as obtaining the same inner hole diameter for different prepattern diameters).

The above analysis assumed that the prepattern was a perfectly cylindrical hole. However, in practice, the shape of the hole often deviates from perfectly cylindrical (barrel-shaped, slanted cylinders, etc.) In the case where the prepattern is written in the resist using UV lithography (probably not directly applicable to PS–PMMA DSA where a hard mask is used, but possible for other polymers), one can simulate the UV process first, predict the shape of the hole, and then import this shape into SCFT as boundary conditions (Reilly et al., 2013). The methodology is shown in Figure 4.16. The specific details of the lithography simulations are beyond the scope of this chapter, and we refer the readers to appropriate publications (Mack, 1987, 2005, 2006, 2008; Robertson et al., 2009; Flagello and Milster, 1997; Flagello et al., 1996), providing here only a brief summary. The full 3D photoresist profile shapes were simulated using the physics-based lithography simulator PROLITH X4.0 (KLA-Tencor). Immersion exposures with a numerical aperture of 1.35, Quasar illumination

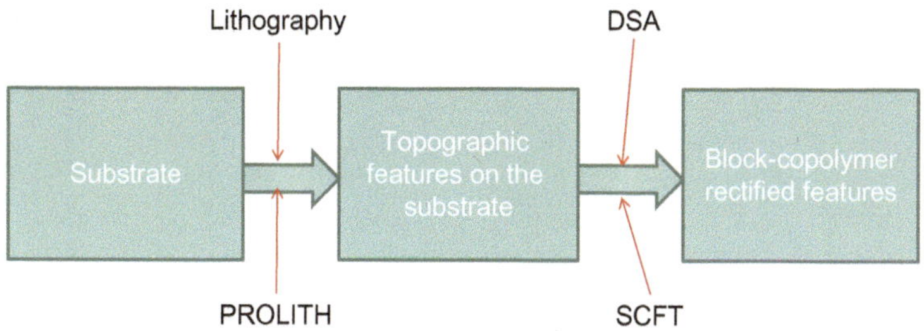

Figure 4.16 Schematic of the combined UV lithography and DSA simulation process. Reprinted from Reilly et al. (2013). Copyright (2013) Society of Photo Optical Instrumentation Engineers.

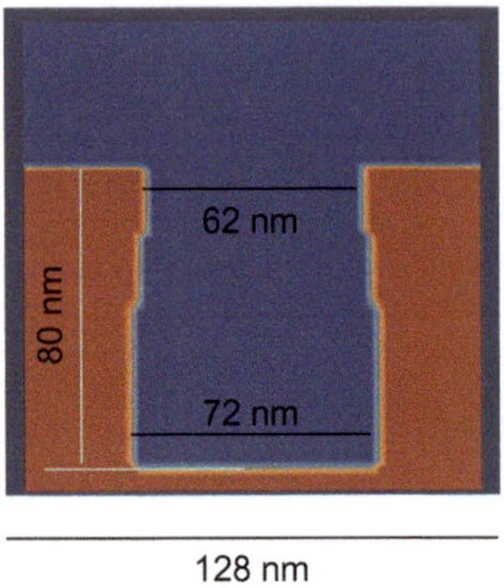

Figure 4.17 Typical shape of the prepattern as calculated using PROLITH software. Red is resist. For more details, see text.
Reprinted from Reilly et al. (2013). Copyright (2013) Society of Photo Optical Instrumentation Engineers.

with outer sigma of 0.7, inner sigma of 0.5, and a blade angle of 40° are assumed. The simulated film stack was 80 nm of Dow^R EPIC-2350^{TM} photoresist on a dual-layer anti-reflection coating. The finite element mesh size was set to 2 nm, total box size was 70×70×41 cells. The dose (exposure) was set at 14.204 mJ/cm^2, and the focus was shifted from the center of the film upward by 0.04 μm (40 nm). The cross-section of the prepattern hole is shown in Figure 4.17.

We now consider contact hole shrink with a PS-*b*-PMMA block copolymer having molecular weight 110 kg/mol and PMMA volume fraction, $f = 0.3$. We assume that at the bake temperature, $\chi = 0.037$, thus, $\chi N = 37.0$. Furthermore, we assume that the air surface is neutral, and the resist surface is slightly PMMA-preferential. The radius of gyration for the diblock is estimated at $R_g = 8.9$ nm, and thus the SCFT mesh $dx = 2\,\text{nm} = 0.22Rg$. In this study, we look at the effect of film thickness on the morphology of the shrink and rectification. The apparent diameter of the central cylinder depends on whether the polymer is level with the hole surface, overfilled, or underfilled (Figure 4.18).

Figure 4.18 Side-view of simulated CH shrink DSA, with thickness $T = 74$ nm (left), 80 nm (center), and 86 nm (right). Here, red is PS, white is PMMA, dark blue is resist, and light blue is air.
Reprinted from Reilly et al. (2013). Copyright (2013) Society of Photo Optical Instrumentation Engineers.

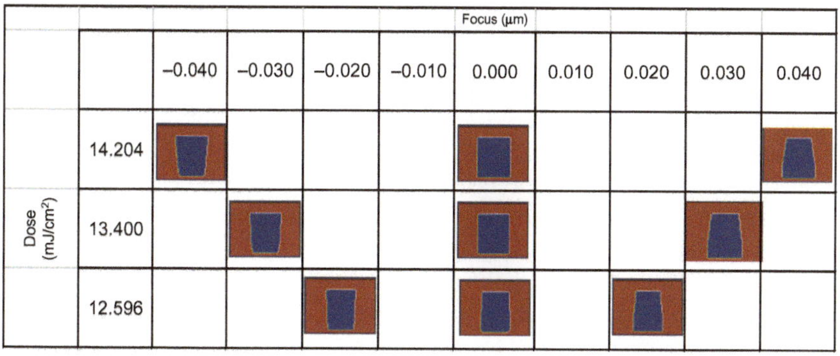

Figure 4.19 Dose-focus map from lithography simulations. Here, blue is the VIA prepattern to be rectified using DSA. Note that the shape of the prepattern is not cylindrical if the focus is shifted up or down from the center of the resist layer.

One can also use this approach to better understand the influence of lithography conditions (such as focus and exposure) on the final DSA. For example, in Figure 4.19, the morphology map for the lithographic prepattern is given as function of dose and focus; in Figure 4.20, we plot the morphologies after DSA, assuming neutral surface conditions and PS-*b*-PMMA diblock with $f = 0.3$, $M = 80$ kg/mol. The polymer is now assumed to be level with the top surface of the resist. Because of the commensurability between the polymer pitch and the prepattern CD, the morphology is favorable (a central cylinder penetrating all the way to the floor) for almost all of the dose-focus conditions. In some cases, it would even seem that the nonuniformity of the initial hole helps stabilize the central cylinder morphology. Overall, the parameter space is very large, and additional studies are ongoing.

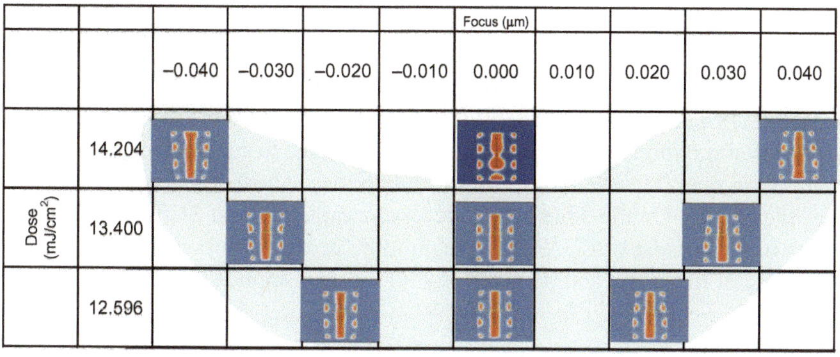

Figure 4.20 Same map after DSA. Now, red depicts PMMA regions. The wall and floor are neutral, and the central PMMA cylinder is formed in a wide process window and reaches all the way from top to bottom.

4.5 Summary and outlook

Today, field-theoretic simulations of DSA, and especially SCFT—and to some extent, dynamic density functional theory—are fast becoming the tool of choice for polymer formulation design and pattern optimization. In this review, we described several examples of how SCFT is used to better understand block copolymer morphologies in both chemoepitaxial and graphoepitaxial applications. Certainly, as DSA becomes more accepted in the advanced nanopatterning industry, SCFT will become part of the product development toolbox in the same fashion as finite element mechanics models became part of the product design in manufacturing a generation ago.

There are still many challenges remaining. Understanding the dynamics of annealing is not yet fully straightforward (though recent studies utilizing the minimum energy pathway approach (Takahashi et al., 2012) appear to make headway in this direction). (It is important to re-emphasize here that SCFT and similar approaches predict thermodynamically equilibrium morphologies and are not very good at describing kinetically trapped or otherwise nonequilibrium states.) Modeling defect free energies and densities depends crucially on our ability to guess which defects are the most dangerous; and this certainly requires both intuition and experimental data. The simulation methods become much more computationally challenging for block copolymers with large segregation strengths ($\chi N > 50$), since in such cases there is a wide separation between the interfacial width scale and the domain period scale. Increasing the pattern dimensions and complexity also make calculations more time-consuming; thus, there are considerable opportunities for development of faster, improved SCFT algorithms and more extensive and optimized parallelization in combination with advanced hardware (e.g., parallel algorithms across multiple GPUs).

In another important development, SCFT simulations can now be integrated in the "inverse design" process. In a typical "forward" or "direct modeling," one prepares (on his/her computer) both the litho prepattern and the block copolymer formulation and then runs SCFT to determine the equilibrium morphology. The "inverse design" approach starts from the desired morphology and attempts to find the prepattern/formulation package that comes close to that objective. There have been several examples of the application of inverse design with SCFT in the DSA context (Hannon et al., 2013; Latypov, 2013), but the methods remain limited in capability at this stage. For detailed discussion of this exciting topic, see Chapter 9.

Overall, DSA technology and field-theoretic simulations seem to be a perfect match. The use of simulations allows more rapid turnaround, faster product development, and ability to quickly "size up" various opportunities (whether a particular approach is feasible or not). On the other hand, successful application of SCFT and other field theories in the context of DSA makes them more accepted in both the chemical and semiconductor industries and prepares ground for their application elsewhere.

References

Albert, J.N.L., Epps, T.H., 2010. Self-assembly of block copolymer thin films. Mater. Today 13, 24–33.

Alexander-Katz, A., Moreira, A.G., Fredrickson, G.H., 2003a. Field-theoretic simulations of confined polymer solutions. J. Chem. Phys. 118, 9030–9036.
Alexander-Katz, A., Moreira, A.G., Fredrickson, G.H., 2003b. Simulations of polymer solutions: a field-theoretic approach. Mesoscale Phenom. Fluid Syst. 861, 279–289.
Audus, D.J., Delaney, K.T., Ceniceros, H.D., Fredrickson, G.H., 2013. Comparison of pseudospectral algorithms for field-theoretic simulations of polymers. Macromolecules 46, 8383–8391.
Bates, F.S., Fredrickson, G.H., 1999. Block copolymers—designer soft materials. Phys. Today 52, 32–38.
Bates, F.S., Hillmyer, M.A., Lodge, T.P., Bates, C.M., Delaney, K.T., Fredrickson, G.H., 2012. Multiblock polymers: panacea or Pandora's box? Science 336, 434–440.
Bencher, C., Smith, J., Miao, L., Cai, C., Chen, Y., Cheng, J.Y., Sanders, D.P., Tjio, M., Truong, H.D., Holmes, S., Hinsberg, W.D., 2011. Self-assembly patterning for sub-15 nm half-pitch: a transition from lab to fab. In: SPIE Advanced Lithography, 2011. International Society for Optics and Photonics, pp. 79700F–79700F-9.
Bita, I., Yang, J.K.W., Jung, Y.S., Ross, C.A., Thomas, E.L., Berggren, K.K., 2008. Graphoepitaxy of self-assembled block copolymers on two-dimensional periodic patterned templates. Science 321, 939–943.
Bosse, A.W., Garcia-Cervera, C.J., Fredrickson, G.H., 2007. Microdomain ordering in laterally confined block copolymer thin films. Macromolecules 40, 9570–9581.
Ceniceros, H.D., Fredrickson, G.H., 2004. Numerical solution of polymer self-consistent field theory. Multiscale Model. Simul. 2, 452–474.
Chang, S.-W., Ginzburg, V.V., Kramer, J.W., Lee, C., Li, M., Murray, D.J., Park, J., Roy, R., Sharma, R., Trefonas III, P., 2013. Opportunities and challenges for directed self-assembly for advanced patterning. J. Photopolym. Sci. Technol. 26, 31–37.
Cheng, J.Y., Ross, C.A., Chan, V.Z.H., Thomas, E.L., Lammertink, R.G.H., Vancso, G.J., 2001. Formation of a cobalt magnetic dot array via block copolymer lithography. Adv. Mater. 13, 1174–1178.
Cheng, J.Y., Ross, C.A., Thomas, E.L., Smith, H.I., Vancso, G.J., 2002. Fabrication of nanostructures with long-range order using block copolymer lithography. Appl. Phys. Lett. 81, 3657–3659.
Cheng, J.Y., Ross, C.A., Thomas, E.L., Smith, H.I., Vancso, G.J., 2003. Templated self-assembly of block copolymers: effect of substrate topography. Adv. Mater. 15, 1599–1602.
Cheng, J.Y., Mayes, A.M., Ross, C.A., 2004. Nanostructure engineering by templated self-assembly of block copolymers. Nat. Mater. 3, 823–828.
Cheng, J.Y., Ross, C.A., Smith, H.I., Thomas, E.L., 2006a. Templated self-assembly of block copolymers: top-down helps bottom-up. Adv. Mater. 18, 2505–2521.
Cheng, J.Y., Zhang, F., Chuang, V.P., Mayes, A.M., Ross, C.A., 2006b. Self-assembled one-dimensional nanostructure arrays. Nano Lett. 6, 2099–2103.
Cheng, J.Y., Zhang, F., Smith, H.I., Vancso, G.J., Ross, C.A., 2006c. Pattern registration between spherical block-copolymer domains and topographical templates. Adv. Mater. 18, 597–601.
Cheng, J.Y., Sanders, D.P., Truong, H.D., Harrer, S., Friz, A., Holmes, S., Colburn, M., Hinsberg, W.D., 2010. Simple and versatile methods to integrate directed self-assembly with optical lithography using a polarity-switched photoresist. ACS Nano 4, 4815–4823.
Chuang, V.P., Cheng, J.Y., Savas, T.A., Ross, C.A., 2006. Three-dimensional self-assembly of spherical block copolymer domains into V-shaped grooves. Nano Lett. 6, 2332–2337.
Chuang, V.P., Ross, C.A., Bilalis, P., Hadjichristidis, N., 2008. Nanoscale rings fabricated using self-assembled triblock terpolymer templates. ACS Nano 2, 2007–2014.
Daoulas, K.C., Muller, M., Stoykovich, M.P., Kang, H., De Pablo, J.J., Nealey, P.F., 2008. Directed copolymer assembly on chemical substrate patterns: a phenomenological and

single-chain-in-mean-field simulations study of the influence of roughness in the substrate pattern. Langmuir 24, 1284–1295.

Delaney, K.T., Fredrickson, G.H., 2013. Polymer field-theory simulations on graphics processing units. Comput. Phys. Commun. 184, 2102–2110.

Detcheverry, F.A., Kang, H.M., Daoulas, K.C., Muller, M., Nealey, P.F., De Pablo, J.J., 2008. Monte Carlo simulations of a coarse grain model for block copolymers and nanocomposites. Macromolecules 41, 4989–5001.

Detcheverry, F.A., Pike, D.Q., Nagpal, U., Nealey, P.F., De Pablo, J.J., 2009a. Theoretically informed coarse grain simulations of block copolymer melts: method and applications. Soft Matter 5, 4858–4865.

Detcheverry, F.A., Pike, D.Q., Nealey, P.F., Muller, M., De Pablo, J.J., 2009b. Monte Carlo simulation of coarse grain polymeric systems. Phys. Rev. Lett. 102, 197801.

Detcheverry, F.A., Liu, G.L., Nealey, P.F., De Pablo, J.J., 2010a. Interpolation in the directed assembly of block copolymers on nanopatterned substrates: simulation and experiments. Macromolecules 43, 3446–3454.

Detcheverry, F.A., Nealey, P.F., De Pablo, J.J., 2010b. Directed assembly of a cylinder-forming diblock copolymer: topographic and chemical patterns. Macromolecules 43, 6495–6504.

Detcheverry, F.A., Pike, D.Q., Nealey, P.F., Muller, M., De Pablo, J.J., 2010c. Simulations of theoretically informed coarse grain models of polymeric systems. Faraday Discuss. 144, 111–125.

Doerk, G.S., Liu, C.-C., Cheng, J.Y., Rettner, C.T., Pitera, J.W., Krupp, L.E., Topuria, T., Arellano, N., Sanders, D.P., 2012. Pattern placement accuracy in block copolymer directed self-assembly based on chemical epitaxy. ACS Nano 7, 276–285.

Drolet, F., Fredrickson, G.H., 1999. Combinatorial screening of complex block copolymer assembly with self-consistent field theory. Phys. Rev. Lett. 83, 4317–4320.

Duchs, D., Ganesan, V., Fredrickson, G.H., Schmid, F., 2003. Fluctuation effects in ternary AB + A + B polymeric emulsions. Macromolecules 36, 9237–9248.

Edwards, S.F., 1965. The statistical mechanics of polymers with excluded volume. Proc. Phys. Soc. 85, 613.

Edwards, E.W., Montague, M.F., Solak, H.H., Hawker, C.J., Nealey, P.F., 2004. Precise control over molecular dimensions of block-copolymer domains using the interfacial energy of chemically nanopatterned substrates. Adv. Mater. 16, 1315–1319.

Edwards, E.W., Muller, M., Stoykovich, M.P., Solak, H.H., De Pablo, J.J., Nealey, P.F., 2007. Dimensions and shapes of block copolymer domains assembled on lithographically defined chemically patterned substrates. Macromolecules 40, 90–96.

Flagello, D.G., Milster, T., 1997. High-numerical-aperture effects in photoresist. Appl. Opt. 36, 8944–8951.

Flagello, D.G., Milster, T., Rosenbluth, A.E., 1996. Theory of high-NA imaging in homogeneous thin films. J. Opt. Soc. Am. A 13, 53–64.

Flory, P.J., 1941. Thermodynamics of high polymer solutions. J. Chem. Phys. 9, 660–661.

Fredrickson, G.H., 2002. Dynamics and rheology of inhomogeneous polymeric fluids: a complex Langevin approach. J. Chem. Phys. 117, 6810–6820.

Fredrickson, G., 2006. The Equilibrium Theory of Inhomogeneous Polymers. Oxford University Press, Oxford.

Fredrickson, G.H., 2007. Computational field theory of polymers: opportunities and challenges. Soft Matter 3, 1329–1334.

Fredrickson, G.H., Ganesan, V., Drolet, F., 2002. Field-theoretic computer simulation methods for polymers and complex fluids. Macromolecules 35, 16–39.

Geisinger, T., Muller, M., Binder, K., 1999a. Symmetric diblock copolymers in thin films. I. Phase stability in self-consistent field calculations and Monte Carlo simulations. J. Chem. Phys. 111, 5241–5250.

Geisinger, T., Muller, M., Binder, K., 1999b. Symmetric diblock copolymers in thin films. II. Comparison of profiles between self-consistent field calculations and Monte Carlo simulations. J. Chem. Phys. 111, 5251–5258.

Ginzburg, V.V., Weinhold, J.D., Hustad, P.D., Trefonas III, P., 2013. Modeling chemoepitaxy of block copolymer thin films using self-consistent field theory. J. Photopolym. Sci. Technol. 26, 817–823.

Ginzburg, V.V., Weinhold, J.D., Trefonas, P., 2015. Computational modeling of block-copolymer directed self-assembly. J. Polym. Sci. B 53, 90–95.

Ham, S., Shin, C., Kim, E., Ryu, D.Y., Jeong, U., Russell, T.P., Hawker, C.J., 2008. Microdomain orientation of PS-*b*-PMMA by controlled interfacial interactions. Macromolecules 41, 6431–6437.

Han, E., Stuen, K.O., Leolukman, M., Liu, C., Nealey, P.F., Gopalan, P., 2009. Perpendicular orientation of domains in cylinder-forming block copolymer thick films by controlled interfacial interactions. Macromolecules 42, 4896–4901.

Hannon, A.F., Gotrik, K.W., Ross, C.A., Alexander-Katz, A., 2013. Inverse design of topographical templates for directed self-assembly of block copolymers. ACS Macro Lett. 2, 251–255.

Huggins, M.L., 1941. Solutions of long chain compounds. J. Chem. Phys. 9, 440.

Hur, S.M., Garcia-Cervera, C.J., Kramer, E.J., Fredrickson, G.H., 2009. SCFT simulations of thin film blends of block copolymer and homopolymer laterally confined in a square well. Macromolecules 42, 5861–5872.

Izumi, K., Laachi, N., Man, X., Delaney, K.T., Fredrickson, G.H., 2014. Self-consistent field theory of directed self-assembly on chemically-prepatterned surfaces. In: SPIE Advanced Lithography, 2014. International Society for Optics and Photonics, pp. 904922–904922-7.

Ji, S.X., Nagpal, U., Liao, W., Liu, C.C., De Pablo, J.J., Nealey, P.F., 2011. Three-dimensional directed assembly of block copolymers together with two-dimensional square and rectangular nanolithography. Adv. Mater. 23, 3692–3697.

Jung, Y.S., Ross, C.A., 2007. Orientation-controlled self-assembled nanolithography using a polystyrene-polydimethylsiloxane block copolymer. Nano Lett. 7, 2046–2050.

Kikuchi, M., Binder, K., 1994. Microphase separation in thin-films of the symmetrical diblock-copolymer melt. J. Chem. Phys. 101, 3367–3377.

Kim, S.-K., 2014. Stochastic simulation studies of line-edge roughness in block copolymer lithography. J. Nanosci. Nanotechnol. 14, 6143–6145.

Kim, S.O., Kim, B.H., Kim, K., Koo, C.M., Stoykovich, M.P., Nealey, P.F., Solak, H.H., 2006. Defect structure in thin films of a lamellar block copolymer self-assembled on neutral homogeneous and chemically nanopatterned surfaces. Macromolecules 39, 5466–5470.

Kim, B., Laachi, N., Fredrickson, G.H., 2013. Defectivity study of directed self-assembly of cylindrical diblock copolymers in laterally confined thin channels. In: SPIE Advanced Lithography, 2013. International Society for Optics and Photonics, pp. 868016–868016-8.

Kim, B., Laachi, N., Delaney, K.T., Carilli, M., Kramer, E.J., Fredrickson, G.H., 2014a. Thermodynamic and kinetic aspects of defectivity in directed self-assembly of cylinder-forming diblock copolymers in laterally confining thin channels. J. Appl. Polym. Sci. 131, 40790.

Kim, B., Laachi, N., Delaney, K.T., Fredrickson, G.H., 2014b. Directed self-assembly of diblock copolymers in laterally confining channels: effect of rough surfaces on line-edge roughness (LER) and defectivity. In: SPIE Advanced Lithography, 2014. International Society for Optics and Photonics, pp. 90491D–90491D-9.

Knoll, A., Horvat, A., Lyakhova, K.S., Krausch, G., Sevink, G.J.A., Zvelindovsky, A.V., Magerle, R., 2002. Phase behavior in thin films of cylinder-forming block copolymers. Phys. Rev. Lett. 89, 035501.

Kodera, K., Maeda, S., Tanaka, S., Mimotogi, S., Seino, Y., Yonemitsu, H., Sato, H., Azuma, T., 2013. Novel error mode analysis method for graphoepitaxial directed self-assembly lithography based on the dissipative particle dynamics method. In: SPIE Advanced Lithography, 2013. International Society for Optics and Photonics, pp. 868015–868015-7.

Laachi, N., Delaney, K.T., Kim, B., Hur, S.-M., Bristol, R., Shykind, D., Weinheimer, C.J., Fredrickson, G.H., 2013. The hole shrink problem: theoretical studies of directed self-assembly in cylindrical confinement. In: SPIE Advanced Lithography, 2013. International Society for Optics and Photonics, pp. 868014–868014-9.

Laachi, N., Delaney, K.T., Kim, B., Hur, S.-M., Bristol, R., Shykind, D., Weinheimer, C.J., Fredrickson, G.H., 2014. Self-consistent field theory investigation of directed self-assembly in cylindrical confinement. J. Polym. Sci. B 53, 142–153.

Latypov, A., 2013. Computational solution of inverse directed self-assembly problem. In: SPIE Advanced Lithography, 2013. International Society for Optics and Photonics, pp. 86800Z–86800Z-10.

Lawson, R.A., Peters, A.J., Ludovice, P.J., Henderson, C.L., 2013. Coarse grained molecular dynamics model of block copolymer directed self-assembly. In: SPIE Advanced Lithography, 2013. International Society for Optics and Photonics, pp. 86801Y–86801Y-11.

Lee, J., Popov, Y.O., Fredrickson, G.H., 2008. Complex coacervation: a field theoretic simulation study of polyelectrolyte complexation. J. Chem. Phys. 128, 224908.

Leibler, L., 1980. Theory of microphase separation in block co-polymers. Macromolecules 13, 1602–1617.

Lennon, E.M., Katsov, K., Fredrickson, G.H., 2008a. Free energy evaluation in field-theoretic polymer simulations. Phys. Rev. Lett. 101, 138302.

Lennon, E.M., Mohler, G.O., Ceniceros, H.D., Garcia-Cervera, C.J., Fredrickson, G.H., 2008b. Numerical solutions of the complex Langevin equations in polymer field theory. Multiscale Model. Simul. 6, 1347–1370.

Liu, C.C., Han, E., Onses, M.S., Thode, C.J., Ji, S.X., Gopalan, P., Nealey, P.F., 2011. Fabrication of lithographically defined chemically patterned polymer brushes and mats. Macromolecules 44, 1876–1885.

Liu, G.L., Detcheverry, F., Ramirez-Hernandez, A., Yoshida, H., Tada, Y., De Pablo, J.J., Nealey, P.F., 2012a. Nonbulk complex structures in thin films of symmetric block copolymers on chemically nanopatterned surfaces. Macromolecules 45, 3986–3992.

Liu, G.L., Ramirez-Hernandez, A., Yoshida, H., Nygard, K., Satapathy, D.K., Bunk, O., De Pablo, J.J., Nealey, P.F., 2012b. Morphology of lamellae-forming block copolymer films between two orthogonal chemically nanopatterned striped surfaces. Phys. Rev. Lett. 108, 065502.

Liu, C.C., Ramirez-Hernandez, A., Han, E., Craig, G.S.W., Tada, Y., Yoshida, H., Kang, H.M., Ji, S.X., Gopalan, P., De Pablo, J.J., Nealey, P.F., 2013. Chemical patterns for directed self-assembly of lamellae-forming block copolymers with density multiplication of features. Macromolecules 46, 1415–1424.

Luo, M., Epps, T.H., 2013. Directed block copolymer thin film self-assembly: emerging trends in nanopattern fabrication. Macromolecules 46, 7567–7579.

Lyakhova, K.S., Sevink, G.J.A., Zvelindovsky, A.V., Horvat, A., Magerle, R., 2004. Role of dissimilar interfaces in thin films of cylinder-forming block copolymers. J. Chem. Phys. 120, 1127–1137.

Mack, C., 1987. Development of positive photoresist. J. Electrochem. Soc. 139, 148–152.

Mack, C.A., 2005. Thirty years of lithography simulation. In: Microlithography 2005. International Society for Optics and Photonics, pp. 1–12.

Mack, C.A., 2006. Field Guide to Optical Lithography. SPIE Press Bellingham, Washington, USA.

Mack, C., 2008. Fundamental Principles of Optical Lithography: The Science of Microfabrication. John Wiley and Sons, West Sussex, England.

Man, X.K., Delaney, K.T., Villet, M.C., Orland, H., Fredrickson, G.H., 2014. Coherent states formulation of polymer field theory. J. Chem. Phys. 140, 024905.

Mansky, P., Liu, Y., Huang, E., Russell, T.P., Hawker, C.J., 1997a. Controlling polymer-surface interactions with random copolymer brushes. Science 275, 1458–1460.

Mansky, P., Russell, T.P., Hawker, C.J., Pitsikalis, M., Mays, J., 1997b. Ordered diblock copolymer films on random copolymer brushes. Macromolecules 30, 6810–6813.

Matsen, M.W., 1997. Thin films of block copolymer. J. Chem. Phys. 106, 7781–7791.

Matsen, M.W., Bates, F.S., 1996a. Origins of complex self-assembly in block copolymers. Macromolecules 29, 7641–7644.

Matsen, M.W., Bates, F.S., 1996b. Unifying weak- and strong-segregation block copolymer theories. Macromolecules 29, 1091–1098.

Matsen, M.W., Schick, M., 1994a. Stable and unstable phases of a diblock copolymer. Melt Phys. Rev. Lett. 72, 2660–2663.

Matsen, M.W., Schick, M., 1994b. Stable and unstable phases of a linear multiblock copolymer melt. Macromolecules 27, 7157–7163.

Mickiewicz, R.A., Yang, J.K.W., Hannon, A.F., Jung, Y.S., Alexander-Katz, A., Berggren, K.K., Ross, C.A., 2010. Enhancing the potential of block copolymer lithography with polymer self-consistent field theory simulations. Macromolecules 43, 8290–8295.

Millward, D.B., Lugani, G.S., Khurana, R., Light, S.L., Niroomand, A., Hustad, P.D., Trefonas, P., Chang, S.-W., Lee, C.N., Quach, D., 2014. A comparison of the pattern transfer of line-space patterns from graphoepitaxial and chemoepitaxial block co-polymer directed self-assembly. In: SPIE Advanced Lithography, 2014. International Society for Optics and Photonics, pp. 90540M–90540M-14.

Mishra, V., Fredrickson, G.H., Kramer, E.J., 2012. Effect of film thickness and domain spacing on defect densities in directed self-assembly of cylindrical morphology block copolymers. ACS Nano 6, 2629–2641.

Morita, H., 2013. Applicable simulation methods for directed self-assembly—advantages and disadvantages of these methods. J. Photopolym. Sci. Technol. 26, 801–807.

Morita, H., 2014. Development of fast DSA simulation method using OCTA system. In: SPIE Advanced Lithography, 2014. International Society for Optics and Photonics, pp. 90492O–90492O-8.

Muramatsu, M., Nakono, T., Tomita, T., Yamamoto, K., Matsuzaki, K., Kitano, T., 2014. Simulation analysis of directed self-assembly for hole multiplication in guide patterns. In: SPIE Advanced Lithography, 2014. International Society for Optics and Photonics, pp. 904921–904921-14

Nagpal, U., Detcheverry, F.A., Nealey, P.F., De Pablo, J.J., 2011a. Morphologies of linear triblock copolymers from Monte Carlo simulations. Macromolecules 44, 5490–5497.

Nagpal, U., Kang, H.M., Craig, G.S.W., Nealey, P.F., De Pablo, J.J., 2011b. Pattern dimensions and feature shapes of ternary blends of block copolymer and low molecular weight homopolymers directed to assemble on chemically nanopatterned surfaces. ACS Nano 5, 5673–5682.

Nagpal, U., Muller, M., Nealey, P.F., De Pablo, J.J., 2012. Free energy of defects in ordered assemblies of block copolymer domains. ACS Macro Lett. 1, 418–422.

Nakano, T., Matsukuma, M., Matsuzaki, K., Muramatsu, M., Tomita, T., Kitano, T., 2013. Dissipative particle dynamics study on directed self-assembly in holes. In: SPIE Advanced Lithography, 2013. International Society for Optics and Photonics, pp. 86801J–86801J-6.

Park, S.M., Stoykovich, M.P., Ruiz, R., Zhang, Y., Black, C.T., Nealey, P.F., 2007. Directed assembly of lamellae- forming block copolymers by using chemically and topographically patterned substrates. Adv. Mater. 19, 607–611.

Pereira, G.G., Williams, D.R.M., 1998. Equilibrium properties of diblock copolymer thin films on a heterogeneous, striped surface. Macromolecules 31, 5904–5915.

Peters, B.L., Ramirez-Hernandez, A., Pike, D.Q., Muller, M., De Pablo, J.J., 2012. Nonequilibrium simulations of lamellae forming block copolymers under steady shear: a comparison of dissipative particle dynamics and Brownian dynamics. Macromolecules 45, 8109–8116.

Pickett, G.T., Balazs, A.C., 1997. Equilibrium orientation of confined diblock copolymer films. Macromolecules 30, 3097–3103.

Ramirez-Hernandez, A., Detcheverry, F.A., De Pablo, J.J., 2010. Numerical simulation of Gaussian chains near hard surfaces. J. Chem. Phys. 133, 064905.

Ramirez-Hernandez, A., Liu, G.L., Nealey, P.F., De Pablo, J.J., 2012. Symmetric diblock copolymers confined by two nanopatterned surfaces. Macromolecules 45, 2588–2596.

Rasmussen, K.O., Kalosakas, G., 2002. Improved numerical algorithm for exploring block copolymer mesophases. J. Polym. Sci. B Polym. Phys. 40, 1777–1783.

Reilly, M., Ginzburg, V., Smith, M.D., 2013. Combining physical resist modeling and self-consistent field theory for pattern simulation in directed self-assembly. In: SPIE Advanced Lithography, 2013. International Society for Optics and Photonics, pp. 86820G–86820G-9.

Robertson, S.A., Reilly, M.T., Graves, T., Biafore, J.J., Smith, M.D., Perret, D., Ivin, V., Potashov, S., Silakov, M., Elistratov, N., 2009. Simulation of optical lithography in the presence of topography and spin-coated films. In: SPIE Advanced Lithography, 2009. International Society for Optics and Photonics, pp. 727340–727340.

Ross, C.A., Cheng, J.Y., 2008. Patterned magnetic media made by self-assembled block-copolymer lithography. MRS Bull. 33, 838–845.

Ross, C.A., Jung, Y.S., Chuang, V.P., Ilievski, F., Yang, J.K.W., Bita, I., Thomas, E.L., Smith, H.I., Berggren, K.K., Vancso, G.J., Cheng, J.Y., 2008. Si-containing block copolymers for self-assembled nanolithography. J. Vac. Sci. Technol. B 26, 2489–2494.

Ruiz, R., Kang, H.M., Detcheverry, F.A., Dobisz, E., Kercher, D.S., Albrecht, T.R., De Pablo, J.J., Nealey, P.F., 2008. Density multiplication and improved lithography by directed block copolymer assembly. Science 321, 936–939.

Sanders, D.P., Cheng, J., Rettner, C.T., Hinsberg, W.D., Kim, H.C., Truong, H., Friz, A., Harrer, S., Holmes, S., Colburn, M., 2010. Integration of directed self-assembly with 193 nm lithography. J. Photopolym. Sci. Technol. 23, 11–18.

Sato, H., Yonemitsu, H., Seino, Y., Kato, H., Kanno, M., Kobayashi, K., Kawanishi, A., Kodera, K., Azuma, T., 2013. Dissipative particle dynamics simulations to optimize contact hole shrink process using graphoepitaxial directed self-assembly. In: SPIE Advanced Lithography, 2013. International Society for Optics and Photonics, pp. 86801K–86801K-6.

Schmid, F., 2011. Theory and simulation of multiphase polymer systems. Handbook of Multiphase Polymer Systems. John Wiley & Sons, Ltd., Chichester, UK.

Segalman, R.A., Yokoyama, H., Kramer, E.J., 2001. Graphoepitaxy of spherical domain block copolymer films. Adv. Mater. 13, 1152–1155.

Semenov, A.N., 1985. Contribution to the theory of microphase layering in block-copolymer melts. Zh. Eksp. Teor. Fiz. 88, 1242–1256.

Sevink, G.J.A., Zvelindovsky, A.V., Kyrylyuk, A.V., Lyakhova, K.A., Fraaije, J.G.E.M., 2001. Dynamic pathways in mesoscale simulations of block copolymer systems. In: Abstracts of Papers of the American Chemical Society, Washington, DC, USA, vol. 222, pp. U364.

Shi, A.C., 2004. Self-consistent field theory of block copolymers. In: Hamley, I.W. (Ed.), Developments in Block Copolymer Science and Technology. Wiley-VCH, Weinheim.

Sides, S.W., Fredrickson, G.H., 2003. Parallel algorithm for numerical self-consistent field theory simulations of block copolymer structure. Polymer 44, 5859–5866.

Stoykovich, M.P., Daoulas, K.C., Muller, M., Kang, H.M., De Pablo, J.J., Nealey, P.F., 2010. Remediation of line edge roughness in chemical nanopatterns by the directed assembly of overlying block copolymer films. Macromolecules 43, 2334–2342.

Suh, H.S., Kang, H.M., Nealey, P.F., Char, K., 2010. Thickness dependence of neutral parameter windows for perpendicularly oriented block copolymer thin films. Macromolecules 43, 4744–4751.

Sundrani, D., Sibener, S.J., 2002. Spontaneous spatial alignment of polymer cylindrical nanodomains on silicon nitride gratings. Macromolecules 35, 8531–8539.

Takahashi, H., Laachi, N., Delaney, K.T., Hur, S.M., Weinheimer, C.J., Shykind, D., Fredrickson, G.H., 2012. Defectivity in laterally confined lamella-forming diblock copolymers: thermodynamic and kinetic aspects. Macromolecules 45, 6253–6265.

Wang, Q., Nath, S.K., Graham, M.D., Nealey, P.F., De Pablo, J.J., 2000a. Symmetric diblock copolymer thin films confined between homogeneous and patterned surfaces: simulations and theory. J. Chem. Phys. 112, 9996–10010.

Wang, Q., Yan, Q.L., Nealey, P.F., De Pablo, J.J., 2000b. Monte Carlo simulations of diblock copolymer thin films confined between chemically heterogeneous hard surfaces. Macromolecules 33, 4512–4525.

Wang, Q., Yan, Q.L., Nealey, P.F., De Pablo, J.J., 2000c. Monte Carlo simulations of diblock copolymer thin films confined between two homogeneous surfaces. J. Chem. Phys. 112, 450–464.

Wang, Q., Nealey, P.F., De Pablo, J.J., 2001. Monte Carlo simulations of asymmetric diblock copolymer thin films confined between two homogeneous surfaces. Macromolecules 34, 3458–3470.

Wang, Q., Nealey, P.F., De Pablo, J.J., 2003. Simulations of the morphology of cylinder-forming asymmetric diblock copolymer thin films on nanopatterned substrates. Macromolecules 36, 1731–1740.

Yang, Y.Z., Qiu, F., Zhang, H.D., Yang, Y.L., 2006. Cylindrical phase of diblock copolymers confined in thin films a real-space self-consistent field theory study. Polymer 47, 2205–2216.

Yoshida, H., Suh, H.S., Ramirez-Hernandez, A., Lee, J.I., Aida, K., Wan, L., Ishida, Y., Tada, Y., Ruiz, R., De Pablo, J., Nealey, P.F., 2013. Topcoat approaches for directed self-assembly of strongly segregating block copolymer thin films. J. Photopolym. Sci. Technol. 26, 55–58.

Zhao, Y., Sivaniah, E., Hashimoto, T., 2008. SAXS analysis of the order-disorder transition and the interaction parameter of polystyrene-block-poly(methyl methacrylate). Macromolecules 41, 9948–9951.

Part Two

Templates and patterning for directed self-assembly

Directed self-oriented self-assembly of block copolymers using topographical surfaces

5

J. Choi, K.R. Carter, T.P. Russell
University of Massachusetts, Amherst, Massachusetts, USA

5.1 Introduction

In the semiconductor industry, device cost and performance are strongly correlated to feature sizes (ITRS, 2013). To produce nanoscale feature sizes, "top-down" lithographic approaches are commonly used. However, further decreasing feature sizes below the 22 nm half-pitch by the lithographic processes is becoming extremely expensive and more challenging (ITRS, 2013). To overcome these limitations, the combination of a "top-down" approach with a "bottom-up" self-assembly has been investigated over the past decade (Hamley, 2003; Hawker and Russell, 2005; Park et al., 2003; Stoykovich and Nealey, 2006).

The self-assembly of BCPs has attracted significant research interest due to its ability to create periodic domains such as lamellae, spheres, gyroids, or cylinders with dimension range from 5 to 50 nm (Almdal et al., 1990; Bates, 1991; Bates and Fredrickson, 1990, 1999; Fasolka and Mayes, 2001; Green and Limary, 2001; Guarini et al., 2002; Matsen and Bates, 1996; Russell, 1996). However, typically, BCP microdomains in thin films on an unpatterned substrate meet intrinsic obstacles, including undesirable domain orientation and short-range lateral order, which limit their capabilities in nanofabrication. To overcome these issues, in most cases, topographically or chemically patterned surfaces are required to guide the self-assembly of BCP in thin films.

In this chapter, we will focus on the directed self-assembly of BCP thin films using various topographical surfaces. In Section 5.2, a brief history on controlling the orientation of BCP microdomains in thin films will be discussed. In Section 5.3, we will discuss the influence of topographical surfaces on the directed self-assembly of BCP thin films. First, the concept of graphoepitaxy and the method to fabricate topographical surfaces will be introduced. Second, we will describe BCP microdomains guided by topographical surfaces based on the geometry of the pattern. In Section 5.4, the application areas of BCPs using topographical surfaces such as lithographic masks, metal dots, and nanowires will be addressed. Finally, in Section 5.5, we will summarize and propose an outlook.

Directed Self-assembly of Block Copolymers for Nano-manufacturing. http://dx.doi.org/10.1016/B978-0-08-100250-6.00005-5

5.2 Control of interfacial interactions

Depending on specific applications such as bit patterned media, membranes, or lithographic masks, different microdomain orientations (in-plane or out-of-plane) are required, so that controlling the orientation of BCP microdomains is crucial for BCP applications. In thin films, interfacial interactions at substrate/copolymer and copolymer/air interfaces and the commensurability between the film thickness and the natural period of BCP (L_0) are primary factors dictating the morphology of BCPs (Cai et al., 1993; Hawker et al., 1996; Huang et al., 1998, 1999, 2000; Husseman et al., 1999; Mansky et al., 1997a,b,c, 1999; Russell et al., 2000; Ryu et al., 2005; Xu et al., 2005). Generally, strong preferential interactions of one component of diblock copolymer with the substrate and/or a lower surface energy of one of blocks lead to a parallel orientation of BCP microdomains to the substrate in thin films because of the connectivity of the blocks. This behavior was investigated using symmetric poly(styrene-*b*-methyl methacrylate) (PS-*b*-PMMA) by neutron reflectivity, secondary ion mass spectrometry, and interference microscopy (Anastasiadis et al., 1989; Coulon et al., 1989; Menelle et al., 1992; Russell et al., 1989). In these studies, after thermal annealing, lamellar microdomains of PS-*b*-PMMA diblock copolymer in thin films were oriented parallel to the substrate due to the strong affinity of PMMA block to the oxide layer on the silicon substrate and the lower surface energy of PS block. However, this preferential orientation of BCP microdomains can be overcome by controlling the interfacial energies precisely. If the substrate is neutral, that is, the interfacial interactions of both blocks are balanced (equally favorable or unfavorable), BCP microdomains can orient normal to the substrate. Under this condition, the incommensurability of the film thickness relative to L_0 also affects the orientation of BCP microdomains.

The most common strategy to control the interfacial interaction is modification of the substrate using random copolymer brushes (Huang et al., 1999, 2000; Mansky et al., 1997a,b,c). Russell and coworkers were first to demonstrate the grafting of hydroxyl terminated PS-*r*-PMMA random copolymers, where the fraction, f, of styrene was varied from 0 to 1 on the oxide layer of the silicon substrate. As shown in Figure 5.1, it was found that the interfacial energies of both PS and PMMA blocks of PS-*r*-PMMA random copolymer brushes were equal when $f \cong 0.57$, which means that the substrate has nonpreferential interactions (neutral condition) with both the PS and PMMA blocks (Mansky et al., 1997a). Although anchoring of PS-*r*-PMMA random copolymer brushes is a simple and effective method to control surface affinities, this approach has disadvantages, such as requiring specific surface chemistries, and the process is time consuming. To circumvent these hurdles, several groups further developed a generalized approach to surface modification by cross-linking reaction (Bang et al., 2007; Han et al., 2007; Ryu et al., 2005, 2007). This was demonstrated with thin films of benzocyclobutene (BCB) functionalized PS-*r*-PMMA (PS-*r*-PBCB-*r*-PMMA) random copolymers (Ryu et al., 2005). In this work, it was found that thermally cross-linked PS-*r*-PBCB-*r*-PMMA ultrathin films can be produced on various solid surfaces such as Al, Si_3N_4, Kapton, and PET substrates with controlled surface energies in a simple, rapid, and robust manner. Interfacial interactions can also be

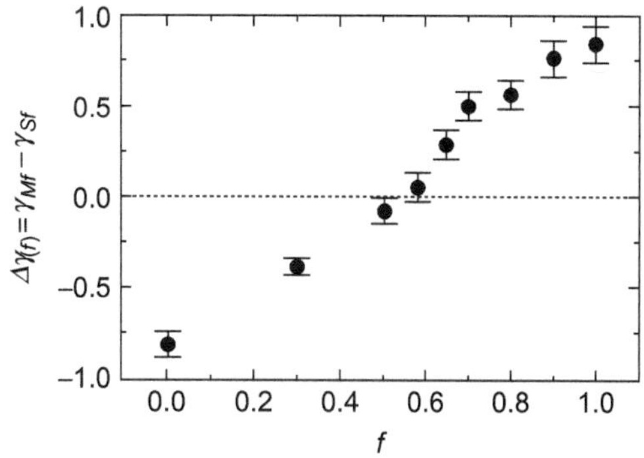

Figure 5.1 $\Delta\gamma_{(f)} = \gamma_{Mf} - \gamma_{Sf}$ on a P(S-*r*-MMA) brush as a function of *f*. From Mansky et al. (1997a). Reprinted with permission from AAAS.

mediated by solvents during solvent vapor annealing (SVA) (Kim and Libera, 1998a,b; Kim et al., 2004a,b; Lin et al., 2002).

The coupling of interfacial interactions and the confinement of BCPs was investigated experimentally and theoretically by confining BCPs between two parallel surfaces (Brown and Chakrabarti, 1995; Huang et al., 1998; Kellogg et al., 1996; Kikuchi and Binder, 1993; Koneripalli et al., 1995; Lambooy et al., 1994; Turner, 1992; Walton et al., 1994). When copolymer films are confined between flat surfaces, the formation of the surface topography such as holes and islands is suppressed and, depending on the wetting conditions at the interfaces of the confining planes and the separation distance between the confining surfaces relative to the equilibrium period of the BCP, the copolymer can be forced into a condition of frustration. Turner (1992) theoretically studied a free energy model that includes the elastic stretching energy of the polymer, the interactions at the A–B interfaces, and the interactions between the polymer and the confining plates for the confined parallel oriented lamellar microdomains of the A–B diblock copolymer in the strong segregation limit. It was found that the number of layers between two parallel plates was changed from n to $n+1/2$ as a function of plate spacing. Walton et al. (1994) extended Turner's model by considering the case of perpendicular lamellar microdomains between two parallel walls. The free energy of parallel and perpendicular lamellar microdomains between walls is expressed as follows:

$$\frac{F_h}{F_0} = \frac{1}{3}\left\{\lambda^2 + \frac{2}{\lambda} + \frac{1}{m\lambda}\left[2\mathbf{r} + \delta\left(\frac{1-(-1)^{2m}}{2}\right)\right]\right\} \tag{5.1}$$

where F_h = free energy of confined parallel lamellar microdomains

F_0 = bulk equilibrium free energy of BCP
$\lambda = L/L_0$ (L = confined lamellar period and L_0 = natural period in bulk)
$\mathbf{r} = \gamma_{AS}/\gamma_{AB}$ (γ_{AS} = A block/substrate interfacial tension and γ_{AB} = A block/B block interfacial tension)

$\delta = \frac{\gamma_{BS} - \gamma_{AS}}{\gamma_{AB}}$ (γ_{BS} = B block/substrate interfacial tension)

$m = n$ for symmetric parallel lamellar microdomains

$m = n + ½$ for antisymmetric parallel lamellar microdomains

$$\frac{F_v}{F_0} = \frac{1}{3}\left[3 + \frac{1}{m\lambda}(2\mathbf{r} + \delta)\right] \tag{5.2}$$

where, F_v = free energy of confined perpendicular lamellar microdomains

In this work, they developed a free energy of the perpendicular lamellar microdomains to the walls and compared it to the symmetric and the antisymmetric parallel lamellar microdomains as a function of wall spacing. The perpendicular lamellar microdomains are favored over the parallel lamellar microdomains when $F_v < F_h$. It was found that the perpendicular lamellar microdomains were favored over in a limited range of wall spacing. Lambooy et al. (1994) used neutron reflectivity to investigate the behavior of symmetric PS-*b*-PMMA diblock copolymer, which was confined by two parallel solid surfaces where the PMMA block preferentially segregated to both surfaces. In this work, the lamellar microdomains showed parallel orientation to the solid surface regardless of the confined film thickness, and copolymer chains were either compressed or expanded to accommodate the imposed frustration. In contrast, Kellogg and coworkers found that when symmetric PS-*b*-PMMA diblock copolymer was confined between parallel substrates that were modified by PS-*r*-PMMA random copolymer brushes, the lamellar microdomains with incommensurate conditions oriented normal to the substrates to accommodate the frustration (Kellogg et al., 1996).

5.3 Graphoepitaxy

We discussed so far strategies to control the orientation of BCP microdomains in thin films by tuning surface energies, but these oriented microdomains on unpatterned substrates show short-range lateral order, which limits their usefulness. Graphoepitaxy, which uses artificial surface relief structures to guide the self-assembly of the BCP in thin films, can enhance the lateral order of BCP microdomains (Geis et al., 1979; Segalman et al., 2001; Smith and Flanders, 1978). In general, the length scale of topographically patterned surfaces for graphoepitaxy is several times greater than L_0, and an annealed BCP material will fill the regions confining the BCP. One advantage of graphoepitaxy is density multiplication. For example, if electron beam lithography (EBL) generates sparse line or post guiding patterns based on L_0, the remaining unpatterned areas will be filled by BCP having a much smaller period, thereby reducing the effective pitch of the patterning so that the writing time of EBL can be significantly reduced. Consequently, this graphoepitaxy, which is the combination of a "top-down" lithographic approach with a "bottom-up" self-assembly of BCPs, provides an opportunity for multilevel controlling of BCP patterns on the nanometer scale on diverse topographically patterned surfaces.

5.3.1 Fabrication of topographical surfaces

Various methods can be employed in the "top-down" lithographic approach to fabricate topographical surfaces for graphoepitaxy. Photolithography is the dominant technique in the microelectronics industry and this method includes serial and parallel processes (Gates et al., 2005). Generally, the feature size is limited by the wavelength of light and the numerical aperture of the lens. By using 193 nm wavelength, photolithography can fabricate patterns with 37 nm feature sizes (Gates et al., 2005). Many topographical surfaces to guide BCP microdomains in thin films were produced by photolithography technique (Black and Bezencenet, 2004; Hammond et al., 2005; Hexemer et al., 2005; Jeong et al., 2009; Segalman et al., 2001).

Interference lithography is a good technique for producing periodic patterns such as 1-D line patterns or 2-D dot/hole arrays over large areas without using the photomask (Farhoud et al., 1999; Savas et al., 1996). Different laser sources such as a Lloyd's mirror system with a 325 nm wavelength He–Cd laser or a 193 nm ArF laser can be used depending on the required feature size. By using shorter wavelengths, interference lithography can extend resolution capability of photolithography. For example, extreme ultraviolet light source with a wavelength of 13.5 nm can achieve a spatial resolution below 10 nm on various photoresists in a single exposure (Auzelyte et al., 2009; Wagner and Harned, 2010). Many researchers have used interference lithography to generate chemically patterned surfaces as well as topographically patterned surfaces (Chang et al., 2009; Cheng et al., 2002; Chuang et al., 2006; Edwards et al., 2006; Jung et al., 2008b, 2010a; Jung and Ross, 2007; La et al., 2005; Liu et al., 2010; Stoykovich et al., 2005).

EBL is a standard technique for writing patterns that does not require a photomask, but this technique has relatively low throughput compared to photolithography (Geissler and Xia, 2004). The resolution of EBL can be below 10 nm, which is determined by several parameters, including the applied voltage, the spot size of the beam, the resist used, and the effect of electron scattering, namely, the proximity effect. PMMA was one of the first resists for EBL and functions as an organic positive-tone resist for high-resolution patterning (Hatzakis, 1969; McCord and Rooks, 1997). The negative-tone resist, hydrogen silsesquioxane (HSQ), and the positive-tone resist, ZEP (Nippon Zeon ZEP-520), have also been developed as high resolution e-beam resists (Grigorescu and Hagen, 2009; Kurihara et al., 1995; Namatsu et al., 1998; Nishida et al., 1992). A great deal of research on the directed self-assembly of BCP in thin films has been done using EBL (Bita et al., 2008; Chai et al., 2007; Chang et al., 2014; Cheng et al., 2004; Hong et al., 2011; Kim et al., 2003; Park et al., 2007; Ruiz et al., 2008; Stoykovich et al., 2007; Sundrani et al., 2004a; Tavakkoli et a,l., 2012; Xiao et al., 2005, 2009; Yang et al., 2009).

Alternative approaches, such as soft lithography, which uses elastomeric molds, thermal nanoimprint lithography, which transfers a surface relief structure from a master mold to a thermoplastic polymer film by heating above glass transition temperature of the polymer, or UV-assisted imprint lithography, which uses a low viscosity photocurable monomer with UV transparent mold, can be employed to produce topographical surfaces (Chou et al., 1996; Long et al., 2007; Marrian and Tennant, 2003; Xia and Whitesides, 1998). Jeong and coworkers used poly(dimethylsiloxane) (PDMS) mold as the topographical pattern and then, by combining the directed self-assembly of BCPs with nanotransfer printing, they obtained ordered sub-10 nm BCP patterns on various substrates such as indium tin oxide, silver, and aluminum beverage can (Jeong et al., 2012). Russell and coworkers fabricated nanoimprinted polymer replica with high fidelity such as polyimide, PDMS, poly(butylene terephthalate), and fluorinated ethylene propylene from a faceted sapphire surface. These replica substrates were used for the directed self-assembly of poly(styrene-*b*-ethylene oxide) (PS-*b*-PEO) diblock copolymer thin films (Hong et al., 2012a; Park et al., 2010).

5.3.2 Geometry with deep patterning

In the directed self-assembly of BCP in thin films, most studies have focused on topographically patterned surfaces with the deep geometry, where the vertical length scale is greater than L_0. As shown in Figure 5.2a, one representative example of deep geometries is a trench pattern. In the deep trench pattern, the depth of the trench is larger than L_0, so that it can easily control BCP microdomains with thickness comparable to L_0 within the confined areas with high spatial resolution and good placement accuracy. However, in this case, the grain size of BCP microdomains is limited by the width of the trench. A reversed case is the shallow trench pattern as shown in Figure 5.2b, that is, the minimal geometry. Here, we will first discuss geometries with deep patterning.

5.3.2.1 Deep trench surfaces

Kramer and coworkers first demonstrated the generation of single crystal films of BCPs over large areas on a trench surface (Segalman et al., 2001). After thermal annealing of the monolayer of spherical microdomains of poly(styrene-*b*-2-vinylpyridine) (PS-*b*-P2VP) diblock copolymer on this surface, they produced at least 4.5 μm single grain hexagonal arrays of spherical microdomains on the mesa as shown in Figure 5.3. In this study, the length scales of mesas and trenches played a crucial

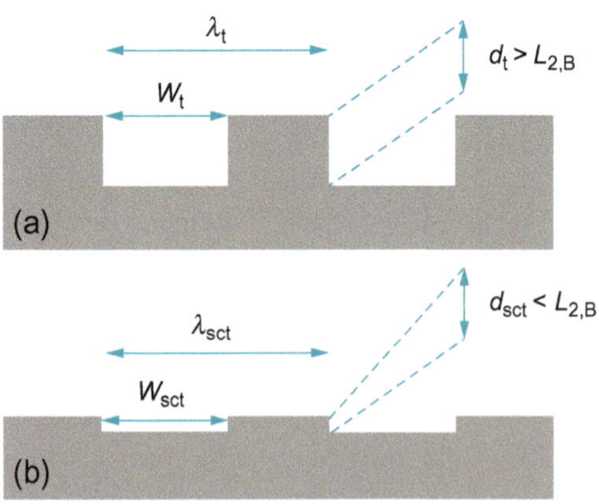

Figure 5.2 Graphical illustrations of (a) the deep trench pattern and (b) the shallow trench pattern, where λ is the pitch of the trench, w is the width of the trench, d is the depth of the trench, and $L_{2,\mathrm{B}}$ is the center-to-center distance between BCP microdomains in the bulk. Adapted with permission from Hong et al. (2011). Copyright 2011 American Chemical Society.

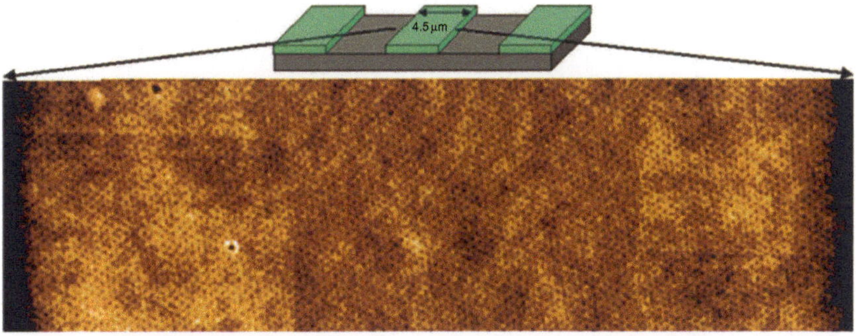

Figure 5.3 SFM of PS-*b*-PVP film on top of a mesa. When the mesa edges are <5 µm apart, a single crystal is formed.
Reproduced with permission from Segalman et al. (2001). Copyright © 2001 Wiley-VCH Verlag GmbH & Co. KGaA, Weinheim.

role in inducing long-range lateral order of spherical microdomains. When the depth of the trench was equal to or greater than L_0 and the width of the mesa was less than 5 µm, they could obtain the single grain of PS-*b*-P2VP microdomains. The same group also investigated the ordering and melting behavior of spherical microdomains of PS-*b*-P2VP on the trench surfaces through a continuous defect generation process (Segalman et al., 2003a). By varying temperatures that corresponded to χN, where χ is the temperature-dependent Flory–Huggins parameter and N is the number of mers in the BCP, they generated a 2-D single crystalline layer with quasi-long-range order with few defects of spherical microdomains of PS-*b*-P2VP when the temperature of thermal annealing was close to but below the temperature of 2-D crystal to hexatic transition ($7.4 < \chi N < 9$). In addition, the same group found that the topographic confining edge provided the spherical microdomains of PS-*b*-P2VP with orientational and translational orders while the influence of the confining edge decayed with the distance from the edge (Segalman et al., 2003b,c). Similarly, the behavior of cylindrical microdomains of BCP in thin films on trench surfaces in terms of orientational order, translational order, and defects were demonstrated by varying temperature (Hammond et al., 2005; Hammond and Kramer, 2006). Furthermore, the grain orientation and the single crystal diffraction of PS-*b*-P2VP microdomains on 2-D hexagonal trench surfaces were characterized using scanning force microscopy (SFM) moiré patterns or grazing incidence small-angle X-ray diffraction by the same group (Hexemer et al., 2005; Stein et al., 2007).

The directed self-assembly of spherical microdomains of poly(styrene-*b*-ferrocenyldimethylsilane) (PS-*b*-PFS) diblock copolymer on trench surfaces was investigated by Ross and coworkers (Cheng et al., 2003, 2004, 2006a,b; Chuang et al., 2006). They used PS-*b*-PFS diblock copolymer because of its high etching selectivity, which is useful for the pattern transfer into an underlying layer or a substrate. To understand confinement effects on the self-assembled array of spherical microdomains, they fabricated the trench pattern of 80 nm depth with various widths by EBL, so that the number of rows of PS-*b*-PFS microdomains was controlled (Figure 5.4a; Cheng et al., 2004). As shown in Figure 5.4b, N rows of spherical microdomains were

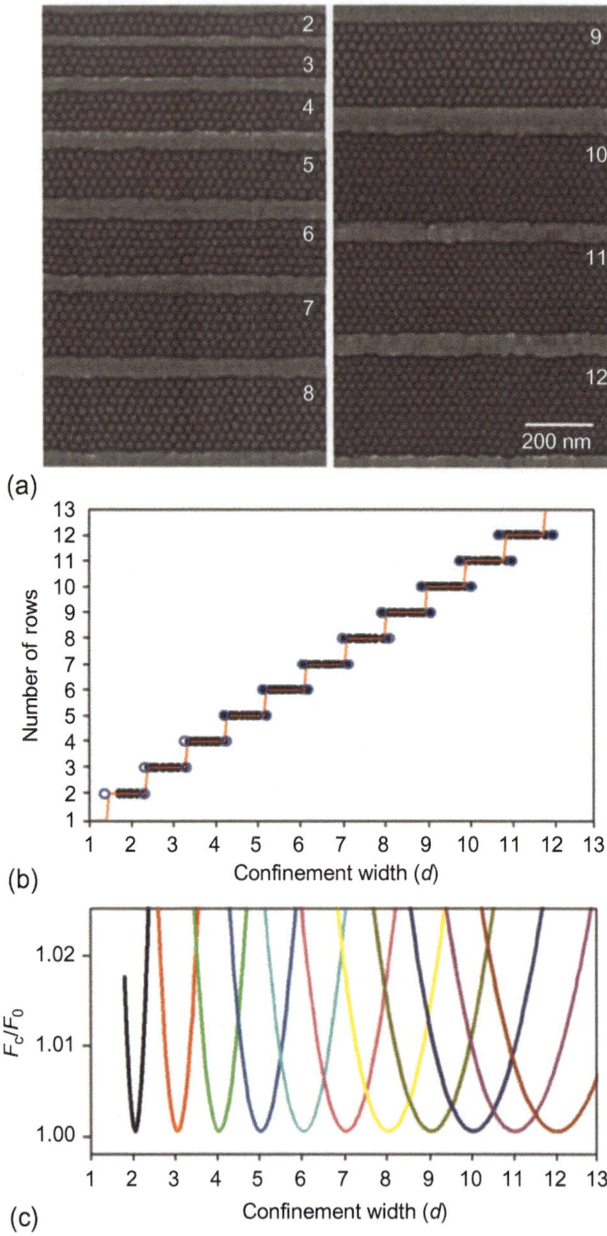

Figure 5.4 The block copolymer PS-*b*-PFS in parallel-sided grooves. (a) Plan-view scanning electron micrograph of ordered arrays of PFS domains with $N = 2–12$ rows. (b) The number of rows in the groove, N, plotted against confinement width, W, showing the widths at which arrays with N rows are stable. The confinement width is expressed in terms of d, the equilibrium row spacing, which is 24.8 nm in this polymer. (c) Energy against confinement width of block copolymer system. The free energy of the confined block copolymer (F_C) is presented relative to the free energy of the bulk block copolymer (F_0).
Reprinted by permission from Macmillan Publishers Ltd: *Nature Materials*, Cheng et al. (2004), copyright 2004.

maintained when $(N-0.5)L_0 < W < (N+0.5)L_0$, where L_0 is the natural period of PS-*b*-PFS and W is the width of the trench. It was found that when there was incommensurability between W and L_0, PS-*b*-PFS microdomains were extended or compressed to comply with the confined trench patterns. To demonstrate this behavior, they developed the free energy model as shown in Figure 5.4c. The free energy shows its local minimum when $W=NL_0$, so that, in this case, spherical microdomains will ideally choose an integer value of N row. When $W \approx (N+0.5)L_0$, however, there will be a transition in the number of rows from N to $N+1$ due to the change in the free energy. Also, the same group generated 2-D topographical patterns having a sharp corner of 60° in order to demonstrate high positional accuracy of spherical microdomains (Cheng et al., 2006b). Compared with 1-D trench patterns, this 2-D pattern could reduce the placement error of hexagonal arrays of PS-*b*-PFS microdomains due to another confinement in a direction of 60°, so that they could achieve the pattern registration with an error of 3 nm, which was comparable to the error by other lithographic techniques.

Graphoepitaxy with poly(styrene-*b*-dimethylsiloxane) (PS-*b*-PDMS) diblock copolymer was also developed by Ross and coworkers (Jung et al., 2008a, 2010a; Jung and Ross, 2007). Since PS-*b*-PDMS has a relatively high Flory–Huggins interaction parameter ($\chi \approx 0.26$ at room temperature), which is related to scalability, smaller L_0 can be achieved in comparison to other BCPs, such as PS-*b*-PMMA ($\chi \approx 0.04$) and PS-*b*-P2VP ($\chi \approx 0.18$). Also, the PDMS block exhibits a high etching contrast to O_2 plasma in comparison to the PS block, due to the presence of Si in the backbone of PDMS, so that it can be used for lithographic masks. However, due to the large difference in the surface tensions between PS ($\gamma = 40.7$ mN/m) and PDMS ($\gamma = 19.9$ mN/m), which cause preferential segregation of PDMS to the air/copolymer interface, the surface modification with a brush layer on topographical surfaces is critical in the graphoepitaxy with PS-*b*-PDMS. They fabricated trench patterns by Lloyd's Mirror interference lithography followed by modification of the surface of the trench by PDMS brush layer to facilitate the self-assembly of cylinder-forming PS-*b*-PDMS diblock copolymer (Jung and Ross, 2007). In this work, they systematically controlled the width ratio (mesa/trench) from 0.18 to 1 with maintaining the trench pitch of 1 μm, the toluene vapor pressure, and the annealing time. Under a high vapor pressure of toluene for 15 h with the narrow mesa ($W_{mesa} = 125$ nm and $W_{trench} = 875$ nm), they obtained in-plain cylindrical microdomains oriented parallel to the trench walls. In contrast, at a lower vapor pressure of toluene for 15 h with the wide mesa ($W_{mesa} = 270$ nm and $W_{trench} = 730$ nm), in-plain cylindrical microdomains were aligned perpendicular to the trench walls.

The same group also produced the concentric ring patterns of PS-*b*-PDMS (Jung et al., 2008a). They fabricated circular trench patterns of 40 nm depth with the diameter, D, from 60 to 350 nm by Lloyd's Mirror interference lithography. The representative ring patterns after the self-assembly of PS-*b*-PDMS are shown in Figure 5.5. They observed only a PDMS sphere when $D = 60$ or 77 nm, a single PDMS ring without an inner sphere when 86 nm $< D <$ 94 nm, reappearance of the inner sphere when 102 nm $< D <$ 134 nm. These periodic changes in the presence and absence of the inner sphere were maintained in the whole range of the trench diameter. Besides the self-assembly of diblock copolymers, Ross and coworkers studied the self-assembly of triblock terpolymer with trench surfaces (Chuang et al., 2009). In this case, to induce

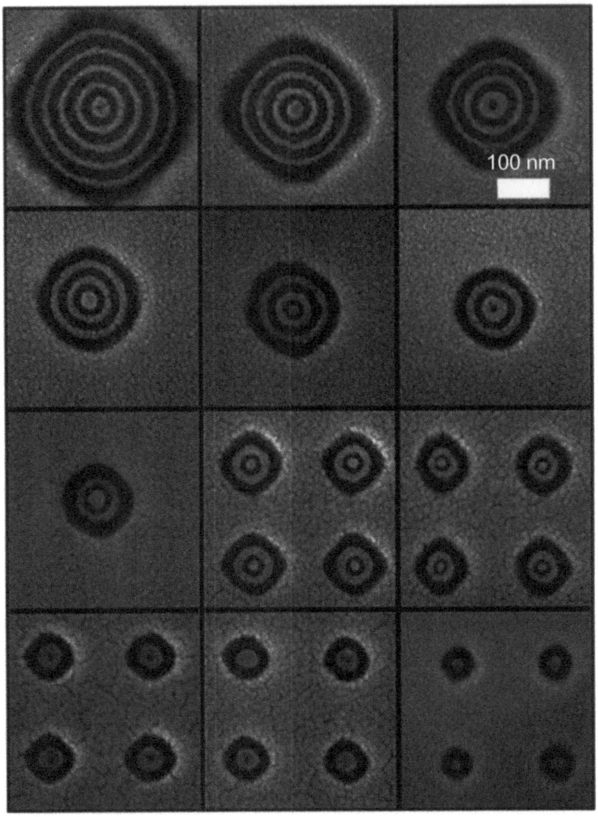

Figure 5.5 Concentric ring patterns in templates with various confinement diameters. For some diameters, the innermost feature is a PDMS sphere (light contrast) and for others, a smaller PS sphere (dark) is formed.
Reprinted with permission from Jung et al. (2008a). Copyright 2008 American Chemical Society.

square symmetry arrays of BCP microdomains, they used the blend of poly(isoprene-*b*-styrene-*b*-ferrocenylsilane) (PI-*b*-PS-*b*-PFS) triblock terpolymer with PS homopolymer. By tuning the surface properties of the trench patterns with various widths, they demonstrated the self-assembly of square-packed triblock terpolymer microdomains.

Sibener and coworkers studied the self-assembly of BCPs using trench surfaces with various widths and depths (Sundrani et al., 2004a,b; Sundrani and Sibener, 2002). It was observed that the trench depths greater than 35 nm induced the alignment of poly(styrene-*b*-(ethylene-*alt*-propylene)) (PS-*b*-PEP) microdomains on the mesas as well as in the trenches by overcoming the confined volumes of the trenches when overfilled (Sundrani et al., 2004a). Also, it was found that if the width of the trench was wider, a longer annealing time was required to align in-plane cylindrical microdomains of PS-*b*-PEP. Buriak and coworkers demonstrated alignment of metallic patterns by employing rectangular, triangular, and circular trenches of 35 nm depth (Chai and Buriak, 2008; Chai et al., 2007). In this work, PS-*b*-P2VP diblock copolymers were chosen to load various metal ions into the P2VP blocks. Recently, the same group investigated the compatibility of microwave annealing with graphoepitaxy (Zhang et al., 2010). By using PS-*b*-P2VP and PS-*b*-PMMA diblock copolymers with microwave annealing in the presence of tetrahydrofuran, the highly ordered parallel cylindrical microdomains were achieved in the trenches having SiO_2 walls.

Trench patterns having chemically homogeneous surfaces—namely, nonpreferential trench bottoms and sidewalls—are useful for guiding sphere- and cylinder-forming BCPs. In contrast, to control the orientation of lamellar-forming BCPs, trench patterns with chemically heterogeneous surfaces, which have nonpreferential trench bottoms and preferential sidewalls, are required. For example, in order to achieve desired orientation of the self-assembled lamellar microdomains of BCPs with graphoepitaxy, Nealey and coworkers controlled the interactions of the blocks of PS-*b*-PMMA with the bottoms and the sidewalls of the trench patterns (Park et al., 2007). As shown in Figure 5.6a, when the bottom of the trench is neutral to both PS and PMMA blocks while the sidewall of the trench selectively interacts with PS blocks, perpendicular lamellar microdomains were aligned parallel with the wall of the trench. In contrast, when both the bottom and the sidewall of the trench were SiO_2, which preferentially interact with the PMMA blocks, parallel lamellar microdomains were observed in the trench pattern (Figure 5.6b). Also, when uniformly modified by neutral brush layers (PS-*r*-PMMA), the trench pattern induced perpendicular lamellar microdomains to both the bottom and the sidewall as shown in Figure 5.6c. In this approach, the

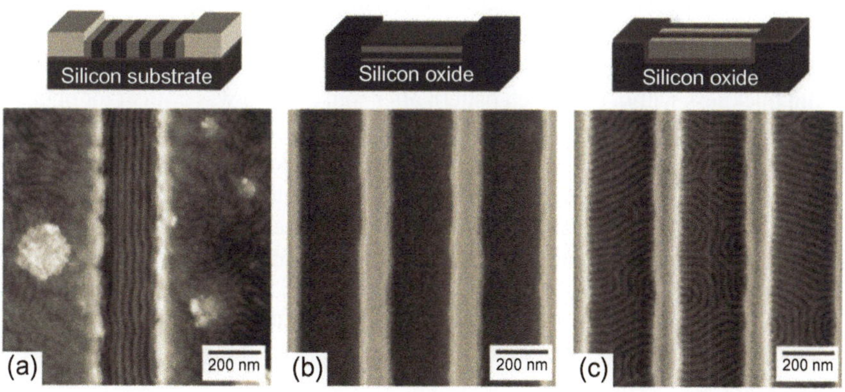

Figure 5.6 Lamellar structures of symmetric PS-*b*-PMMA (36 kg mol^{-1}) on topographically patterned substrates. (a) The sidewalls of the gold patterns (right and left) are preferentially wetted by the PS block. The bottom of the groove (middle) exhibits neutral wetting behavior to both blocks. (b) Lamellar structures are oriented parallel to the SiO_2 bottom of the grooves (dark) within a patterned SiO_2 substrate (light). (c) The lamellar domains are oriented perpendicular to both the substrate and the groove sidewalls when the topographic pattern in (b) is uniformly treated with a neutral brush.

optimization of experimental conditions such as the film thickness of BCP, the molecular weight of BCP, and the commensurability between the width of the trench and L_0, is important to obtain high degree of alignment of lamellar microdomains. Similarly, Yamaguchi and coworkers demonstrated perpendicular lamellar microdomains of PS-*b*-PMMA diblock copolymer in the line, rectangular, and hexagonal trench patterns with heterogeneous surfaces produced by the combination of the neutralization of the bottom surface with HSQ resist patterns (Yamaguchi and Yamaguchi, 2006, 2008). Recently, Gopalan and coworkers studied not only symmetric BCP microdomains, but also asymmetric BCP microdomains in the trench patterns by simultaneously controlling the surface chemical composition and the BCP film thickness (Han et al., 2010; Kim et al., 2013b).

Kim and coworkers examined the self-assembly of BCPs by using various trench geometries generated by EBL (Cheng et al., 2007; Kim et al., 2008; Park et al., 2009a,b). By using the hybrid mixture of PS-*b*-PEO and polymethylsiloxane resin, they produced wave-like structure of lamellar microdomains perpendicular to the trench direction because of large difference in the bending properties of aligned microdomains along two axes (Kim et al., 2008). In this case, however, the same trench patterns were not successful for lamellar microdomains of PS-*b*-PMMA diblock copolymer. Recently, the same group extended a previous approach to align lamellar microdomains of PS-*b*-PMMA by using broader range of length scales of the trench patterns (Park et al., 2009a). It was found that the alignment behavior of lamellar microdomains of PS-*b*-PMMA on trench surfaces was dependent on not only the film thickness of BCP, but also on the underlying pattern dimension such as the depth and the pitch.

Black and coworkers systematically demonstrated pattern registration and alignment of PS-*b*-PMMA diblock copolymers by using square, circular, and general trench patterns (Black and Bezencenet, 2004). It was found that the initial variation of the film thickness within the trench pattern after spin coating was critical for obtaining well-ordered cylindrical microdomains with parallel orientation in the trenches. As shown in Figure 5.7a, when the film thickness was thicker near a step edge than that of a middle of the trench, the microdomains were first nucleated along the step edges and then, propagated to the middle of the trench. However, the thickness variations within the trench pattern were dependent on the pattern dimensions, such as a narrow width (Figure 5.7b) and a wide width (Figure 5.7c), so that the length scales of the pattern should be optimized to achieve registration of BCP microdomains. Ruiz and

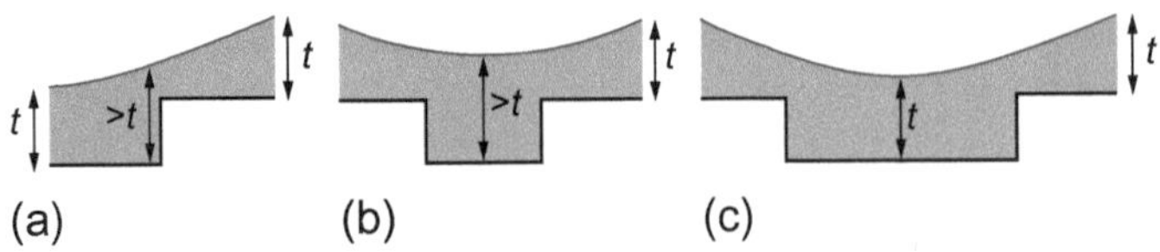

Figure 5.7 (a) Schematic of diblock copolymer film coverage of lithographic step edge. (b) and (c) Schematics of polymer film coverage of small (b) and large (c) lithographic features, illustrating thicker effective film in smaller structures.

coworkers investigated the local defectivity of parallel lamellar microdomains of PS-*b*-PMMA confined in a tapered trench geometry (Ruiz et al., 2007). In this geometry, the narrow region imposes a higher free energy penalty for defect formation than that of the wider areas. By using this approach, they produced defect-free microdomains areas in the narrow trench regions up to the width of 0.64 μm ($20L_0$) and the length of 5 μm ($150L_0$).

The influence of the line-edge roughness (LER) of the trench pattern on the self-assembly of BCP films was studied (Welander et al., 2008; Yang et al., 2004). Yang and coworkers reported that the ordering of cylindrical microdomains of PS-*b*-PMMA oriented normal to the substrate was strongly affected by the LER of the trench pattern (Yang et al., 2004). Similarly, Nealey and coworkers investigated the influence of the trench width roughness (TWR) on defectivity of cylindrical microdomains of PS-*b*-PMMA (Welander et al., 2008). In this case, defect formation was strongly dependent on commensurability between L_0 and the width of the trench while TWR did not strongly impact defectivity.

Kim and coworkers demonstrated the lateral ordering of BCP microdomains by graphoepitaxy with a photoresist pattern (Jeong et al., 2009, 2010a,b). One of the advantages using this method is that the photoresist confinement can be completely removed by a simple cleaning after pattern transfer, so that further overlay processes can be facilitated. By employing symmetric and asymmetric PS-*b*-PMMA thin films with a SU8 photoresist trench pattern, they obtained highly ordered BCP microdomains (Jeong et al., 2009). In this case, since the sidewall of the photoresist trench preferentially interacted with the PS blocks, perpendicular orientation of BCP microdomains was induced when interactions between the blocks of the copolymer and the bottom of the trench were balanced. They demonstrated that lateral order of lamellar microdomains oriented normal to the film surface could be achieved over large areas (17.5 μm by 12.5 μm) by combining graphoepitaxy with epitaxial self-assembly (Jeong et al., 2010a).

5.3.2.2 Post surfaces

To control the self-assembly of spherical microdomains of PS-*b*-PDMS thin films, Ross and coworkers generated 2-D arrays of HSQ posts by EBL (Bita et al., 2008). In contrast to previous trench surfaces, here, the sparse HSQ posts were fabricated over the surface, interacting both locally and globally with spherical microdomains of the BCP. As shown in Figure 5.8a, the HSQ posts and the native oxide on the silicon wafer were functionalized by a short PDMS brush layer, so that posts can be used as surrogates for minor PDMS sphere in the close-packed array. These HSQ posts also can be modified by a PS brush layer. When there were no guiding posts on the surface, spherical microdomains of PS-*b*-PDMS thin film showed short-range lateral order (Figure 5.8b). In contrast, when brush-coated HSQ guiding posts were employed, a single grain of spherical microdomains was produced, as shown in Figure 5.8c and d. Also, they found that the orientation and the domain spacing of the spherical microdomains on post surfaces were dictated by the commensurability between the equilibrium spacing of the BCP on a flat surface (L_0) and the period of HSQ post (L_{post}), which is supported by a free energy model as following:

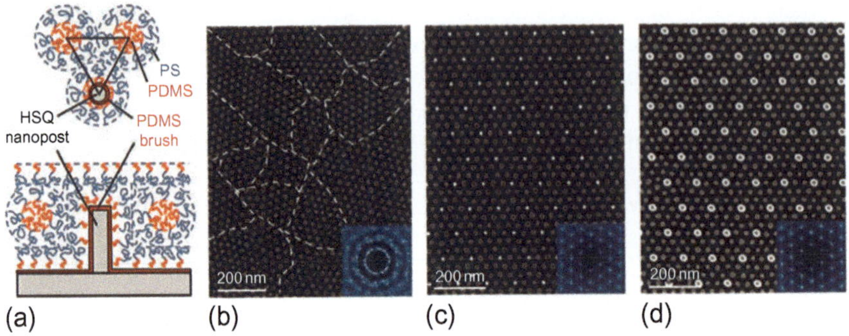

Figure 5.8 (a) Top-down and side-view schematics showing the arrangement of PS-*b*-PDMS block copolymer molecules in the region surrounding a single post made from cross-linked HSQ resist. The post and substrate surfaces have been chemically functionalized by a monolayer of short-chain PDMS brush. (b) SEM images of a poorly ordered monolayer of BCP spherical domains formed on a flat surface, that is, without templating. The boundaries between different grain orientations are indicated with dashed lines. The inset is a 2-D Fourier transform of the domain positions that shows the absence of long-range order. (c and d) SEM images of ordered BCP spheres formed within a sparse 2-D lattice of HSQ posts (brighter dots). The substrate and post surfaces were functionalized with a PDMS brush layer in (c), which corresponds to the schematic in (a), and with a PS brush layer in (d). The insets show the 2-D Fourier transforms in which the low-frequency components originate from the post lattice.
From Bita et al. (2008). Reprinted with permission from AAAS.

$$\frac{\Delta F_{\text{chain}}}{kT} = \frac{2Mb}{L}\sqrt{\frac{\chi_{\text{AB}}}{6}} + \frac{1}{2}\left(\frac{L^2}{4Mb^2} + \frac{4b\sqrt{M}}{L} - 3\right) \tag{5.3}$$

where, ΔF_{chain} is the free energy per BCP chain, k is Boltzmann's constant, T is the temperature, M is the number of statistical segments of the BCP chain, b is the statistical segment length, L is the strained BCP spacing required for the BCP lattice to be commensurate with the template for the considered $\langle i\ \ j\rangle$ configuration, and χ_{AB} is the Flory–Huggins interaction parameter. When they plotted this free energy curve as a function of L_{post}/L_0, the free energy showed a minimum at L_{post}/L_0 values where the HSQ post lattice was commensurate with a lattice of unstrained BCP microdomain. This theoretical prediction was closely aligned with the obtained experimental results, so that they could observe all predicted BCP microdomains lattice types, such as $\langle 1\ \ 1\rangle$, $\langle 2\ \ 0\rangle$, $\langle 2\ \ 1\rangle$, $\langle 3\ \ 0\rangle$, $\langle 2\ \ 2\rangle$, $\langle 3\ \ 1\rangle$, $\langle 4\ \ 0\rangle$, $\langle 3\ \ 2\rangle$, and $\langle 4\ \ 1\rangle$ by varying L_{post}. Also, a single BCP lattice orientation or two or more different BCP lattice orientations were observed depending on a value of L_{post}/L_0 due to changing width and number of the potential wells in the free energy curve. Furthermore, Ross and coworkers reported more complex self-assembled BCP patterns by using HSQ post surface (Yang et al., 2010). By changing the post lattice parameters L_x and L_y, where L_x and L_y are the periodicities in the x and y directions, respectively, the lattice orientation angles of BCP were controlled. When they varied the geometry of the HSQ posts and their motif, more complex BCP patterns, including bends, junctions, and zigzag patterns could be produced.

Berggren and coworkers reported the directed self-assembly of cylindrical microdomains of PS-*b*-PDMS with a sub-10-nm period by using rectangular arrays of posts (Chang et al., 2012). In this study, every possible commensurate case to have the orientations of PDMS cylinders from 0° to 90° with respect to the *x*-axis was investigated by varying parameters L_x and L_y. Also, the effect of length scales of HSQ posts on the self-assembly of BCPs was demonstrated. It was found that the larger diameter and greater height of the HSQ posts enhanced the *y*-axis orientation similar to 1-D trench patterns while the smaller diameter and shorter height of the posts provided poor ordering. Recently, Ross and coworkers generated 3-D multilevel structures of BCPs by using a bilayer of in-plane PDMS cylinders in a PS matrix with chemically functionalized HSQ posts (Tavakkoli et al., 2012). By varying the post array periodicity in the *x* and *y* directions, the orientation and morphology of BCP microdomains in each layer were controlled, so that various structures could be achieved such as perpendicular and angled mesh-shaped structures, different periodic superstructures, cylinders on top of spheres, cylinders on top of square symmetry perforated lamellae, and cylinders on top of ellipsoids, which were described by self-consistent field theory simulations. They demonstrated that this strategy for 3-D routing of dense BCP microdomains could be a new route to fabricate nanoscale device manufacturing. Further, Alexander-Katz and coworkers studied the optimization of post templates for the target BCP patterns having complex features by using inverse design simulations (Hannon et al., 2014). In this work, they first performed simulations to predict the pattern of HSQ posts necessary to guide PS-*b*-PDMS diblock copolymers. By using these simulated posts, they produced the target patterns such as junctions, lines, and terminations.

5.3.2.3 Other surfaces

Kim and coworkers investigated the self-assembly of symmetric PS-*b*-PMMA diblock copolymers on wrinkled surfaces having balanced interfacial interactions (Kim et al., 2014). They fabricated a wrinkle pattern by using Norland Optical Adhesive (NOA71) UV-curable epoxy thin film that was biaxially strained using oxygen plasma. After curing with UV irradiation, they obtained randomly oriented wrinkle patterns having a pitch of 4 μm and an amplitude of 800 nm due to the uniform equi-biaxial stress on NOA71 surface. After the directed self-assembly of lamellar microdomains of PS-*b*-PMMA on this wrinkled surface, it was found that lamellar microdomains in the center of the wrinkle were aligned normal to the wrinkle direction, due to the thickness gradient by geometric anchoring. They also studied the wrinkle-directed self-assembly of BCPs on various straight and periodic wrinkle surfaces and on graphene-wrinkled surfaces.

On 1-D or 2-D planar patterned surfaces, such as trench and post patterns, the commensurability between the pitch of the pattern and L_0 and the preferential interaction between the one block of BCP and the substrate play a key role in dictating morphology of BCP in thin films. However, BCP morphologies on nonplanar templates, such as cylindrical or spherical, can be affected by both the commensurability and curvature of the template (Dobriyal et al., 2009; He et al., 2001; Sevink et al., 2001; Shin et al., 2004; Wu et al., 2004; Xiang et al., 2004, 2005a,b). Russell and coworkers studied poly(styrene-*b*-butadiene) (PS-*b*-PBD) diblock copolymer confined in cylindrical nanoporous alumina membranes (Shin et al., 2004). In the bulk, these PS-*b*-PBD diblock copolymers showed lamellar morphology, while when PS-*b*-PBD was confined in cylindrical pore with diameter, *d*, of 45 nm ($d/L_0 \sim 2.6$), it showed a stacked-disc or toroidal-type structures, which deviated from the bulk morphology because of the imposed high degree of curvature and incommensurability.

5.3.3 Geometry with minimal patterning

Although topographical surfaces with the deep geometry can control orientation and/or position of BCP microdomains with lateral order, there is limitation in the final grain size of BCP microdomains owing to the confined width. Minimal perturbation patterning can be a route to overcome this limitation. In general, the confining depth of the BCP guide pattern is smaller than L_0 so that on surfaces with minimal topographical patterning, grain sizes several orders of magnitude greater than that of BCPs on deep topographical surfaces can be achieved by overcoming the width of the confinement.

5.3.3.1 Faceted surfaces

Perhaps the simplest means of topographically patterning a surface is to reconstruct the surface of a single crystalline material, like Si or sapphire, that is miscut, namely, cut along a crystal plane that is stable at room temperature but unstable at elevated temperatures (Gabai et al., 2007; Huth et al., 2002; Song and Mochrie, 1995; Song et al., 1995). The facets have a depth of several nanometers with a pitch of tens of nanometers, so that it can be used to guide the self-assembly of BCPs. Fasolka et al. (1997) studied symmetric diblock copolymer thin films with thickness below L_0 on the faceted Si surface. They found that BCP microdomains were decorated along the facet ridges due to the variation in film thickness on the faceted surface. Rockford et al. (1999) investigated lamellar forming PS-*b*-PMMA diblock copolymer on the faceted surface. In this case, they prepared periodic heterogeneous surfaces of polar (silicon oxide) and nonpolar (gold) by the glancing angle evaporation of metals on the miscut Si wafer having a sawtooth pattern with an amplitude of several nanometers. In this case, the silicon oxide preferentially interacts with PMMA block while the gold interacts with PS block. It was found that the orientation and the ordering of lamellar microdomains of PS-*b*-PMMA thin films on the faceted surface were strongly affected by the commensurability between L_0 and the period of the faceted substrate. Also, they reported a surface-directed morphology of lamellar forming PS-*b*-PMMA diblock copolymer in thick films on the Si faceted surface with alternating stripes of silicon oxide and gold (Rockford et al., 2001).

Recently, Russell and coworkers demonstrated an ultradense single grain array of cylindrical microdomains of BCPs on the sapphire faceted surface (Park et al., 2009). To produce the faceted surface, the miscut single crystal sapphire surface (Figure 5.9a) was thermally annealed in air from 1300 to 1500 °C, which produced patterns with pitches from 24 to 160 nm and amplitudes from 3 to 20 nm (Figure 5.9b). These sawtooth patterns were employed to guide cylindrical microdomains of PS-*b*-PEO diblock copolymer with different molecular weights. After producing PS-*b*-PEO thin films

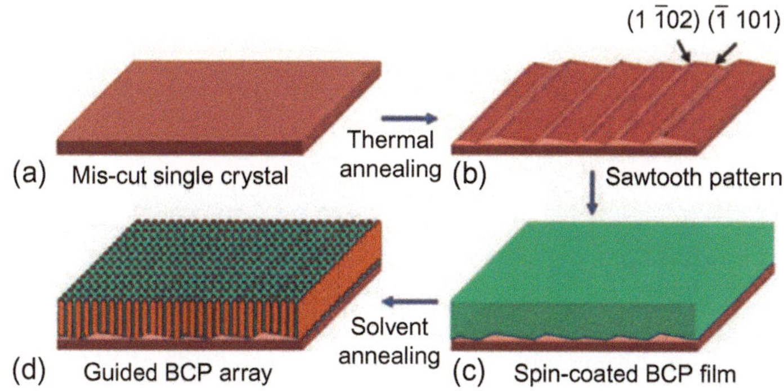

Figure 5.9 Schematic illustration of the strategy used for generating BCP cylindrical microdomains on highly oriented crystalline facets on a single crystal surface. Here, the M-plane sapphire (a) was annealed in air at 1300–1500 °C for 24 h to form sawtooth patterns. (b) Upon thermal annealing, the flat surface of the M-plane sapphire reconstructs into facets with the $(\bar{1}101)$ and $(1\bar{1}02)$ planes on opposite edges. (c) Smooth PS-*b*-PEO (Mn = 43.0 kg/mol) thin films were spin-coated onto these surfaces that, upon annealing in *o*-xylene vapors (d), produced highly ordered cylindrical microdomains oriented normal to the film surface.
From Park et al. (2009). Reprinted with permission from AAAS.

(Figure 5.9c) followed by SVA, they produced a single hexagonal array of cylindrical microdomains with perpendicular orientation having areal densities of 0.74–10.5 Tbit/in.^{2} on a 1.5 cm × 1.5 cm faceted pattern (Figure 5.9d). Grazing incidence small-angle X-ray scattering (GISAXS) was used to characterize the single grain of PS-*b*-PEO thin film on the sapphire sawtooth pattern by rotating the sample stage from 0° to 90°. Even though there were defects in the underlying faceted pattern and variations in the pitches and the amplitudes of the pattern, the generated single grain of BCP microdomains showed a perfect orientational order and a quasi-long-range crystalline order. Furthermore, they extended this method to align cylindrical microdomains of PS-*b*-PEO with perpendicular orientation over large areas on various flexible substrates, which were replicated from the faceted sapphire surface (Park et al., 2010).

Line patterns of BCP thin films on imprinted polymer replicas as well as faceted sapphire surfaces were investigated by Russell and coworkers (Hong et al., 2012a). They produced single grain line patterns of PS-*b*-PEO diblock copolymer on the sapphire sawtooth surface with SVA, which was driven by entropic penalties of chain packing at the substrate. These line patterns were investigated by GISAXS as a function of the rotation angle. With increasing the rotation angle of the sample from 0°, where the X-ray beam was parallel with (1 0) lattice line, to 4°, the (1 0) reflection was broadened and the intensity was significantly decreased. Based on these GISAXS data, the orientation parameter, f, where $f = 1.0$ for perfect orientation of the microdomains, and $f = 0$ for random orientation, was calculated. In this study, f was more than 0.97, which was characteristic of unidirectionally aligned line patterns of cylindrical microdomains with a single grain. Furthermore, the same group demonstrated the line pattern of PS-*b*-PEO thin films on the faceted Si pattern, which showed more asymmetric geometry than that of the faceted sapphire pattern (Hong et al., 2012b). In this case, by rotating PS-*b*-PEO thin films during GISAXS experiments, the orientation and the lateral ordering of PS-*b*-PEO microdomains over macroscopic areas were characterized.

5.3.3.2 Shallow trench surfaces

Russell and coworkers studied the self-assembly of poly(styrene-*b*-4-vinylpyridine) (PS-*b*-P4VP) diblock copolymer with SVA on the shallow trench patterns of two different depths, 5 and 14 nm, respectively (Park et al., 2008). They varied the widths of the mesa from 50 to 120 nm with maintaining the trench depth of 14 nm. In this case, all obtained hexagonal arrays of PS-*b*-P4VP microdomains showed long-range order in the confined trenches, so that the grain size was determined by the width of the trench. In contrast, when they used the trench depth of 5 nm, they produced a single grain of at least 5 μm × 5 μm of hexagonal array of PS-*b*-P4VP microdomains on patterned areas with a high degree of lateral order after SVA for 6 h. To characterize the orientation and the order of BCP microdomains, moirè pattern analyses were used by SFM.

Circular line patterns of BCPs with areal density multiplication on circular shallow trench patterns 8 nm deep were generated by Russell and coworkers (Hong et al., 2011). As shown in Figure 5.10, they fabricated three different pattern dimensions such as $I_{29,80}$ (trench width = 29 nm and trench pitch = 80 nm), $M_{31,99}$ (trench width = 31 nm and trench pitch = 99 nm), and $O_{34,121}$ (trench width = 34 nm and trench width = 121 nm) using EBL. Of these three patterns, parallel cylindrical microdomains of PS-*b*-PEO thin film were guided by only the $M_{31,99}$ pattern due to commensurability, showing long-range lateral order with areal density multiplication by a factor of 3. To demonstrate the generality of this approach, they also used a thin film of cylinder-forming PS-*b*-P2VP diblock copolymer. In this case, highly oriented and ordered PS-*b*-P2VP microdomains were observed only on the $I_{29,80}$ pattern because of commensurability. Since the depth of the trench pattern was 8 nm, which was much smaller than L_0, the surfaces of both PS-*b*-PEO and PS-*b*-P2VP microdomains were flat over large areas, which corresponded to a single grain of circular pattern by overcoming the limitation from the width of the trench.

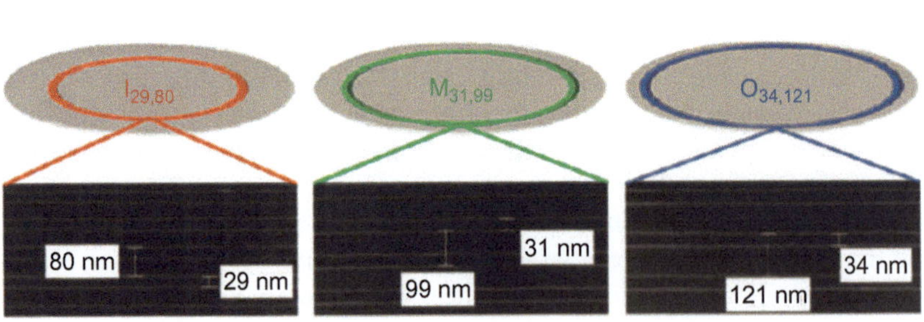

Figure 5.10 Graphical illustrations and the SEM images of three different kinds of circularly patterned areas and the magnifications of inner ($I_{29,80}$), middle ($M_{31,99}$), and outer ($O_{34,121}$) rim regions.

Reprinted with permission from Hong et al. (2011). Copyright 2011 American Chemical Society.

5.4 Application of BCPs guided by topographical surfaces

BCP lithographic masks with graphoepitaxy have been investigated for pattern transfer (Tsai et al., 2012; Tseng et al., 2011). Tseng and coworkers demonstrated pattern transfer directly from PS-*b*-PMMA films to silicon by the combination of plasma etching with topographic pattern (Tseng et al., 2011). In this study, after obtaining well-ordered line patterns of PS-*b*-PMMA in trench patterns, they used sequential infiltration synthesis to enhance the etch contrast between PS and PMMA blocks. By applying plasma etching with HBr and O_2, PMMA cylinders were directly transferred into the underlying silicon, so that a high aspect ratio pattern transfer of the line patterns could be achieved. Also, pattern transfer of BCP line patterns generated by graphoepitaxy into dielectric materials, such as silicon nitride and silicon oxides, was systematically investigated in terms of the evolution of critical dimension, LER, and line-width roughness (Tsai et al., 2012). It was found that the final LER value can be varied depending on the choice of the hard mask material. Also, they demonstrated that the applied stresses in BCPs, due to the strength of the confinement in the trench pattern, can have an impact on final LER value.

Metal dot arrays can be produced by the directed self-assembly of BCP in thin films with graphoepitaxy (Naito et al., 2002; Shin et al., 2013; Xiao et al., 2005; Yang et al., 2004). Nealey and coworkers fabricated high-quality patterned magnetic dot arrays using the trench pattern (Xiao et al., 2005). The trench pattern was first modified by PS-*r*-PMMA random copolymer to induce cylindrical microdomains of PS-*b*-PMMA diblock copolymer normal to the substrate. After producing hexagonally packed PMMA cylinders in the PS matrix in the trench pattern with thermal annealing, PMMA blocks were selectively removed by UV irradiation followed by rinsing with acetic acid, so that nanoporous PS templates could be deposited with Co. It was found that the defect-free hexagonal arrays of Co dots with long-range lateral order in the trench pattern could be achieved when they used a dry etching for the lift-off process as opposed to a wet etching. Recently, Kim and coworkers reported the multicomponent nanopatterns, including Pt nanowire-Pd nanodot and Au–Pt nanodot arrays by the combination of the self-assembly of BCPs with graphoepitaxy (Shin et al., 2013).

The fabrication of nanowire using the directed self-assembly of BCPs with graphoepitaxy has been explored (Black, 2005, 2007; Chang et al., 2009; Jeong et al., 2009, 2010b; Jung et al., 2008b, 2010b; Kim et al., 2013a; Moon et al., 2014). Black demonstrated the generation of silicon field effect transistor (FET) with 6, 8, 10, and 16 nanowires using BCP lithography (Black, 2005). Since the number of parallel nanowires can be controlled by changing the width of the trench, the applied self-assembly approach helps to control the scalability of the nanowire-based devices. In this work, the silicon nanowires showed current drives of ~5 μA/wire and on-current to off-current ratio of nearly ~10^5. Ross and coworkers prepared conducting poly(3,4-ethylenedioxythiophene)-poly(styrenesulfonate) (PEDOT:PSS) nanowires by using cylinder-forming PS-*b*-PDMS diblock copolymers in topographical patterns (Jung et al., 2008b). The in-plane PDMS cylinders in a PS matrix within the trench pattern were transferred to the underlying PEDOT:PSS by plasma etching, so that the generated nanowires could be used for gas sensing. Due to the high surface-to-volume ratio

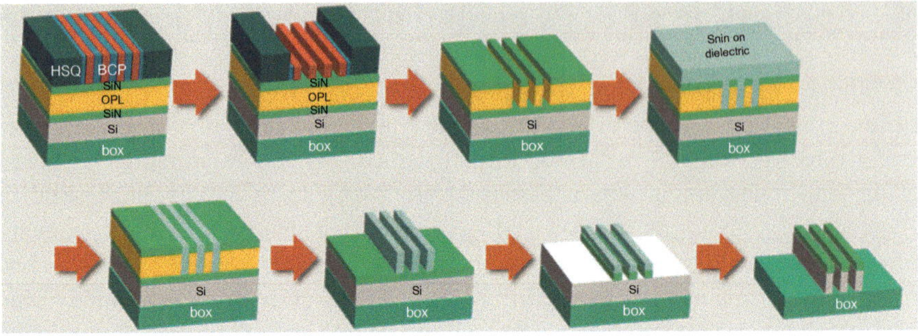

Figure 5.11 Process flow for forming groups of silicon on insulator (SOI) fins with a controlled fin spacing. A tone inversion operation is applied after etch transferring the DSA pattern into an organic planarizing layer (OPL). The tone inverted image is then transferred into a fin hard mask layer and 20 nm of SOI.
Reprinted with permission from Tsai et al. (2014). Copyright 2014 American Chemical Society.

and the short diffusion length, this 1-D nanowire sensor showed higher sensitivity to ethanol vapor than the unpatterned film with the same thickness. The same group also developed the fabrication of various metallic nanowires such as Ti, W, Pt, Co, Ni, Ta, Au, and Al with a high degree of ordering within the trench pattern by using cylindrical microdomains of PS-*b*-PDMS with parallel orientation (Jung et al., 2010b). More recently, Tsai and coworkers fabricated a fin field effect transistor (FinFET) by using graphoepitaxy of lamellar-forming PS-*b*-PMMA diblock copolymers (Tsai et al., 2014). Using HSQ trench patterns, they controlled the lateral order of the BCP micorodomains, so that these aligned microdomains were transferred into the underlying layers by the series of plasma etching steps to fabricate fins, as shown in Figure 5.11.

5.5 Summary and outlook

In this chapter, we reviewed the control of interfacial interactions at the substrate/copolymer interface using random copolymer brushes to generate a neutral substrate. This controlling method can be used to fabricate chemically controlled surfaces on topographic patterns, so that a desired orientation of BCP microdomains can be achieved. We also discussed several standard lithography techniques to fabricate topographic patterns such as photolithography, interference lithography, EBL, soft lithography, and nanoimprint lithography. We largely categorized topographical surfaces for guiding the self-assembly of BCPs into two geometric types. The first type is the geometry with deep patterning, where the vertical length scale is greater than L_0, so that it can easily control BCP microdomains within the confined areas with high spatial resolution and good placement accuracy. However, in this case, the grain size of BCP microdomains is limited by the confined width. The second type is the geometry with minimal patterning. One representative example of these minimal geometries is a faceted pattern, where the confining depth of the topographic pattern is smaller than L_0 so that on surfaces with the faceted pattern, grain size several orders of magnitude greater than that of BCPs on deep topographical surfaces can be achieved by overcoming the confined width. Also, the applications of topographically guided BCP microdomains, including lithographic masks, metal dot arrays, and nanowire arrays, were reviewed.

There has been enormous progress in understanding and developing the self-assembly of BCP thin films using topographical surfaces over the past decade. However, to date, most efforts have focused on the fabrication of periodic nanoscopic materials by using diblock copolymers with topographical surfaces as templates and scaffolds. There has been little effort to fabricate aperiodic patterns such as isolated lines, jogs, T-junctions, isolated spots, and sharp bends using the self-assembly of BCPs with topographical surfaces because of the complexity. The use of other types of BCPs such as miktoarm star terpolymers could widen the processing window for the fabrication of these complex patterns. Also, theoretical research on complex patterns are required to gain insight for graphoepitaxy, so that one can reduce the number of trial and error to sift through a relatively large parameter space. Lastly, generating 3-D hierarchical structures with the self-assembly of BCPs is still challenging. Further efforts in this area can lead us to unprecedented materials that have not yet been achieved using 2-D BCP films.

References

Almdal, K., Rosedale, J.H., Bates, F.S., Wignall, G.D., Fredrickson, G.H., 1990. Gaussian-to stretched-coil transition in block copolymer melts. Phys. Rev. Lett. 65, 1112–1115.

Anastasiadis, S.H., Russell, T.P., Satija, S.K., Majkrzak, C.F., 1989. Neutron reflectivity studies of the surface-induced ordering of diblock copolymer films. Phys. Rev. Lett. 62, 1852–1855.

Auzelyte, V., Dais, C., Farquet, P., Grützmacher, D., Heyderman, L.J., Luo, F., Olliges, S., Padeste, C., Sahoo, P.K., Thomson, T., Turchanin, A., David, C., Solak, H.H., 2009. Extreme ultraviolet interference lithography at the Paul Scherrer Institut. J. Micro/Nanolith. MEMS MOEMS 8, 021204–021210.

Bang, J., Bae, J., Löwenhielm, P., Spiessberger, C., Given-Beck, S.A., Russell, T.P., Hawker, C.J., 2007. Facile routes to patterned surface neutralization layers for block copolymer lithography. Adv. Mater. 19, 4552–4557.

Bates, F.S., 1991. Polymer-polymer phase behavior. Science 251, 898–905.

Bates, F.S., Fredrickson, G.H., 1990. Block copolymer thermodynamics: theory and experiment. Annu. Rev. Phys. Chem. 41, 525–557.

Bates, F.S., Fredrickson, G.H., 1999. Block copolymers-designer soft materials. Phys. Today 52, 32–38.

Bita, I., Yang, J.K.W., Jung, Y.S., Ross, C.A., Thomas, E.L., Berggren, K.K., 2008. Graphoepitaxy of self-assembled block copolymers on two-dimensional periodic patterned templates. Science 321, 939–943.

Black, C.T., 2005. Self-aligned self assembly of multi-nanowire silicon field effect transistors. Appl. Phys. Lett. 87, 163116.

Black, C.T., 2007. Block copolymers: nanowire arrays build themselves. Nat. Nanotechnol. 2, 464–465.

Black, C.T., Bezencenet, O., 2004. Nanometer-scale pattern registration and alignment by directed diblock copolymer self-assembly. IEEE Trans. Nanotechnol. 3, 412–415.

Brown, G., Chakrabarti, A., 1995. Ordering of block copolymer melts in confined geometry. J. Chem. Phys. 102, 1440–1448.

Cai, Z.h., Huang, K., Montano, P.A., Russell, T.P., Bai, J.M., Zajac, G.W., 1993. Experimental study of the surface structure of diblock copolymer films using microscopy and X-ray scattering. J. Chem. Phys. 98, 2376–2386.

Chai, J., Buriak, J.M., 2008. Using cylindrical domains of block copolymers to self-assemble and align metallic nanowires. ACS Nano 2, 489–501.

Chai, J., Wang, D., Fan, X., Buriak, J.M., 2007. Assembly of aligned linear metallic patterns on silicon. Nat. Nanotechnol. 2, 500–506.

Chang, S.-W., Chuang, V.P., Boles, S.T., Ross, C.A., Thompson, C.V., 2009. Densely packed arrays of ultra-high-aspect-ratio silicon nanowires fabricated using block-copolymer lithography and metal-assisted etching. Adv. Funct. Mater. 19, 2495–2500.

Chang, J.-B., Son, J.G., Hannon, A.F., Alexander-Katz, A., Ross, C.A., Berggren, K.K., 2012. Aligned sub-10-nm block copolymer patterns templated by post arrays. ACS Nano 6, 2071–2077.

Chang, J.-B., Choi, H.K., Hannon, A.F., Alexander-Katz, A., Ross, C.A., Berggren, K.K., 2014. Design rules for self-assembled block copolymer patterns using tiled templates. Nat. Commun. 5, 1–9, Article number: 3305.

Cheng, J.Y., Ross, C.A., Thomas, E.L., Smith, H.I., Vancso, G.J., 2002. Fabrication of nanostructures with long-range order using block copolymer lithography. Appl. Phys. Lett. 81, 3657–3659.

Cheng, J.Y., Ross, C.A., Thomas, E.L., Smith, H.I., Vancso, G.J., 2003. Templated self-assembly of block copolymers: effect of substrate topography. Adv. Mater. 15, 1599–1602.

Cheng, J.Y., Mayes, A.M., Ross, C.A., 2004. Nanostructure engineering by templated self-assembly of block copolymers. Nat. Mater. 3, 823–828.

Cheng, J.Y., Zhang, F., Chuang, V.P., Mayes, A.M., Ross, C.A., 2006a. Self-assembled one-dimensional nanostructure arrays. Nano Lett. 6, 2099–2103.

Cheng, J.Y., Zhang, F., Smith, H.I., Vancso, G.J., Ross, C.A., 2006b. Pattern registration between spherical block-copolymer domains and topographical templates. Adv. Mater. 18, 597–601.

Cheng, J.Y., Pitera, J., Park, O.-H., Flickner, M., Ruiz, R., Black, C.T., Kim, H.-C., 2007. Rapid directed self assembly of lamellar microdomains from a block copolymer containing hybrid. Appl. Phys. Lett. 91, 143106-1–143106-3.

Chou, S.Y., Krauss, P.R., Renstrom, P.J., 1996. Nanoimprint lithography. J. Vac. Sci. Technol. B 14, 4129–4133.

Chuang, V.P., Cheng, J.Y., Savas, T.A., Ross, C.A., 2006. Three-dimensional self-assembly of spherical block copolymer domains into V-shaped grooves. Nano Lett. 6, 2332–2337.

Chuang, V.P., Gwyther, J., Mickiewicz, R.A., Manners, I., Ross, C.A., 2009. Templated self-assembly of square symmetry arrays from an ABC triblock terpolymer. Nano Lett. 9, 4364–4369.

Coulon, G., Russell, T.P., Deline, V.R., Green, P.F., 1989. Surface-induced orientation of symmetric, diblock copolymers: a secondary ion mass-spectrometry study. Macromolecules 22, 2581–2589.

Dobriyal, P., Xiang, H., Kazuyuki, M., Chen, J.-T., Jinnai, H., Russell, T.P., 2009. Cylindrically confined diblock copolymers. Macromolecules 42, 9082–9088.

Edwards, E.W., Stoykovich, M.P., Solak, H.H., Nealey, P.F., 2006. Long-range order and orientation of cylinder-forming block copolymers on chemically nanopatterned striped surfaces. Macromolecules 39, 3598–3607.

Farhoud, M., Ferrera, J., Lochtefeld, A.J., Murphy, T.E., Schattenburg, M.L., Carter, J., Ross, C.A., Smith, H.I., 1999. Fabrication of 200 nm period nanomagnet arrays using interference lithography and a negative resist. J. Vac. Sci. Technol. B 17, 3182–3185.

Fasolka, M.J., Mayes, A.M., 2001. Block copolymer thin films: physics and applications. Annu. Rev. Mater. Res. 31, 323–355.

Fasolka, M.J., Harris, D.J., Mayes, A.M., Yoon, M., Mochrie, S.G.J., 1997. Observed substrate topography-mediated lateral patterning of diblock copolymer films. Phys. Rev. Lett. 79, 3018–3021.

Gabai, R., Ismach, A., Joselevich, E., 2007. Nanofacet lithography: a new bottom-up approach to nanopatterning and nanofabrication by soft replication of spontaneously faceted crystal surfaces. Adv. Mater. 19, 1325–1330.

Gates, B.D., Xu, Q., Stewart, M., Ryan, D., Willson, C.G., Whitesides, G.M., 2005. New approaches to nanofabrication: molding, printing, and other techniques. Chem. Rev. 105, 1171–1196.

Geis, M.W., Flanders, D.C., Smith, H.I., 1979. Crystallographic orientation of silicon on an amorphous substrate using an artificial surface-relief grating and laser crystallization. Appl. Phys. Lett. 35, 71–74.

Geissler, M., Xia, Y., 2004. Patterning: principles and some new developments. Adv. Mater. 16, 1249–1269.

Green, P.F., Limary, R., 2001. Block copolymer thin films: pattern formation and phase behavior. Adv. Colloid Interf. Sci. 94, 53–81.

Grigorescu, A.E., Hagen, C.W., 2009. Resists for sub-20-nm electron beam lithography with a focus on HSQ: state of the art. Nanotechnology 20, 292001.

Guarini, K.W., Black, C.T., Yeung, S.H.I., 2002. Optimization of diblock copolymer thin film self assembly. Adv. Mater. 14, 1290–1294.

Hamley, I.W., 2003. Nanostructure fabrication using block copolymers. Nanotechnology 14, R39–R54.

Hammond, M.R., Kramer, E.J., 2006. Edge effects on thermal disorder in laterally confined diblock copolymer cylinder monolayers. Macromolecules 39, 1538–1544.

Hammond, M.R., Cochran, E., Fredrickson, G.H., Kramer, E.J., 2005. Temperature dependence of order, disorder, and defects in laterally confined diblock copolymer cylinder monolayers. Macromolecules 38, 6575–6585.

Han, E., In, I., Park, S.M., La, Y.H., Wang, Y., Nealey, P.F., Gopalan, P., 2007. Photopatternable imaging layers for controlling block copolymer microdomain orientation. Adv. Mater. 19, 4448–4452.

Han, E., Kang, H., Liu, C.-C., Nealey, P.F., Gopalan, P., 2010. Graphoepitaxial assembly of symmetric block copolymers on weakly preferential substrates. Adv. Mater. 22, 4325–4329.

Hannon, A.F., Ding, Y., Bai, W., Ross, C.A., Alexander-Katz, A., 2014. Optimizing topographical templates for directed self-assembly of block copolymers via inverse design simulations. Nano Lett. 14, 318–325.

Hatzakis, M., 1969. Electron resists for microcircuit and mask production. J. Electrochem. Soc. 116, 1033–1037.

Hawker, C.J., Russell, T.P., 2005. Block copolymer lithography: merging "bottom-up" with "top-down" processes. MRS Bull. 30, 952–966.

Hawker, C.J., Elce, E., Dao, J., Volksen, W., Russell, T.P., Barclay, G.G., 1996. Well-defined random copolymers by a "living" free-radical polymerization process. Macromolecules 29, 2686–2688.

He, X., Song, M., Liang, H., Pan, C., 2001. Self-assembly of the symmetric diblock copolymer in a confined state: Monte Carlo simulation. J. Chem. Phys. 114, 10510–10513.

Hexemer, A., Stein, G.E., Kramer, E.J., Magonov, S., 2005. Block copolymer monolayer structure measured with scanning force microscopy moiré patterns. Macromolecules 38, 7083–7089.

Hong, S.W., Gu, X., Huh, J., Xiao, S., Russell, T.P., 2011. Circular nanopatterns over large areas from the self-assembly of block copolymers guided by shallow trenches. ACS Nano 5, 2855–2860.

Hong, S.W., Huh, J., Gu, X., Lee, D.H., Jo, W.H., Park, S., Xu, T., Russell, T.P., 2012a. Unidirectionally aligned line patterns driven by entropic effects on faceted surfaces. Proc. Natl. Acad. Sci. U. S. A. 109, 1402–1406.

Hong, S.W., Voronov, D.L., Lee, D.H., Hexemer, A., Padmore, H.A., Xu, T., Russell, T.P., 2012b. Controlled orientation of block copolymers on defect-free faceted surfaces. Adv. Mater. 24, 4278–4283.

Huang, E., Russell, T.P., Harrison, C., Chaikin, P.M., Register, R.A., Hawker, C.J., Mays, J., 1998. Using surface active random copolymers to control the domain orientation in diblock copolymer thin films. Macromolecules 31, 7641–7650.

Huang, E., Pruzinsky, S., Russell, T.P., Mays, J., Hawker, C.J., 1999. Neutrality conditions for block copolymer systems on random copolymer brush surfaces. Macromolecules 32, 5299–5303.

Huang, E., Mansky, P., Russell, T.P., Harrison, C., Chaikin, P.M., Register, R.A., Hawker, C.J., Mays, J., 2000. Mixed lamellar films: evolution, commensurability effects, and preferential defect formation. Macromolecules 33, 80–88.

Husseman, M., Malmström, E.E., McNamara, M., Mate, M., Mecerreyes, D., Benoit, D.G., Hedrick, J.L., Mansky, P., Huang, E., Russell, T.P., Hawker, C.J., 1999. Controlled synthesis of polymer brushes by "living" free radical polymerization techniques. Macromolecules 32, 1424–1431.

Huth, M., Ritley, K.A., Oster, J., Dosch, H., Adrian, H., 2002. Highly ordered Fe and Nb stripe arrays on facetted α-Al_2O_3 (1010). Adv. Funct. Mater. 12, 333–338.

ITRS, 2013. International Technology Roadmap for Semiconductors (ITRS) 2013 Edition. http://www.itrs.net.

Jeong, S.-J., Kim, J.E., Moon, H.-S., Kim, B.H., Kim, S.M., Kim, J.B., Kim, S.O., 2009. Soft graphoepitaxy of block copolymer assembly with disposable photoresist confinement. Nano Lett. 9, 2300–2305.

Jeong, S.-J., Moon, H.-S., Kim, B.H., Kim, J.Y., Yu, J., Lee, S., Lee, M.G., Choi, H., Kim, S.O., 2010a. Ultralarge-area block copolymer lithography enabled by disposable photoresist prepatterning. ACS Nano 4, 5181–5186.

Jeong, S.-J., Moon, H.-S., Shin, J., Kim, B.H., Shin, D.O., Kim, J.Y., Lee, Y.-H., Kim, J.U., Kim, S.O., 2010b. One-dimensional metal nanowire assembly via block copolymer soft graphoepitaxy. Nano Lett. 10, 3500–3505.

Jeong, J.W., Park, W.I., Do, L.-M., Park, J.-H., Kim, T.-H., Chae, G., Jung, Y.S., 2012. Nanotransfer printing with sub-10nm resolution realized using directed self-assembly. Adv. Mater. 24, 3526–3531.

Jung, Y.S., Ross, C.A., 2007. Orientation-controlled self-assembled nanolithography using a polystyrene-polydimethylsiloxane block copolymer. Nano Lett. 7, 2046–2050.

Jung, Y.S., Jung, W., Ross, C.A., 2008a. Nanofabricated concentric ring structures by templated self-assembly of a diblock copolymer. Nano Lett. 8, 2975–2981.

Jung, Y.S., Jung, W., Tuller, H.L., Ross, C.A., 2008b. Nanowire conductive polymer gas sensor patterned using self-assembled block copolymer lithography. Nano Lett. 8, 3776–3780.

Jung, Y.S., Chang, J.B., Verploegen, E., Berggren, K.K., Ross, C.A., 2010a. A path to ultranarrow patterns using self-assembled lithography. Nano Lett. 10, 1000–1005.

Jung, Y.S., Lee, J.H., Lee, J.Y., Ross, C.A., 2010b. Fabrication of diverse metallic nanowire arrays based on block copolymer self-assembly. Nano Lett. 10, 3722–3726.

Kellogg, G.J., Walton, D.G., Mayes, A.M., Lambooy, P., Russell, T.P., Gallagher, P.D., Satija, S.K., 1996. Observed surface energy effects in confined diblock copolymers. Phys. Rev. Lett. 76, 2503–2506.

Kikuchi, M., Binder, K., 1993. Monte carlo study of thin films of the symmetric diblock-copolymer melt. Europhys. Lett. 21, 427.

Kim, G., Libera, M., 1998a. Morphological development in solvent-cast polystyrene-polybutadiene-polystyrene (SBS) triblock copolymer thin films. Macromolecules 31, 2569–2577.

Kim, G., Libera, M., 1998b. Kinetic constraints on the development of surface microstructure in sbs thin films. Macromolecules 31, 2670–2672.

Kim, S.O., Solak, H.H., Stoykovich, M.P., Ferrier, N.J., de Pablo, J.J., Nealey, P.F., 2003. Epitaxial self-assembly of block copolymers on lithographically defined nanopatterned substrates. Nature 424, 411–414.

Kim, S.H., Misner, M.J., Russell, T.P., 2004a. Solvent-induced ordering in thin film diblock copolymer/homopolymer mixtures. Adv. Mater. 16, 2119–2123.

Kim, S.H., Misner, M.J., Xu, T., Kimura, M., Russell, T.P., 2004b. Highly oriented and ordered arrays from block copolymers via solvent evaporation. Adv. Mater. 16, 226–231.

Kim, H.-C., Rettner, C.T., Sundström, L., 2008. Fabrication of 20nm half-pitch gratings by corrugation-directed self-assembly. Nanotechnology 19, 235301-1–235301-5.

Kim, J.Y., Kim, B.H., Hwang, J.O., Jeong, S.-J., Shin, D.O., Mun, J.H., Choi, Y.J., Jin, H.M., Kim, S.O., 2013a. Flexible and transferrable self-assembled nanopatterning on chemically modified graphene. Adv. Mater. 25, 1331–1335.

Kim, M., Han, E., Sweat, D.P., Gopalan, P., 2013b. Interplay of surface chemical composition and film thickness on graphoepitaxial assembly of asymmetric block copolymers. Soft Matter 9, 6135–6141.

Kim, B.H., Choi, Y., Kim, J.Y., Shin, H., Kim, S., Son, S.-W., Kim, S.O., Kim, P., 2014. Wrinkle-directed self-assembly of block copolymers for aligning of nanowire arrays. Adv. Mater. 26, 4665–4670.

Koneripalli, N., Singh, N., Levicky, R., Bates, F.S., Gallagher, P.D., Satija, S.K., 1995. Confined block copolymer thin films. Macromolecules 28, 2897–2904.

Kurihara, K., Iwadate, K., Namatsu, H., Nagase, M., Takenaka, H., Murase, K., 1995. An electron beam nanolithography system and its application to Si nanofabrication. Jpn. J. Appl. Phys. 34, 6940.

La, Y.-H., Edwards, E.W., Park, S.-M., Nealey, P.F., 2005. Directed assembly of cylinder-forming block copolymer films and thermochemically induced cylinder to sphere transition: a hierarchical route to linear arrays of nanodots. Nano Lett. 5, 1379–1384.

Lambooy, P., Russell, T.P., Kellogg, G.J., Mayes, A.M., Gallagher, P.D., Satija, S.K., 1994. Observed frustration in confined block copolymers. Phys. Rev. Lett. 72, 2899–2902.

Lin, Z.Q., Kim, D.H., Wu, X.D., Boosahda, L., Stone, D., LaRose, L., Russell, T.P., 2002. A rapid route to arrays of nanostructures in thin films. Adv. Mater. 14, 1373–1376.

Liu, C.-C., Nealey, P.F., Raub, A.K., Hakeem, P.J., Brueck, S.R.J., Han, E., Gopalan, P., 2010. Integration of block copolymer directed assembly with 193 immersion lithography. J. Vac. Sci. Technol. B 28, C6B30–C6B34.

Long, B.K., Keitz, B.K., Willson, C.G., 2007. Materials for step and flash imprint lithography (S-FIL®). J. Mater. Chem. 17, 3575–3580.

Mansky, P., Liu, Y., Huang, E., Russell, T.P., Hawker, C., 1997a. Controlling polymer-surface interactions with random copolymer brushes. Science 275, 1458–1460.

Mansky, P., Russell, T.P., Hawker, C.J., Mays, J., Cook, D.C., Satija, S.K., 1997b. Interfacial segregation in disordered block copolymers: effect of tunable surface potentials. Phys. Rev. Lett. 79, 237–240.

Mansky, P., Russell, T.P., Hawker, C.J., Pitsikalis, M., Mays, J., 1997c. Ordered diblock copolymer films on random copolymer brushes. Macromolecules 30, 6810–6813.

Mansky, P., Tsui, O.K.C., Russell, T.P., Gallot, Y., 1999. Phase coherence and microphase separation transitions in diblock copolymer thin films. Macromolecules 32, 4832–4837.

Marrian, C.R.K., Tennant, D.M., 2003. Nanofabrication. J. Vac. Sci. Technol. A 21, S207–S215.

Matsen, M.W., Bates, F.S., 1996. Unifying weak- and strong-segregation block copolymer theories. Macromolecules 29, 1091–1098.

McCord, M.A., Rooks, M.J., 1997. In: Handbook of microlithography, micromachining, and microfabrication. Microlithography, vol. 1. SPIE, Bellingham.

Menelle, A., Russell, T.P., Anastasiadis, S.H., Satija, S.K., Majkrzak, C.F., 1992. Ordering of thin diblock copolymer films. Phys. Rev. Lett. 68, 67–70.

Moon, H.-S., Kim, J.Y., Jin, H.M., Lee, W.J., Choi, H.J., Mun, J.H., Choi, Y.J., Cha, S.K., Kwon, S.H., Kim, S.O., 2014. Atomic layer deposition assisted pattern multiplication of block copolymer lithography for 5 nm scale nanopatterning. Adv. Funct. Mater. 24, 4343–4348.

Naito, K., Hieda, H., Sakurai, M., Kamata, Y., Asakawa, K., 2002. 2.5-inch disk patterned media prepared by an artificially assisted self-assembling method. IEEE Trans. Magn. 38, 1949–1951.

Namatsu, H., Takahashi, Y., Yamazaki, K., Yamaguchi, T., Nagase, M., Kurihara, K., 1998. Three-dimensional siloxane resist for the formation of nanopatterns with minimum line-width fluctuations. J. Vac. Sci. Technol. B 16, 69–76.

Nishida, T., Notomi, M., Iga, R., Tamamura, T., 1992. Quantum wire fabrication by e-beam elithography using high-resolution and high-sensitivity e-beam resist ZEP-520. Jpn. J. Appl. Phys. 31, 4508–4514.

Park, C., Yoon, J., Thomas, E.L., 2003. Enabling nanotechnology with self assembled block copolymer patterns. Polymer 44, 6725–6760.

Park, S.M., Stoykovich, M.P., Ruiz, R., Zhang, Y., Black, C.T., Nealey, P.F., 2007. Directed assembly of lamellae-forming block copolymers by using chemically and topographically patterned substrates. Adv. Mater. 19, 607–611.

Park, S., Kim, B., Yavuzcetin, O., Tuominen, M.T., Russell, T.P., 2008. Ordering of PS-b-P4VP on patterned silicon surfaces. ACS Nano 2, 1363–1370.

Park, S.-M., Berry, B.C., Dobisz, E., Kim, H.-C., 2009a. Observation of surface corrugation-induced alignment of lamellar microdomains in PS-b-PMMA thin films. Soft Matter 5, 957–961.

Park, S.-M., Rettner, C.T., Pitera, J.W., Kim, H.-C., 2009b. Directed self-assembly of lamellar microdomains of block copolymers using topographic guiding patterns. Macromolecules 42, 5895–5899.

Park, S., Lee, D.H., Xu, J., Kim, B., Hong, S.W., Jeong, U., Xu, T., Russell, T.P., 2009. Macroscopic 10-terabit-per-square-inch arrays from block copolymers with lateral order. Science 323, 1030–1033.

Park, S., Lee, D.H., Russell, T.P., 2010. Self-assembly of block copolymers on flexible substrates. Adv. Mater. 22, 1882–1884.

Rockford, L., Liu, Y., Mansky, P., Russell, T.P., Yoon, M., Mochrie, S.G.J., 1999. Polymers on nanoperiodic, heterogeneous surfaces. Phys. Rev. Lett. 82, 2602–2605.

Rockford, L., Mochrie, S.G.J., Russell, T.P., 2001. Propagation of nanopatterned substrate templated ordering of block copolymers in thick films. Macromolecules 34, 1487–1492.

Ruiz, R., Ruiz, N., Zhang, Y., Sandstrom, R.L., Black, C.T., 2007. Local defectivity control of 2D self-assembled block copolymer patterns. Adv. Mater. 19, 2157–2162.

Ruiz, R., Kang, H., Detcheverry, F.A., Dobisz, E., Kercher, D.S., Albrecht, T.R., de Pablo, J.J., Nealey, P.F., 2008. Density multiplication and improved lithography by directed block copolymer assembly. Science 321, 936–939.

Russell, T.P., 1996. Copolymers at surfaces and interfaces. Curr. Opin. Colloid Interface Sci. 1, 107–115.

Russell, T.P., Coulon, G., Deline, V.R., Miller, D.C., 1989. Characteristics of the surface-induced orientation for symmetric diblock PS/PMMA copolymers. Macromolecules 22, 4600–4606.

Russell, T.P., Thurn-Albrecht, T., Tuominen, M., Huang, E., Hawker, C.J., 2000. Block copolymers as nanoscopic templates. Macromol. Symp. 159, 77–88.

Ryu, D.Y., Shin, K., Drockenmuller, E., Hawker, C.J., Russell, T.P., 2005. A generalized approach to the modification of solid surfaces. Science 308, 236–239.

Ryu, D.Y., Wang, J.-Y., Lavery, K.A., Drockenmuller, E., Satija, S.K., Hawker, C.J., Russell, T.P., 2007. Surface modification with cross-linked random copolymers: minimum effective thickness. Macromolecules 40, 4296–4300.

Savas, T.A., Schattenburg, M.L., Carter, J.M., Smith, H.I., 1996. Large-area achromatic interferometric lithography for 100 nm period gratings and grids. J. Vac. Sci. Technol. B 14, 4167–4170.

Segalman, R.A., Yokoyama, H., Kramer, E.J., 2001. Graphoepitaxy of spherical domain block copolymer films. Adv. Mater. 13, 1152–1155.

Segalman, R.A., Hexemer, A., Hayward, R.C., Kramer, E.J., 2003a. Ordering and melting of block copolymer spherical domains in 2 and 3 dimensions. Macromolecules 36, 3272–3288.

Segalman, R.A., Hexemer, A., Kramer, E.J., 2003b. Edge effects on the order and freezing of a 2D array of block copolymer spheres. Phys. Rev. Lett. 91, 196101.

Segalman, R.A., Hexemer, A., Kramer, E.J., 2003c. Effects of lateral confinement on order in spherical domain block copolymer thin films. Macromolecules 36, 6831–6839.

Sevink, G.J.A., Zvelindovsky, A.V., Fraaije, J.G.E.M., Huinink, H.P., 2001. Morphology of symmetric block copolymer in a cylindrical pore. J. Chem. Phys. 115, 8226–8230.

Shin, K., Xiang, H., Moon, S.I., Kim, T., McCarthy, T.J., Russell, T.P., 2004. Curving and frustrating flatland. Science 306, 76.

Shin, D.O., Mun, J.H., Hwang, G.-T., Yoon, J.M., Kim, J.Y., Yun, J.M., Yang, Y.-B., Oh, Y., Lee, J.Y., Shin, J., Lee, K.J., Park, S., Kim, J.U., Kim, S.O., 2013. Multicomponent nanopatterns by directed block copolymer self-assembly. ACS Nano 7, 8899–8907.

Smith, H.I., Flanders, D.C., 1978. Oriented crystal growth on amorphous substrates using artificial surface-relief gratings. Appl. Phys. Lett. 32, 349–350.

Song, S., Mochrie, S.G.J., 1995. Attractive step-step interactions, tricriticality, and faceting in the orientational phase diagram of silicon surfaces between [113] and [114]. Phys. Rev. B 51, 10068–10084.

Song, S., Mochrie, S.G.J., Stephenson, G.B., 1995. Faceting kinetics of stepped Si(113) surfaces: a time-resolved X-ray scattering study. Phys. Rev. Lett. 74, 5240–5243.

Stein, G.E., Kramer, E.J., Li, X., Wang, J., 2007. Single-crystal diffraction from two-dimensional block copolymer arrays. Phys. Rev. Lett. 98, 086101.

Stoykovich, M.P., Nealey, P.F., 2006. Block copolymers and conventional lithography. Mater. Today 9, 20–29.

Stoykovich, M.P., Müller, M., Kim, S.O., Solak, H.H., Edwards, E.W., de Pablo, J.J., Nealey, P.F., 2005. Directed assembly of block copolymer blends into nonregular device-oriented structures. Science 308, 1442–1446.

Stoykovich, M.P., Kang, H., Daoulas, K.C., Liu, G., Liu, C.-C., de Pablo, J.J., Müller, M., Nealey, P.F., 2007. Directed self-assembly of block copolymers for nanolithography: fabrication of isolated features and essential integrated circuit geometries. ACS Nano 1, 168–175.

Sundrani, D., Sibener, S.J., 2002. Spontaneous spatial alignment of polymer cylindrical nanodomains on silicon nitride gratings. Macromolecules 35, 8531–8539.

Sundrani, D., Darling, S.B., Sibener, S.J., 2004a. Guiding polymers to perfection: macroscopic alignment of nanoscale domains. Nano Lett. 4, 273–276.

Sundrani, D., Darling, S.B., Sibener, S.J., 2004b. Hierarchical assembly and compliance of aligned nanoscale polymer cylinders in confinement. Langmuir 20, 5091–5099.

Tavakkoli, K.G.A., Gotrik, K.W., Hannon, A.F., Alexander-Katz, A., Ross, C.A., Berggren, K.K., 2012. Templating three-dimensional self-assembled structures in bilayer block copolymer films. Science 336, 1294–1298.

Tsai, H.-Y., Miyazoe, H., Engelmann, S., To, B., Sikorski, E., Bucchignano, J., Klaus, D., Liu, C.-C., Cheng, J., Sanders, D., Fuller, N., Guillorn, M., 2012. Sub-30nm pitch line-space patterning of semiconductor and dielectric materials using directed self-assembly. J. Vac. Sci. Technol. B 30, 06F205-1–06F205-6.

Tsai, H., Pitera, J.W., Miyazoe, H., Bangsaruntip, S., Engelmann, S.U., Liu, C.-C., Cheng, J.Y., Bucchignano, J.J., Klaus, D.P., Joseph, E.A., Sanders, D.P., Colburn, M.E., Guillorn, M.A., 2014. Two-dimensional pattern formation using graphoepitaxy of PS-b-PMMA block copolymers for advanced FinFET device and circuit fabrication. ACS Nano 8, 5227–5232.

Tseng, Y.-C., Peng, Q., Ocola, L.E., Elam, J.W., Darling, S.B., 2011. Enhanced block copolymer lithography using sequential infiltration synthesis. J. Phys. Chem. C 115, 17725–17729.

Turner, M.S., 1992. Equilibrium properties of a diblock copolymer lamellar phase confined between flat plates. Phys. Rev. Lett. 69, 1788–1791.

Wagner, C., Harned, N., 2010. EUV lithography: lithography gets extreme. Nat. Photonics 4, 24–26.

Walton, D.G., Kellogg, G.J., Mayes, A.M., Lambooy, P., Russell, T.P., 1994. A free energy model for confined diblock copolymers. Macromolecules 27, 6225–6228.

Welander, A.M., Nealey, P.F., Cao, H., Bristol, R., 2008. Impact of trench width roughness on the graphoepitaxial assembly of block copolymers. J. Vac. Sci. Technol. B 26, 2484–2488.

Wu, Y., Cheng, G., Katsov, K., Sides, S.W., Wang, J., Tang, J., Fredrickson, G.H., Moskovits, M., Stucky, G.D., 2004. Composite mesostructures by nano-confinement. Nat. Mater. 3, 816–822.

Xia, Y., Whitesides, G.M., 1998. Soft lithography. Annu. Rev. Mater. Sci. 28, 153–184.

Xiang, H., Shin, K., Kim, T., Moon, S.I., McCarthy, T.J., Russell, T.P., 2004. Block copolymers under cylindrical confinement. Macromolecules 37, 5660–5664.

Xiang, H., Shin, K., Kim, T., Moon, S., McCarthy, T.J., Russell, T.P., 2005a. The influence of confinement and curvature on the morphology of block copolymers. J. Polym. Sci., Part B: Polym. Phys 43, 3377–3383.

Xiang, H., Shin, K., Kim, T., Moon, S.I., McCarthy, T.J., Russell, T.P., 2005b. From cylinders to helices upon confinement. Macromolecules 38, 1055–1056.

Xiao, S., Yang, X., Edwards, E.W., La, Y.-H., Nealey, P.F., 2005. Graphoepitaxy of cylinder-forming block copolymers for use as templates to pattern magnetic metal dot arrays. Nanotechnology 16, S324–S329.

Xiao, S., Yang, X., Park, S., Weller, D., Russell, T.P., 2009. A novel approach to addressable 4 teradot/in.2 patterned media. Adv. Mater. 21, 2516–2519.

Xu, T., Hawker, C.J., Russell, T.P., 2005. Interfacial interaction dependence of microdomain orientation in diblock copolymer thin films. Macromolecules 38, 2802–2805.

Yamaguchi, T., Yamaguchi, H., 2006. Resist-pattern guided self-assembly of symmetric diblock copolymer. J. Photopolym. Sci. Technol. 19, 385–388.

Yamaguchi, T., Yamaguchi, H., 2008. Two-dimensional patterning of flexible designs with high half-pitch resolution by using block copolymer lithography. Adv. Mater. 20, 1684–1689.

Yang, X., Xiao, S., Liu, C., Pelhos, K., Minor, K., 2004. Nanoscopic templates using self-assembled cylindrical diblock copolymers for patterned media. J. Vac. Sci. Technol. B 22, 3331–3334.

Yang, X., Wan, L., Xiao, S., Xu, Y., Weller, D.K., 2009. Directed block copolymer assembly versus electron beam lithography for bit-patterned media with areal density of 1 terabit/inch^2 and beyond. ACS Nano 3, 1844–1858.

Yang, J.K.W., Jung, Y.S., Chang, J.-B., Mickiewicz, R.A., Alexander Katz, A., Ross, C.A., Berggren, K.K., 2010. Complex self-assembled patterns using sparse commensurate templates with locally varying motifs. Nat. Nanotechnol. 5, 256–260.

Zhang, X., Harris, K.D., Wu, N.L.Y., Murphy, J.N., Buriak, J.M., 2010. Fast assembly of ordered block copolymer nanostructures through microwave annealing. ACS Nano 4, 7021–7029.

Directed self-oriented self-assembly of block copolymers using chemically modified surfaces

6

R. Seidel, L. Williamson*, T. Segal-Peretz†, G. Wu*, H.S. Suh†, C. Zhou*, S. Xiong†, G.S.W. Craig*, P.F. Nealey**

*University of Chicago, Chicago, IL, USA; †Division of Materials Science, Argonne, IL, USA

6.1 Introduction

Directed self-assembly (DSA) of thin films of block copolymer (BCP) on chemically patterned substrates is an effective method to pattern dense arrays of features with dimensions in the nanoscale. DSA combines the ability of BCPs to self-assemble into nanoscale features with the use of lithographic tools and materials to create chemical patterns that guide the assembly of BCP domains into desired structures. DSA has been used to assemble BCP domains in a collection of individual and arrayed device-oriented geometries (Stoykovich et al., 2007). In addition, DSA can be used to form nonregular geometric patterns (Ruiz et al., 2008; Cheng et al., 2008, 2010; Park et al., 2010; Jeong et al., 2010; Hinsberg et al., 2010; Bencher et al., 2011). Also, DSA can multiplicatively increase the feature density and resolution of underlying chemical patterns (Ruiz et al., 2008; Cheng et al., 2008, 2010). In addition to its pattern formation capabilities, DSA of BCPs on chemical patterns offers the possibility of improving the capability of traditional lithographic processes (Bates and Fredrickson, 1990, 1999; Hamley, 1998; Black et al., 2007; Black, 2007; Ross and Cheng, 2008) because of the ability of the BCP or blend to transfer the important features, such as pattern period, feature shape, and registration (Stoykovich et al., 2005, 2007; Liu et al., 2010a; Park et al., 2007; Wilmes et al., 2006), from the chemical pattern to its free surface with a higher etching aspect ratio. Because DSA relies on the self-assembly of the BCP, and, therefore, the BCP's inherent physical properties, DSA can reduce positional fluctuation from the perfect lattice caused by the imperfection in the underlying, lithographically defined guiding patterns (Ruiz et al., 2008; Cheng et al., 2008; Liu et al., 2010b), while at the same time forming a pattern that enhances other aspects of the underlying chemical pattern, such as critical dimension (CD) control (Stuen et al., 2009; Liu et al., 2009), line-edge roughness (Daoulas et al., 2008; Stoykovich et al., 2010), pattern perfection, and resolution (Ruiz et al., 2008; Cheng et al., 2008; Xiao et al., 2009; Tada et al., 2008, 2009a,b). Indeed, DSA on chemical patterns has shown the ability to accommodate up to ~±10% of mismatch between the natural period of the BCPs (L_0) and the period of underlying guiding patterns (L_S) (Edwards et al., 2004). These attributes of DSA make it a potentially viable patterning technique for magnetic storage media (Ruiz et al., 2008; Xiao et al., 2009) and semiconductor devices (Guarini et al., 2001; Hong et al., 2010).

Directed Self-assembly of Block Copolymers for Nano-manufacturing. http://dx.doi.org/10.1016/B978-0-08-100250-6.00006-7

An overview of the DSA process is presented in Figure 6.1 for a lamellae-forming BCP on striped chemical patterns, with L_S values suitable for either 1:1 or 2:1 DSA. 2:1 DSA is an example of density multiplication, in which the density of features in the assembled BCP film is an integer multiple of the density of features on the chemical pattern. Regardless of whether DSA incorporates density multiplication or not, the DSA process starts with the formation of a chemical pattern on the substrate, which will be presented in more detail in the following section. Using advanced lithographic techniques such as electron beam lithography (Ruiz et al., 2008; Stoykovich et al., 2005; Xiao et al., 2009), 193-nm immersion lithography (Liu et al., 2010c, 2011a), or extreme ultraviolet interference lithography (Edwards et al., 2005; Kim et al., 2003), regions with a strong preference to be wet by a specific block of the copolymer (*guiding* regions) are defined, while the remaining regions (*background* regions) can be modified to be either preferentially wet by the other block or be non-preferential to the two blocks. A thin ($\leq$100 nm) film of BCP is coated onto the chemical pattern. Thermal annealing of the BCP film causes it to equilibrate in the presence of the chemical pattern. The selective wetting of the BCP domains (blue and red in Figure 6.1) on the different regions of the chemical pattern lead to the registration, orientation, and ordering of the BCP domains (Stoykovich et al., 2005; Kim et al., 2003). For technological applications, one of the blocks can be selectively removed, leaving the remaining block to serve as a template. Unlike the formation of the BCP morphology in the bulk, the domain orientation and ordering is driven primarily by the chemical pattern and less strongly by neighboring domains.

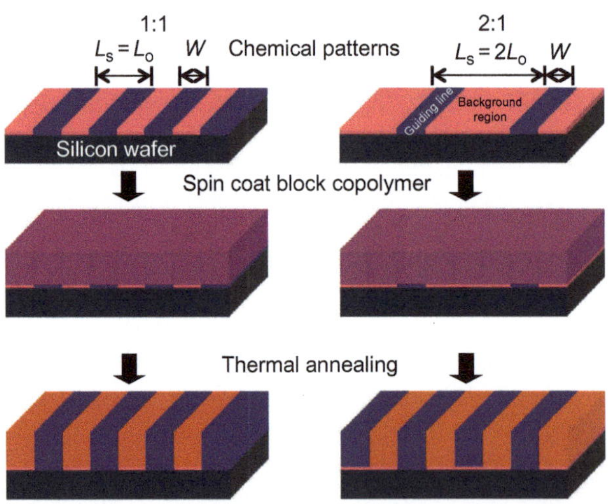

Figure 6.1 Simple schematic of directed self-assembly for 1:1 and 2:1 assemblies of a block copolymer film on a chemical pattern. The blue and pink regions represent the guiding lines (width *W*) and background regions of the chemical pattern and are selectively wet by the blue and red domains of the block copolymer, respectively. In the 1:1 assembly, the pattern period L_S equals the bulk period of the block copolymer L_0. In the 2:1 assembly, corresponding to DSA with density multiplication, $L_S = 2L_0$.

6.2 Fabrication of chemical patterns

Various materials have been used to form chemical patterns, including self-assembled monolayers (Kim et al., 2003), HSQ resist (Cheng et al., 2008), cross-linked polymer mats (Liu et al., 2011a, 2013a), and end-grafted homopolymers or random copolymer brushes (Ruiz et al., 2008; Stoykovich et al., 2005; Liu et al., 2011a, 2013a). Usually, these materials served as the guiding regions, and etching away sections of the guiding regions created the background regions. The resulting background material, whether it was exposed substrate or oxidized organic material, provided sufficient chemical contrast with the guiding regions to enable DSA studies. However, to achieve independent control over both the guiding and the background regions chemistries, it is necessary to apply the background material independently of the guiding lines.

Such a chemical pattern fabrication technique was developed by Cheng and coworkers at IBM (Cheng et al., 2010). In their approach, the photoresist is patterned directly on the substrate and mask areas of the substrate during the grafting of the background regions. Subsequent lift-off of the photoresist (Cheng et al., 2010; Somervell et al., 2012), or photoresist pull process (Liu et al., 2013b), exposes the guiding regions on the substrate. While this process is straightforward, there is a concern that the geometry of the chemical pattern cannot be precisely controlled due to the lift-off process (Somervell et al., 2012).

The fabrication of chemical patterns for the chemoepitaxial DSA process, with precise control over the chemistries and geometry of the chemical pattern, was achieved in the Liu-Nealey flow (LiNe flow), shown in Figure 6.2 (Liu et al., 2011a). The process starts with the deposition of a cross-linked polymer mat on the substrate, which is lithographically patterned and subsequently etched to expose the substrate in the desired locations; for a periodic pattern, this step will determine the chemical pattern period, L_S. Importantly, the polymer mat regions can be further etched in an anisotropic etch process (trim etch) to achieve sublithographic dimensions and to control the width W of the guiding regions. After photoresist removal, the interspatial regions are functionalized with polymer brushes with the desired chemistry. The polymer mat is cross-linked sufficiently such that the grafting of the polymer brush does not affect the surface chemistry of the mat (Liu et al., 2011b, 2013a). The control over the chemistry of the background regions is especially important when directing assembly with density multiplication because domains of each block will need to assemble over the background region for perpendicular orientation to be obtained. Thus, end-grafted random copolymers brushes, for which the interfacial energy can be tuned by controlling the random copolymer composition, are used to create the background region.

One of the advantages of the LiNe flow is its fab compatibility. By introducing a thin silicon nitride layer as an antireflective coating, Liu et al. demonstrated the compatibility of the LiNe flow with 193-nm immersion lithography (Liu et al., 2010c). In order to transfer the process from the lab to the fab, additional fab compatibility issues were addressed, including solvent toxicity, throughput, process latitude, and uniformity. Moving to a fab-friendly solvent such as PGMEA, and implementing a high-temperature bake (250° C) under nitrogen environment to increase throughput (Liu et al., 2011b), enabled the successful implementation of 300-mm DSA pilot lines at both

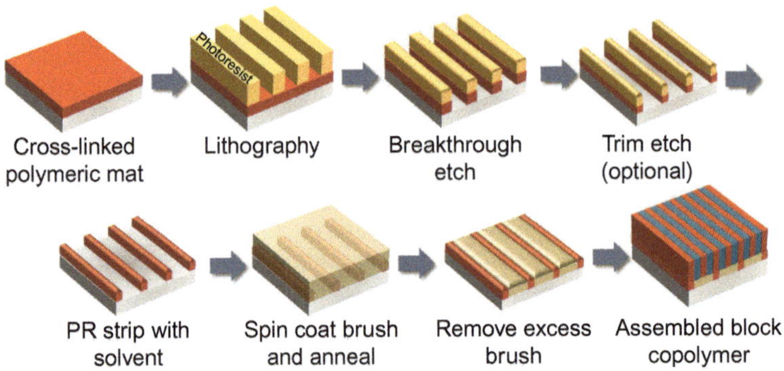

Figure 6.2 Schematic of the chemical pattern formation process (LiNe flow), followed by BCP-directed assembly. A film of cross-linkable polymeric mat is cross-linked, patterned, and etched. Subsequently, polymer brush (such as P(S-*r*-MMA)-OH) with the desired chemistry is grafted onto the exposed spaces of the substrate to complete the chemical pattern. As a demonstration of DSA on the chemical pattern in the schematic, a film of BCP is directed to assemble on the chemical pattern with 2× density multiplication.
Reprinted with permission from Liu et al. (2011a). Copyright © 2011, ACS Publishing Group.

IMEC and IBM (Bencher et al., 2011; Somervell et al., 2012; Liu et al., 2013b; Rincon Delgadillo et al., 2012). These pilot lines have demonstrated large process latitude and high uniformity of the chemical pattern dimensions across the wafer, enabling detailed research on the kinetic, thermodynamics, and origin of defects in the DSA process.

6.3 Thermodynamics of thin film BCP assembly on chemical patterns

6.3.1 Overview

Any discussion of thermodynamics should commence by recognizing the key concepts involved: free energy, kinetics, and equilibrium. A system tends toward states that minimize its free energy, which itself contains information about the enthalpy and entropy of the particular state. The system explores configuration space through fluctuations as it adopts various states, spending more time in those with lower free energy according to Boltzmann statistics. This move through different states and the rate at which such a move occurs, comprise the kinetics of the system as it tends toward equilibrium—the state with the lowest free energy.

The free energy F_{total} of a thin film of BCP takes into account several different contributions (Figure 6.3): the conformational energy free energy of each block of copolymer, the interfacial energy between domains of blocks, the surface energy at the top surface of the film, and the interfacial energy component between the bottom surface and the blocks of the BCP with which it is in contact. With a chemical pattern, the equation is further complicated by the respective, relative interaction of the guiding and background regions of the pattern with the blocks of the BCP in the thin film.

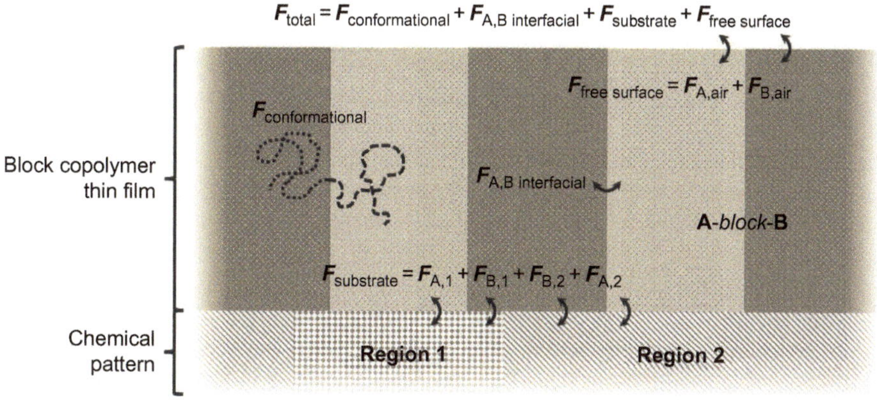

Figure 6.3 The components of the free-energy equation for a block copolymer thin film on a chemically patterned surface. The terms of the form "$F_{i,j}$" denote the free energy of interaction between two phases. Conformational free energy is also sometimes referred to as the entropic term, as it captures the entropic nature of the individual polymer configuration.

To simplify this section, unless stated explicitly otherwise, we will initially limit our discussion to considering incompressible thin films of monodisperse, symmetric (having approximately equal block lengths and therefore lamellae-forming) A-B diblock copolymers with a free (exposed to air) surface. We will further simplify the discussion by assuming that the bulk BCP forms discrete A and B domains at equilibrium and that the surface energies of the two domains are similar ($\gamma_A \approx \gamma_B$) at the annealing temperature. These assumptions are true for our example system of polystyrene-*block*-poly(methyl methacrylate) (PS-*b*-PMMA).

Consider a thin film of BCP extending infinitely across an unpatterned surface that is not preferential to either block of the BCP. The nonpreferential nature of the substrate, coupled with our assumption in the preceding paragraph of equal surface energies of the two blocks, means that there is no enthalpic preference for any particular configuration, and, instead, entropy is the only contributing factor. The result is lamellae oriented perpendicularly to the substrate, which is similar to the lamellae in Figure 6.3. If the lamellae lack any long-range lateral order, at the free surface they form "fingerprint" structures (Park et al., 2007). At equilibrium, our hypothetical film would consist of very large grains of lamellae with extremely long-range order. However, these grains would be randomly oriented and also randomly positioned on the substrate. Such an assembled film—with no positional or orientation control of the BCP domains—is not useful for applications where such control is necessary. The addition of a chemically patterned substrate, in which the guiding and background regions of the substrate are preferential to different blocks of the BCP, modifies the free energy of the system. When the film is in contact with a chemical pattern, there is an enthalpic gain to have similar chemistries in contact (and a corresponding penalty when dissimilar chemistries are adjacent), thereby directing the assembly to assume a specific geometry or pattern.

6.3.2 Chemical patterns

To further explore the effect of chemical patterns on the equilibration of a thin film of BCP, consider the BCP described above on an ideal chemical pattern (perfectly patterned with areas of surface chemistry exactly matching that of the respective BCP domains and at the exact size and spacing of said domains ($L_S = L_0$)). A BCP film on top of such a chemical pattern would, at equilibrium, seek to minimize its free energy by minimizing the amount of dissimilar interactions (minimizing enthalpy) while maximizing entropy. The resulting equilibrium structure would consist of semi-infinite parallel lines of alternating composition, aligned perfectly with the regions of the underlying chemical pattern. The alignment of the BCP domains to the regions of the chemical pattern causes the BCP morphology to be registered to the chemical pattern. These lines would only be broken by occasional high (but finite) free-energy defect structures in the registration and lateral order, such as the dislocations or disclinations shown in Figure 6.4. According to Boltzmann statistics, these defects would be present at equilibrium, but at vanishingly small numbers; current estimates put their equilibrium concentration at smaller than one per $10^{100}\,m^2$ of registered film (Nagpal et al., 2012). (For comparison, the area of a disk the size of our galaxy is around $10^{40}\,m^2$, so this concentration is so small as to be zero in any practical consideration; these structures are mentioned only to preserve the fidelity of this hypothetical system.)

Here, we have touched briefly on a simple but powerful concept, central to the idea of bottom-up manufacturing: the free-energy landscape that a self-assembling system

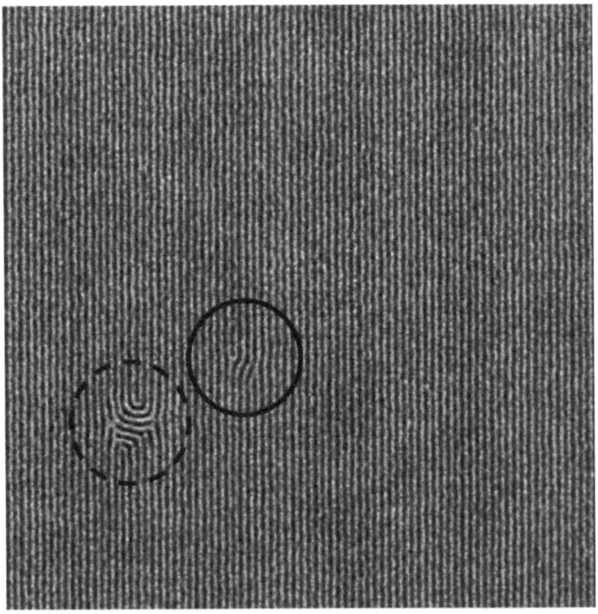

Figure 6.4 Top-down SEM of a large field of registered lamellae on a chemical pattern. The defect within the solid circle is a dislocation pair; that inside the dashed circle is a disclination cluster. Both defect types are present at effectively zero concentration at equilibrium.

explores can be modified to encourage that system to adopt a particular configuration. By introducing chemical boundary conditions, we create regions that have (for example) a lower enthalpy of interaction with one of the blocks, causing that block to preferentially wet that particular region. In more abstract thermodynamics terms, this has the effect of deepening certain free-energy valleys or shifting or eliminating others, shaping the landscape to change its local and global minima as well as the pathways between them. The deeper the energy wells, the more time the system spends in the corresponding configuration; the steeper the slopes of the energy pathways, the faster the system moves to the lower energy state. Conversely, moving away from the ideal chemical pattern described earlier modifies the free-energy valleys. Shifting these valleys often corresponds to moving equilibrium toward frustrated nonbulk morphologies, while making the valleys shallower can cause a decrease in the degree of registration until the film becomes a randomly oriented fingerprint. There are many ways to modify that ideal pattern, changing either the geometric or chemical parameters. The figure of merit typically used to analyze the effect of this modification is the equilibrium (or quasi-equilibrium) degree of registration. While the exact distinction between full versus quasi-equilibrium structures can be challenging to discern for some systems, that distinction is not necessary in the scope of this discussion, and we will for the sake of brevity refer to both collectively as simply "equilibrium."

6.3.3 Nonideal chemical patterns

As mentioned above, an ideal chemical pattern will have $L_S = L_0$ (see Figure 6.1). If L_S is slightly larger or smaller than L_0, the system can choose to ignore the chemical pattern and relax to its unperturbed L_0, but then there is an enthalpic penalty as the BCP domains sit on dissimilar chemical pattern regions. Alternately, the individual chains can stretch or compress to fit the chemical pattern precisely, but that of course brings an entropic cost. Ultimately, a BCP film on such a substrate will choose the configuration that minimizes the overall free energy, accepting a degree of stretching/compressing but with increasing loss of registration as the chemical pattern moves further from the "ideal" geometry. Outside a certain range, registration is lost entirely and the film adopts a fingerprint (or otherwise unregistered) configuration; this range is about $\pm 10\%$ of the ideal L_S for lamellae on 1:1 chemical patterns (Edwards et al., 2004). Similarly, changing the size (for example, W in Figure 6.1) of the chemically patterned regions relative to the BCP domains (rather than their spacing) can be tolerated but only up to a point.

The previous paragraph discusses changing the geometry of a chemical pattern, but it is also worthwhile to consider the effect of modifying the guiding chemistry. Intuitively, one would expect that the stronger the interaction between the guiding region and the domain of the BCP, the deeper the free-energy well, and the better the registration of the BCP film to the chemical pattern at equilibrium. Conversely, we expect the geometric incommensurability tolerance to decrease with decreasing strength of that interfacial interaction (Edwards et al., 2004). Figure 6.5 shows an experimental demonstration of such a system, in which both L_S and the strength of the interaction of the guiding region were systematically varied.

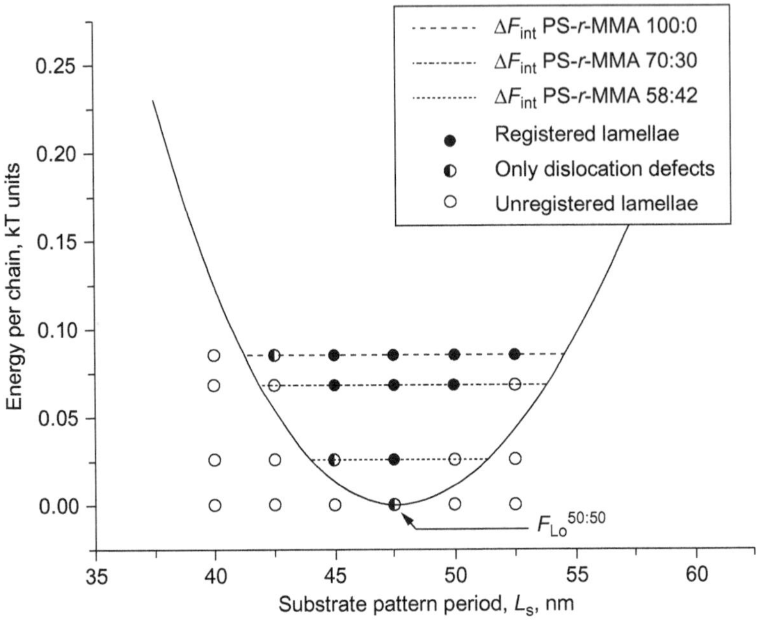

Figure 6.5 Differences in the free energy of a PS-*b*-PMMA block copolymer film self-assembled on a chemical pattern having varying pattern period L_S and chemical composition, which is represented by the interaction energy per chain on the *y*-axis. Registered assembly of the domains was predicted to occur at a value of L_S if the change in interfacial energy (dashed lines) was greater than the change in energy associated with stretching or compressing the lamellar period from its bulk value of 47.5 nm (solid line). The open, half-open, and closed circles represent the corresponding experimental systems in which unregistered lamellae, only dislocation defects, or registered lamellae, respectively, were observed.
Reprinted from Edwards et al. (2004) with permission. Copyright © 2004, WILEY-VCH Verlag GmbH & Co. KGaA, Weinheim.

6.3.4 Density multiplication (DSA on sparse chemical patterns)

The ability of BCPs to self-assemble on sparse chemical patterns ($L_S = nL_0$, n an integer >1), with some domains interpolating between guiding regions on the substrate, thereby achieving feature density multiplication and a corresponding resolution enhancement, increases the utility of DSA of BCPs for nanodevice fabrication. As the degree of density multiplication increases, however, fewer domains of the BCP film are directly guided by the chemical pattern, and the tolerance for the L_S/L_0 incommensurability decreases. Additionally, for sparse chemical patterns, there is a fundamental difference, compared to 1:1 DSA, in the enthalpic nature of the interaction between thin film and substrate. On 1:1 chemical patterns, each region is wet by only a single BCP domain. For density multiplication, however, multiple domains assemble above each background region. The more similar that region is to the one block, the more dissimilar it must necessarily be to the other, and, therefore, even at optimal dimensions, these regions contribute an enthalpic penalty to the free-energy equation. For a surface that exhibits wetting behavior that is a linear combination of the two blocks, the quantity f can be defined as the percent similarity of the background region of the substrate to the "A" rather than the "B" block of the BCP (so $f=1$ corresponds to a surface exactly like A, while a value of f approaching 0 means the surface is similar to B). An approximation of the normalized BCP-substrate interfacial free energy follows:

$$F_{\text{substrate}} = \frac{1}{2n}\left(F_{\text{A,guide}} + (n-1)\cdot(1-f)\cdot\gamma_{\text{A,B}} + n\cdot f\cdot\gamma_{\text{A,B}}\right),$$

where $F_{\text{A,guide}}$ is the free energy of the A domain on the guided stripe, n is the degree of multiplication, and $\gamma_{\text{A,B}}$ is the interfacial energy between the two blocks (Liu et al., 2013a). It is readily apparent that the free energy of the system is a function of how similar the nonguiding region is to the unguided domain. A more subtle observation is that at different density multiplication factors; the minimum free-energy state can be found at differing degrees of similarity (values of f). In an experimental setup using random copolymer brushes in the background regions, this prediction translates into a different optimum brush composition for different density multiplication factors, as shown in Figure 6.6. The higher the density multiplication, the more nonpreferential the brush composition should be to achieve the lowest free-energy structure.

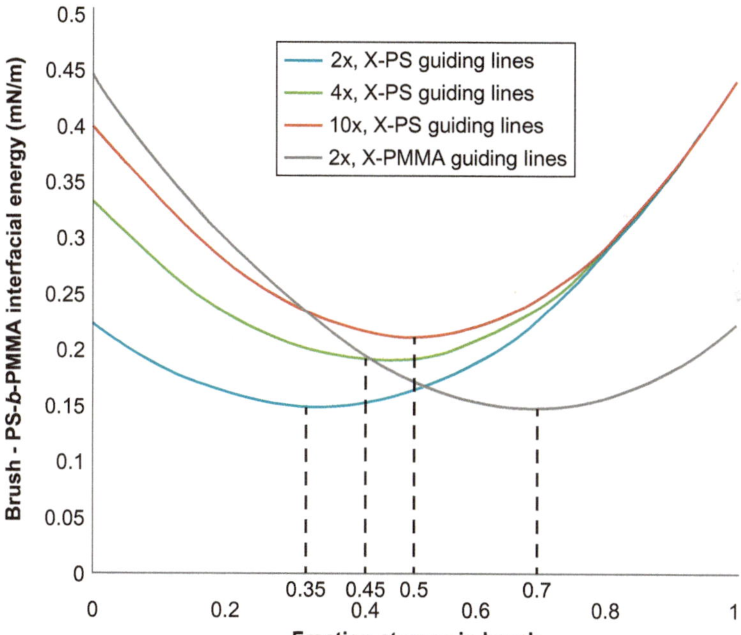

Figure 6.6 Effect of density multiplication factor and chemical pattern composition on the calculated interfacial energy of the block polymer–background interface of assembled PS-*b*-PMMA films. The chemical pattern composition was varied by changing the fraction of styrene in a poly(styrene-*r*-methyl methacrylate) random copolymer brush used in the background region of the chemical pattern. The higher the density multiplication factor, the closer the corresponding optimum composition is to the standard nonpreferential point. Reprinted with permission from Liu et al. (2013a). Copyright © 2013, ACS Publishing Group.

In addition to research that has focused on the behavior of DSA of BCPs at or near optimum conditions, a large body of work has also been devoted to far more complicated systems. The confinement effects in thin films can exacerbate conflicting terms in the overall free-energy equation, leading to frustrated morphologies reflective of the strong, localized interplay between enthalpy and entropy in these systems. Unexpected, bizarre structures have been theoretically predicted and experimentally demonstrated in cases of strong chemical patterns at highly exaggerated geometries, as shown in Figure 6.7 (Detcheverry et al., 2010). Areas of high interfacial curvature, normally not seen because of their strong free-energy penalty, are possible under these nonstandard conditions.

The confinement effects can be further complicated if a second chemically patterned substrate is introduced at the top surface of the BCP thin film (Ramirez-Hernandez et al., 2012), as shown in Figure 6.8. While the rigorous solution to the free-energy minimization equation for each of these systems is too complicated to derive in detail, the fundamental concepts remain the same. The (quasi-) equilibrium structure adopted

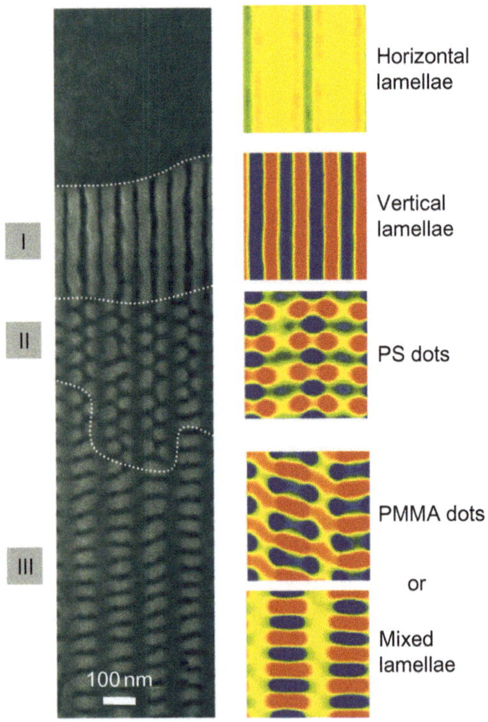

Figure 6.7 Different nonbulk morphologies of assembled PS-*b*-PMMA theoretically predicted (right) and experimentally demonstrated (left), assembled on a range of chemical pattern dimensions. (left) SEM micrograph showing four distinct morphologies. PS-rich domains appear in light gray. (right) Color maps of simulations. PS-rich domains appear in red and PMMA-rich domains in blue. These morphologies are frustrated configurations that represent free-energy valleys significantly altered by the presence of strong chemical patterns. Reprinted with permission from Detcheverry et al. (2010). Copyright © 2010, ACS Publishing Group.

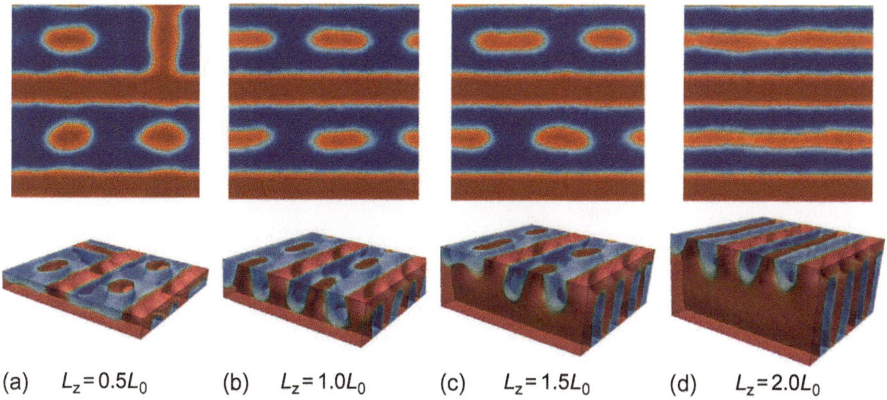

Figure 6.8 (top) Top-down views and (bottom) three-dimensional snapshots of block copolymer morphologies between two chemical patterns at various film thicknesses obtained from Monte Carlo simulations.
Reprinted with permission from Ramirez-Hernandez et al. (2012). Copyright © 2012, ACS Publishing Group.

by the BCP thin film is a fragilely balanced compromise in the tug-of-war between the different free-energy terms.

To fully understand the thermodynamics at play in the very complicated systems described above, it is sometimes easier to treat exhaustively the thermodynamics of more elementary cases. Instead of analyzing the highly tortuous bicontinuous structures shown in Figure 6.8, we can study the less curved yet still frustrated morphologies of programmed bends on more standard chemical patterns (Liu et al., 2010a; Stoykovich et al., 2005). Note that in these cases (Figure 6.9) the "defect" structures are actually part of the chemical pattern, so a well-registered film would look as if it had jogs or other nonequilibrium morphologies. The formation and persistence of these features provide insight into BCP thermodynamics in general but also, more specifically, a parallel way of studying the same structures on ideal chemical patterns. Furthermore, these nonstandard structures are an avenue to realization of the complex shapes required to engineer functional nanoelectronic or other devices through BCP patterning.

6.4 Kinetics of thin film BCP assembly on chemical patterns

6.4.1 Overview

In addition to understanding the equilibrium and quasi-equilibrium structures, it is critical to consider the time required or kinetic pathways taken to reach those structures. Investigating the kinetics of the DSA system not only improves our understanding of the physics involved but also helps to identify industry-compatible production steps, where process time is a key metric. We will start this section on kinetics by describing the kinetic process undergone by a BCP thin film as it goes from disorder toward equilibrium.

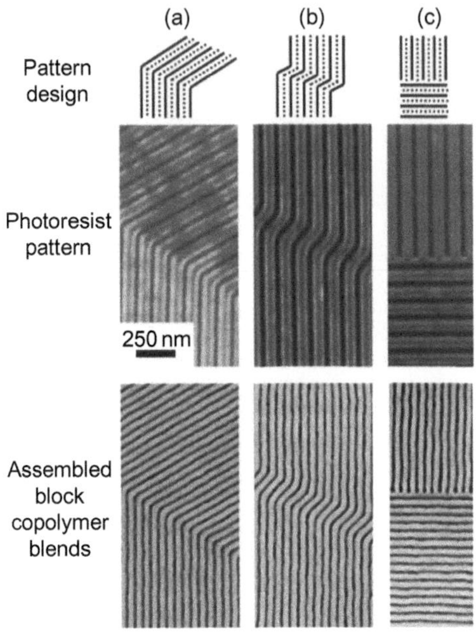

Figure 6.9 Directed assembly of block copolymer/homopolymer blends to achieve nonregular, device-oriented geometries. The schematic chemical pattern design (top row), SEM images of the photoreist pattern (middle row), and the corresponding assembled block copolymer blends (bottom row) directed to assemble into device-oriented structures, such as (a) 120° bends, (b) jogs, (c) T-junctions and terminations on the chemical patterns, with 2× density multiplication. These BCP morphologies may look like nonequilibrium structures, but they actually represent perfect registration over the underlying chemical pattern.
Reprinted with permission from Liu et al. (2010a). Copyright © 2010, WILEY-VCH Verlag GmbH & Co. KGaA, Weinheim.

One distinction that is important to make is the difference between the terms "ordered" and "disordered" as they apply to BCP physics on one hand, and the patterns formed by a BCP assembled on a chemical pattern on the other hand. From a BCP physics point of view, a disordered BCP system is one that has not undergone the order–disorder transition (ODT) that results in microphase separation and the formation of distinct microdomains; a disordered BCP is a single phase and exhibits no scattering peaks. Once the film is brought through the ODT—either through cooling or, in the case of spin coating from solution, removal of solvent—the film immediately becomes ordered; that is, it undergoes spinodal decomposition and the domains microphase separate. Often, the term "ordered" when applied to DSA of BCPs refers to the long-range order of the pattern formed by the assembled BCP. A newly ordered (i.e., microphase separated) film initially has very poor long-range order, though it still exhibits a distinctive first-order peak in scattering experiments. For example, in Figure 6.10, the SEM images and corresponding Fourier transforms are presented for a BCP film at various annealing times (Stuen et al., 2008). Before annealing ($t=0$), the BCP has microphase separated and, therefore, "ordered," but the pattern of the domains is disordered.

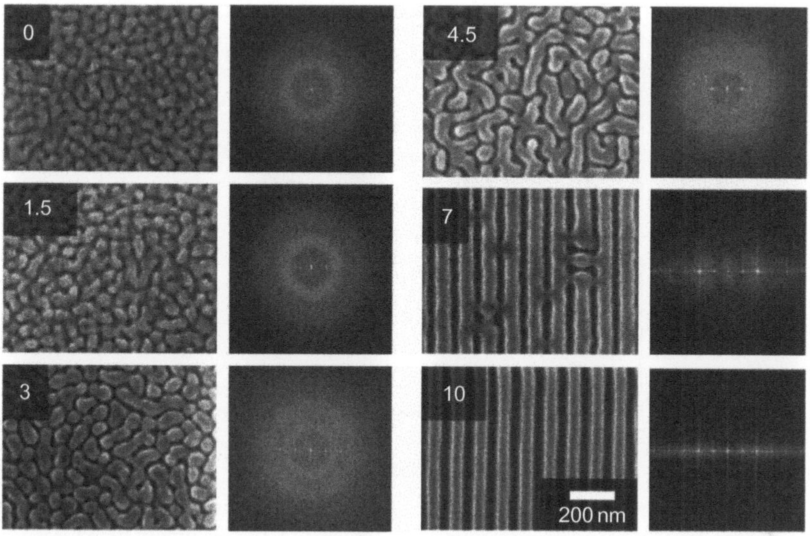

Figure 6.10 Top-down SEM images of block copolymer films on chemically patterned surfaces with their magnified Fourier transforms. The numbers denote annealing times in minutes. The scale bar is 200 nm in length. A spin cast BCP film crosses ODT to become ordered immediately upon solvent evaporation but has no long-range order (top left, time = 0). As the film matures, the polymer chains move until they eventually reach their equilibrium configuration (bottom right, time = 10 min).
Reprinted with permission from Stuen et al. (2008). Copyright © 2008, AVS.

As the film equilibrates, the individual chains move and explore configuration space, diffusing through and along the ever-growing phase-separated domain interfaces. In so doing, they adopt more energetically favorable states, and the film develops toward its equilibrium configuration. In free-energy space, the system moves downhill along the steepest gradient possible at a rate proportional to the slope along that gradient. This means that in the absence of stabilizing impurities, areas of higher interfacial curvature such as disclinations tend to disappear first due to the steep free-energy penalty of these structures. Lower free-energy structures such as dislocations can persist longer but eventually move to the lowest energy state found at equilibrium. When assembling on a chemical pattern, the polymer movement is of course also strongly influenced by the presence of the various substrate regions, which modifies the free-energy landscape and can provide a location for seeding growth of the specific BCP domains.

6.4.2 Time to equilibration

The time scales for the formation of equilibrium structures of polymer systems often can be nontrivial, as can be expected for a process involving macromolecular diffusion, and in fact, sometimes can last indefinitely as a particular state becomes trapped with no way for entangled polymers to disentangle and alter their configuration or position. These states correspond to free-energy wells with either very steep sidewalls

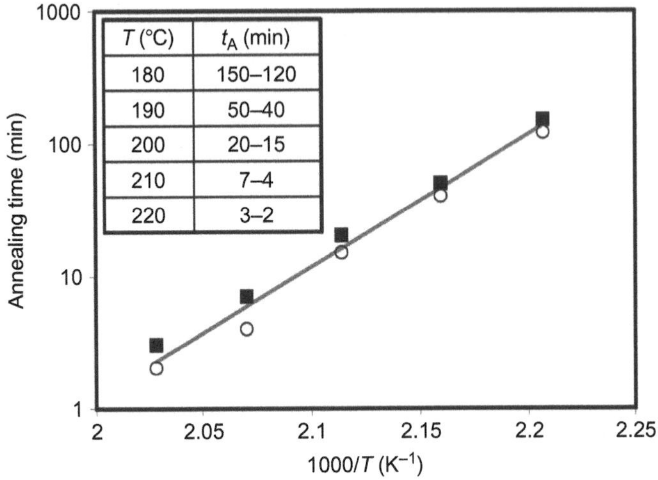

Figure 6.11 Semi-log plot of annealing time as a function of inverse temperature for (1) the minimum time observed for defect-free assembly (■) and (2) the maximum time for which abundant defects still persisted (O). The line represents a linear regression of all data points, where the slope $= \Delta E_A/R$. Annealing exhibits an Arrhenius dependence on temperature; at higher temperatures, annealing is exponentially faster.
Reprinted with permission from Welander et al. (2008). Copyright © 2008, ACS Publishing Group.

or very long pathways to lower free-energy states. However, the time scales can be shortened and the trapped states avoided by judicial selection of processing materials and conditions. Because diffusivity exhibits an Arrhenius dependence on temperature,

$$D = D_0 \cdot e^{-(E_A/RT)}$$

one simple way to shorten maturation times is simply by elevating the temperature (Welander et al., 2008). This approach, known as thermal annealing, has been thoroughly explored and is the method of choice for many processes (Figure 6.11). The temperature for thermal annealing must be carefully chosen so as not to exceed ODT or cause chemical modification of the polymer system.

As mentioned above, the presence of a chemical pattern can strongly influence the kinetics of assembly by modifying the free-energy landscape. Often, this behavior is exhibited in the form of much slower assembly on a nonideal chemical pattern. For values of L_S incommensurate with L_0, even when the equilibrium morphology consists of arrays free of registration defects, the assembly still takes much longer than when $L_S = L_0$ (Delgadillo, 2014). Similarly, when comparing stronger to weaker interactions in different chemical patterns, the shallower free-energy wells corresponding to the weaker chemical interactions led to slower assembly (Figure 6.12) (Edwards et al., 2005). Sparse chemical patterns (used for density multiplication) are another example of weaker chemical patterns and exhibit similarly slower assembly kinetics when compared to the behavior of films on 1:1 patterns (Figure 6.13) (Liu et al., 2010d).

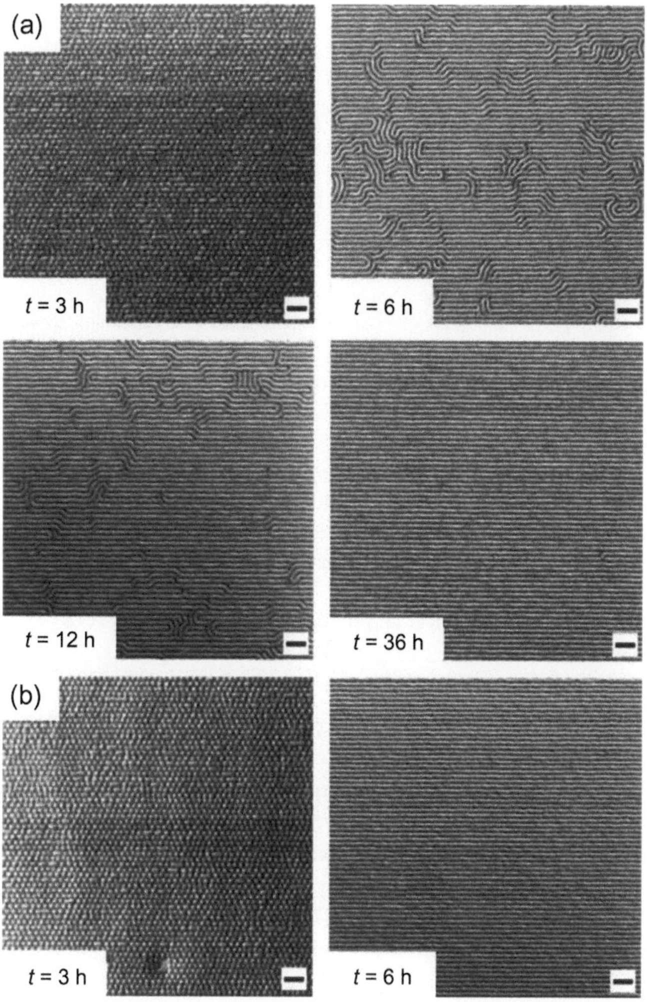

Figure 6.12 Plan-view scanning electron micrographs of PS-*b*-PMMA domains on chemically nanopatterned surface with (a) weaker and (b) stronger chemistry annealed for various times. Assembly on the weaker chemical pattern (with shallower free-energy valleys) takes much longer to equilibrate (36 h) in comparison with a strong chemical pattern (6 h). Reprinted with permission from Edwards et al. (2005). Copyright © 2005, WILEY-VCH Verlag GmbH & Co. KGaA, Weinheim.

6.4.3 Kinetic pathways and trapped states (defects)

One topic of particular interest is mapping out the exact mechanism for how the equilibrium structure is achieved, especially the elimination of the final trapped defect states often seen in line-space patterns. These dislocation pairs have been calculated through simulations to be very high energy states, as they are areas of high interfacial

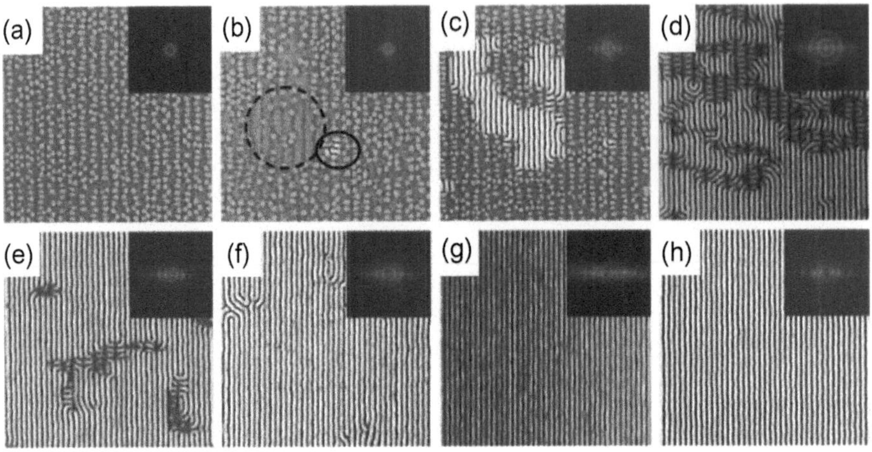

Figure 6.13 Top-down SEM images of thin films of PS-*b*-PMMA annealed at 240 °C for various times. Panels (a)–(g) show images of block copolymer annealed for 10 min, 30 min, 1, 2, 4, 16, and 64 h, respectively, on chemical patterns designed for density multiplication and (h) for 10 min on chemical patterns without density multiplication. Assembly is much slower on the sparse chemical pattern.
Reprinted with permission from Liu et al. (2010d). Copyright © 2010, AVS.

curvature and extra interfacial area, and yet they are observed frequently even after long, high-temperature maturation processes. Ultimately, their persistence is not due to their being equilibrium or quasi-equilibrium states, but rather the very long free-energy path between these states and the equilibrium state. Because these dislocations always occur in pairs, they have to independently diffuse toward each other through the registered film until they finally annihilate upon encountering one another (Figure 6.14) (Tong and Sibener, 2013). The diffusivity of these structures has been determined experimentally and found to have widely differing values depending on whether the diffusion is happening along or across the direction of the line-space pattern. One can understand the factors involved in the persistence of nonequilibrium states in terms of the rearrangement of the polymer chains. In order to move across the line-space array, many polymers have to simultaneously rearrange, temporarily creating a much larger interfacial area between the blocks. In contrast, moving along a domain interface only requires diffusion of the polymer chains along that interface without any additional enthalpic penalty.

Ultimately, the kinetics of BCP assembly are dictated by the accessible free-energy pathways that are available as a system tries to move from one configuration to another. Even when the final state has a much lower free energy than the initial state, assembly can still be slowed indefinitely by the presence of pathway barriers. These barriers can be in the form of higher energy states along the configuration space path between the initial and final states, such as the bumps shown in the free-energy curves in Figure 6.15, or they can be simply due to a deepening of the initial state well through the presence of some sort of stabilizing nucleation material.

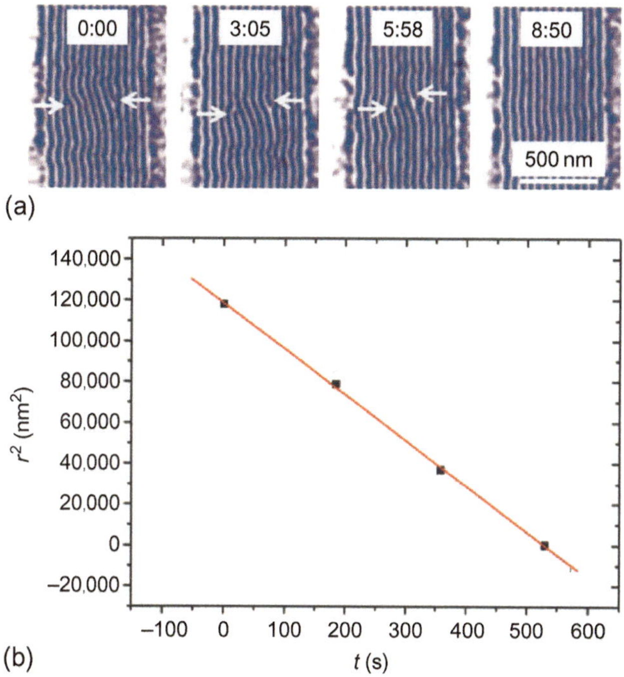

Figure 6.14 (a) Time sequence of AFM phase images showing glide motion of two dislocations of opposite Burgers vectors. The dislocation cores are indicated by the arrows. The number at the top of each image indicates the time when the image was taken. The images are taken at 240 °C. (b) The square of the distance between the two dislocation cores is plotted to the time for the images shown in (a).

Reprinted with permission from Tong and Sibener (2013). Copyright © 2013, ACS Publishing Group.

6.5 High-resolution patterning

6.5.1 Overview

The main driving force behind high-resolution patterning is the continuous pursuit of smaller, cheaper, and more powerful integrated circuitry by the semiconductor industry. As the feature dimensions continue to shrink to sub-22 nm and smaller, standard photolithography is approaching its physical limit and alternative technologies are highly desired to solve the grand challenge facing the semiconductor industry. Strategies including extreme ultraviolet lithography, nanoimprint lithography, interference lithography, and DSA of BCPs have been considered and listed as potential lithographic solutions for the fabrication of sub-22-nm node in the International Technology Roadmap for Semiconductors (ITRS) (International Technology Roadmap for Semiconductors, 2011). Among those strategies, DSA of BCPs on chemical patterns combines advantages of thermodynamics-governed "bottom-up" self-assembly of BCPs and "top-down" photolithography and offers opportunities to create

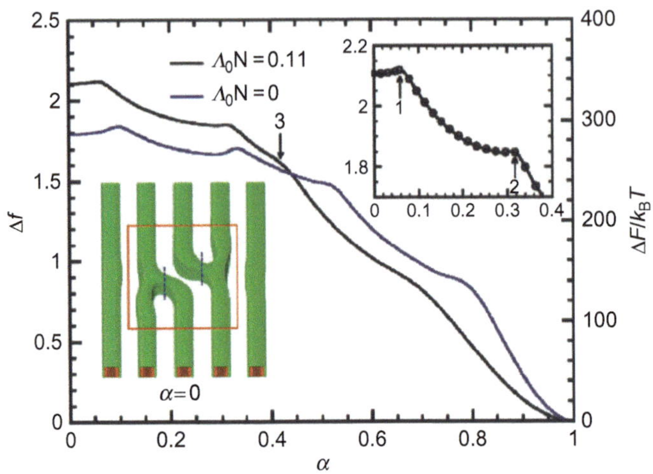

Figure 6.15 Minimum free-energy paths for block copolymer defect annihilation for χN = 30 and Λ_0N = 0 and 0.11. The initial defect, at $\alpha = 0$, is shown in the simulation image in the inset. Bumps along the free-energy paths denote three intermediate morphologies. For Λ_0N = 0.11, the intermediate morphologies are located at the barrier, $\alpha \approx 0.059$, and the two shoulders, $\alpha \approx 0.318$ and 0.420, respectively. These energetic barriers represent kinetically trapped states; the height of the barriers defines how long the system statistically resides in that state.

periodic/aperiodic nanostructures with sub-16-nm half-pitch node for the fabrication of integrated circuits and other devices with high degrees of perfection, accuracy, registration, and complexity. The following subsections will review three approaches to achieve high-resolution patterning with DSA of BCPs.

6.5.2 Low delta surface energy BCPs

Lithography with "high-χ" BCPs is a promising methodology to get sub-10 nm features for next-generation lithographic applications. However, the large mismatch in surface tensions of the blocks in the copolymer result in domain orientation issues, with the lower-surface-energy block coating the free surface, thereby effectively erasing any pattern formed by the interaction of the BCP with the underlying chemical pattern. Kim et al. demonstrated a strategy to decouple the factors that create a large χ value from those that lead to surface energy differences between the blocks of the BCP. They introduced a random copolymer molecular architecture in one block of a diblock copolymer (A-*b*-(B-*r*-C)) to balance the surface energy at the relevant annealing temperatures. With this arrangement, they achieved through-film perpendicular domains with sub-10-nm feature sizes (Kim et al., 2012). The diblock copolymer they used was polystyrene-*block*-polyisoprene (PS-*b*-PI), where the PI block was partially epoxidized, and thus a new kind of diblock copolymer—namely, polystyrene-*block*-poly(isoprene-*ran*-epoxyisoprene) was formed as shown in Figure 6.16. The epoxidation reaction occurs at random positions along the polymer chains, and the extent of epoxidation could be controlled by the amount of the oxidant. It was found that the free surface energy of the PS and PI blocks are nearly

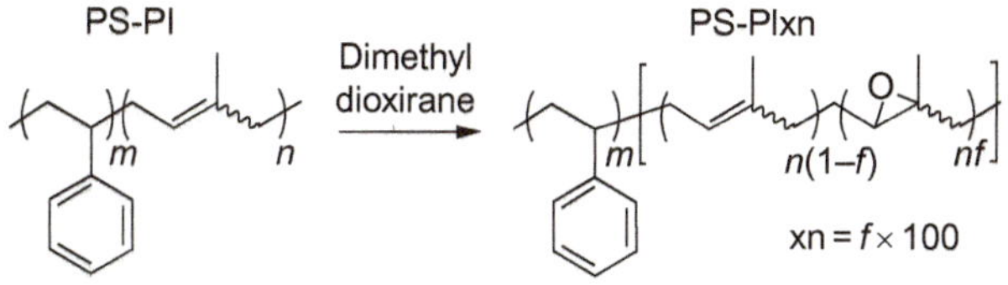

Figure 6.16 Epoxidation of PS-*b*-PI diblock copolymer. The partially epoxidized PS-*b*-PI is denoted as PS-PIxn, while *f* and xn (%) represent the mole fraction and degree of epoxidation, respectively.

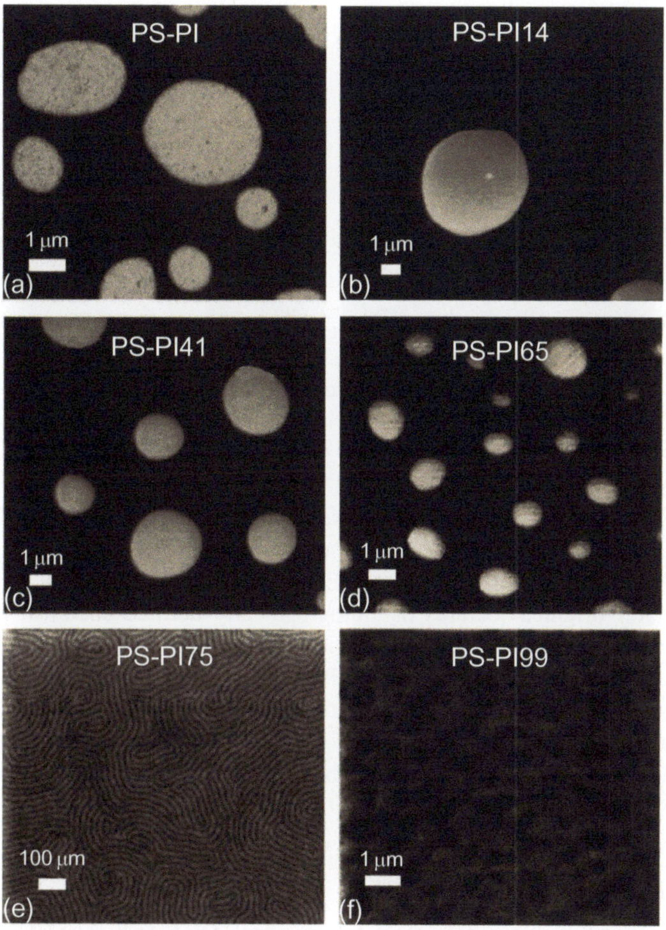

Figure 6.17 Top-down SEM images of PS-*b*-PIxn (xn representing different extents of epoxidation) thin films above a cross-linked P(S-*r*-MMA) brush.

equal when the degrees of partial epoxidation (xn) was ~75%, and defect-free assembly of lamellar domains was realized on end-grafted PS-*r*-PMMA random copolymer brushes (Figure 6.17). The effective interaction parameter (χ_{eff}) depends on χ_{AB}, χ_{AC}, and χ_{BC} interactions, and thus decouples the bulk and surface thermodynamics.

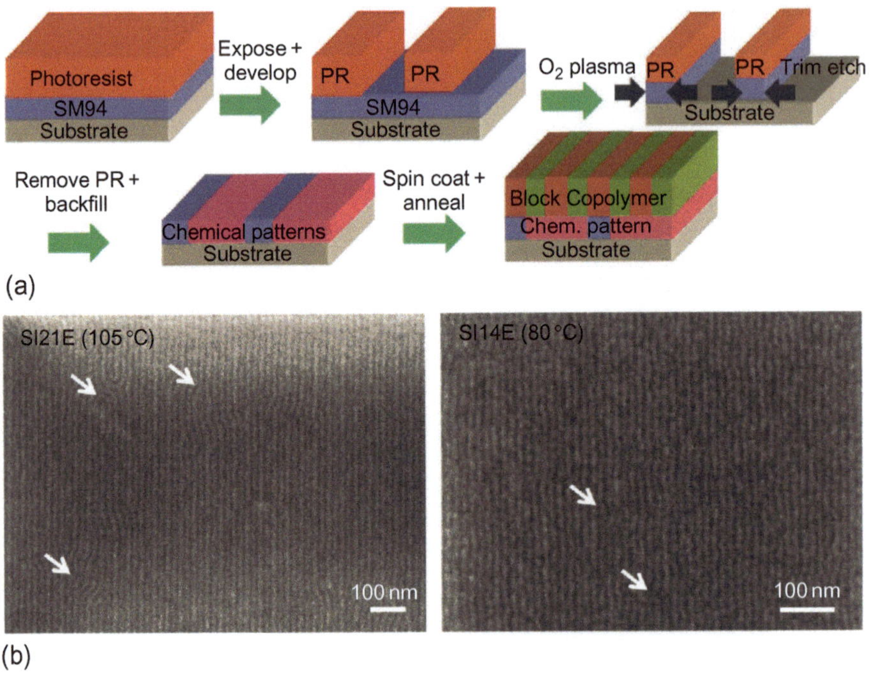

Figure 6.18 (a) Schematic illustrating the process for chemical pattern fabrication. (b) Top-down SEM images of directed self-assembly of polystyrene-*block*-poly(isoprene-*ran*-epoxyisoprene) with $L_0 = 19.1$ nm at 105 °C and $L_0 = 14.5$ nm at 80 °C, respectively.

The χ_{eff} between PS and partially epoxidized PI blocks in PS-*b*-PI_{75} diblock copolymer was ~0.063, which is much higher than the χ of PS-*b*-PMMA, which has been extensively studied in DSA. As anticipated, directed assembly of through-film, perpendicularly oriented features on lithographically defined chemical patterns by simple thermal annealing is therefore feasible at a resolution below 10 nm as shown in Figure 6.18 (Kim et al., 2014). Two BCPs, with L_0 of 19.1 and 14.6 nm, underwent DSA on chemical patterns. Despite some minor defects observed in the SEM images, the significant degree of alignment of these BCPs indicates that the molecularly engineered block chemistry enables simple thermal annealing and a top, free surface to achieve desirable architectures with sub-10 nm features.

6.5.3 Application of topcoats

Another strategy to achieve sub-10 nm features with DSA of high-χ BCP films is to use a topcoat on the BCP film. A nonpreferential topcoat can be selected such that its interaction energy with each block of the BCP is effectively equal, thereby preventing one block from coating over the top surface. As shown in Figure 6.19, when the top surface of an A-*b*-B BCP film is covered by an additional topcoat polymer C, the difference in interfacial energies between the topcoat and each block of the BCP (γ_{AC} vs. γ_{BC}) contributes to the free energy of the BCP film and takes the place of the surface

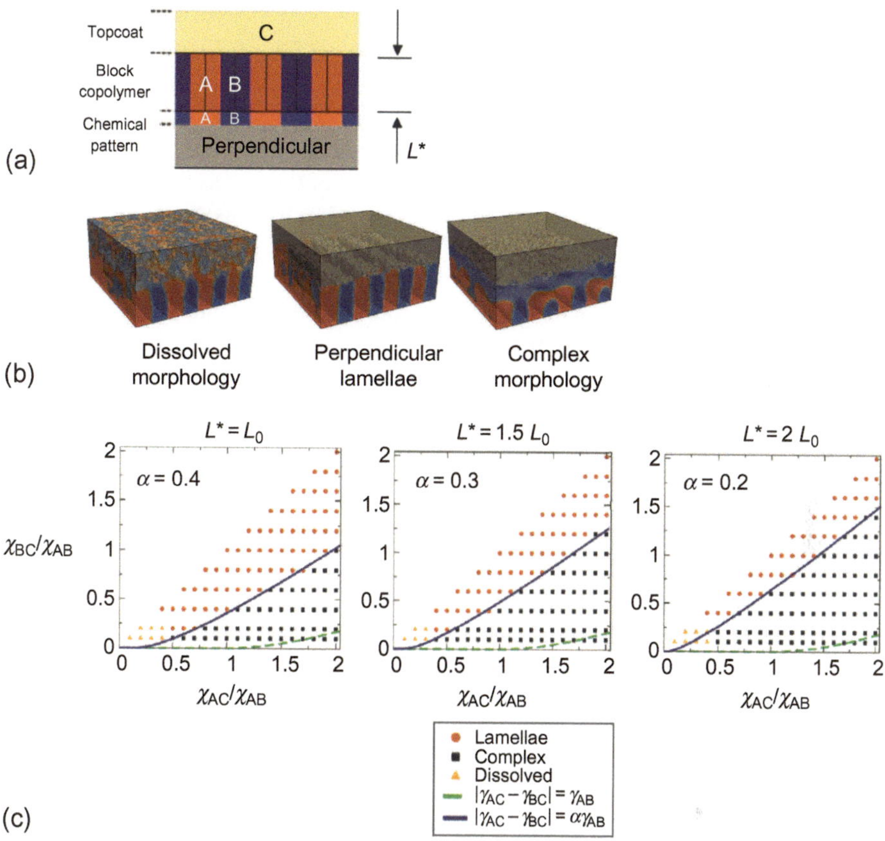

Figure 6.19 (a) Schematic of topcoat strategy for DSA. Symmetric A-*b*-B block copolymer is assembled between a chemical pattern and topcoat polymer C (red for A block, blue for B block, and yellow for polymer C). (b) Representative morphologies of BCP films obtained from Monte Carlo simulations. Topcoat polymer is shown as gray in color. (c) Phase diagram in the plane χ_{BC} versus χ_{AC} for different film thicknesses L_*, as shown above each graph. The phase boundaries depicted in the diagram were generated by the expression $|\gamma_{AC} - \gamma_{BC}| = \alpha\gamma_{AB}$. Reprinted with permission from Ramirez-Hernandez et al. (2014). Copyright © 2014, ACS Publishing Group.

energy of the two blocks (Ramirez-Hernandez et al., 2014). If an appropriate topcoat polymer is applied to balance this interfacial interaction, thermodynamically favorable boundary conditions for DSA with perpendicular orientation can be prepared.

Ramirez-Hernandez et al. used Monte Carlo simulations of a Theoretically Informed Coarse-Grain (TICG) model to explore the properties of topcoat materials for DSA of BCP (Ramirez-Hernandez et al., 2014). In their model, a symmetric A-*b*-B diblock copolymer was assembled between a 1:1 chemical pattern and a homogeneous topcoat polymer C. The set of different morphologies of BCP films predicted by Monte Carlo simulations was classified into three cases: dissolved morphology, perpendicular lamellae, and complex morphology. Examples of these three morphologies are shown in Figure 6.19b. The outcomes of simulations are summarized in the phase diagram in the plane χ_{AC} and χ_{BC}. Figure 6.19c shows the phase diagrams for different thicknesses of BCP films. The perpendicular lamellae was observed around the line where $\chi_{AC}=\chi_{BC}$. This condition corresponds to a nonpreferential boundary condition on the top of BCP films. The perpendicular lamellae became complex morphologies as a result of preferential wetting of one block at the surface of topcoat as the topcoat moved away from the nonpreferential condition. The boundary of this transition is shown with a solid blue line in the phase diagrams, and it was found to be described by the expression $|\gamma_{AC}-\gamma_{BC}|=\alpha\gamma_{AB}$, with α being a fitting parameter. This means one can obtain the perpendicularly oriented domain if the difference in interfacial energies at top surface is smaller than some portion (α) of the interfacial energy between blocks of BCP. Therefore, the α value is defining the process window of a topcoat system. A larger α value indicates a wider selection window for topcoat materials. Figure 6.19c shows examples of how the topcoat process window is affected by the thickness of the BCP film. As thickness increases from L_0 to $2L_0$, the α value decreases from 0.4 to 0.2, indicating the orientation becomes more sensitive to the interfacial interaction of the BCP with the topcoat material. This numerical approach provides useful guidelines for selection of appropriate topcoat materials for any BCP system.

The topcoat approach that enables control of the orientation of high-χ BCP films was also demonstrated experimentally. Yoshida et al. controlled the orientation of polystyrene-*block*-poly(2-vinylpyridene) (PS-*b*-P2VP) films on chemical patterns by using several topcoat materials (Yoshida et al., 2013). When the top surface of a PS-*b*-P2VP film was covered by a cross-linked PS or P2VP mat, which do not fall within the process window predicted by the simulation study, complex mixed morphologies were observed after thermal annealing (Figure 6.20a). In contrast, when cross-linked, nonpreferential poly(styrene-*ran*-2-vinylpyridene) (P(S-*r*-2VP)) was used as a topcoat, the PS-*b*-P2VP domains oriented perpendicularly to the substrate and the top surface. This desired through-film orientation was also obtained when the PMMA was used as a topcoat for P2VP-*b*-PS-*b*-P2VP film on a nonpreferential substrate. Note that PMMA was predicted to be within the process window for the topcoat of a PS-*b*-P2VP system. As shown in Figure 6.20b, the parallel orientation of triblock copolymer domain was observed even on the neutral substrate when being annealed without topcoat because of the strong PS–preferential interaction on the top surface. However, as shown in Figure 6.20c, perpendicularly oriented domains with $L_0=15$ nm (7.5 nm half-pitch) were clearly observed when the triblock copolymer was annealed under a PMMA topcoat. This result strongly suggested that DSA of BCP films with sub-10 nm features is feasible by thermal annealing with thermodynamically favorable boundary conditions, including the appropriate underlying chemical pattern and topcoat polymer. For implementation of this topcoat strategy, application and removal methods of the topcoat layer will need to be developed to be compatible with the DSA process and manufacturing systems.

6.5.4 Solvent annealing

Solvent annealing provides an alternative strategy for self-assembly of BCPs to realize sub-10-nm resolution. Compared with thermal annealing, solvent annealing can be

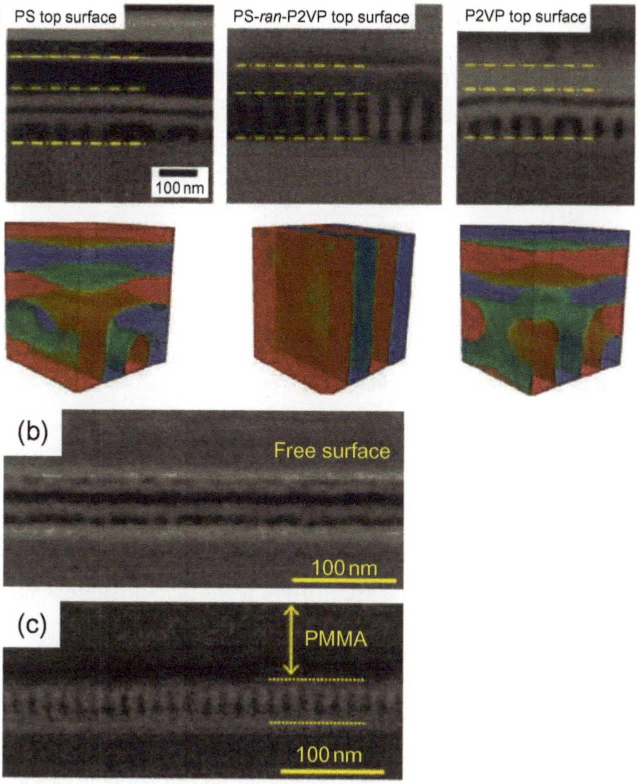

Figure 6.20 (a) Cross-sectional SEM and corresponding 3D Monte Carlo simulation images of PS-*b*-P2VP films assembled between chemical pattern and cross-linked polymeric topcoats. Cross-sectional SEM images of P2VP-*b*-PS-*b*-P2VP film on neutral substrate under (b) free surface and (c) PMMA topcoat.
Reprinted with permission from Yoshida et al. (2013).

applied for the DSA of BCP systems that show very dissimilar surface energy between the blocks and works efficiently for BCPs with low thermal degradation temperature.

Solvent vapor has multiple effects during annealing of BCP films. First, the solvent vapor can mediate surface energy differences between the blocks of the BCP at the free surface of the film to generate perpendicular domain orientation (Kim et al., 2004a,b; Xuan et al., 2004; Ho et al., 2005; Vayer et al., 2010; Cavicchi et al., 2005). Second, the absorbed solvent can effectively lower the glass transition temperatures (T_g) of the blocks and increase the polymer mobility (Cavicchi et al., 2005; Albalak et al., 1998; Kim and Libera, 1998). The increased mobility leads to faster processing times. Kim et al. reported that solvent-vapor annealing significantly facilitated the mass transport for high molecular weight PS-*b*-PMMA BCP, which is hardly achievable using thermal annealing alone (Kim et al., 2013). Third, the solvent in the film alters the interactions between blocks (χ_{eff}, Naughton and Matsen, 2002), which results in changes in morphology (Xuan et al., 2004; Jung and Ross, 2009; Peng et al., 2004) and period of the BCP domains (Kim et al., 2004a; Banaszak and Whitmore, 1992;

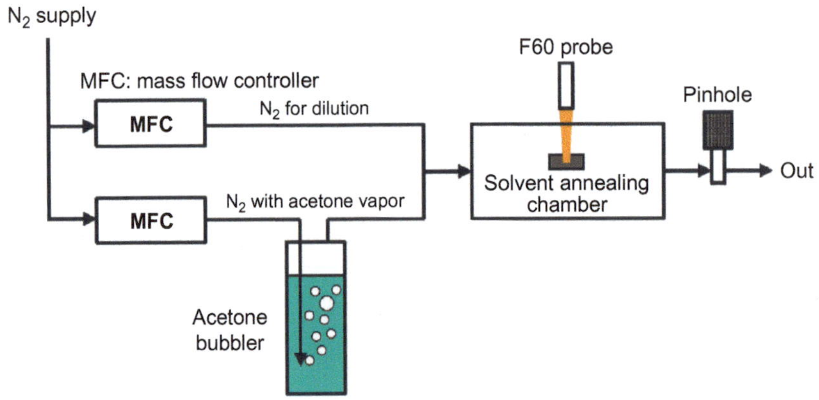

Figure 6.21 Continuous flow process for solvent annealing.

Whitmore and Vavasour, 1992; Di et al., 2010). Cushen et al. demonstrated that by using a combination of domain-selective solvent mixtures, a transition from a disordered to an ordered BCP occurred (Cushen et al., 2014). Additionally, the solvent interactions may lead to dewetting of the BCP films from the substrate's surface (Cavicchi et al., 2005; Harant and Bowman, 2005; Peng et al., 2005).

When exposed to the solvent vapors, the dry BCP film will absorb the solvents and swell. The swelling is driven by the entropy of mixing and reaches the interphase equilibrium only when the chemical potential of the solvent in the film approaches that of the solvent in the vapor phase. The thickness of the swollen film and the relative concentration of polymer and solvent are controlled by the solvent vapor pressure under ambient condition (Sinturel et al., 2013). In the traditional setup for solvent annealing, the BCP film was set in a closed jar with a reservoir of solvent placed beside it for a period of time, and the ratio of solvent surface area to chamber volume was used to control the swelling ratio (Kim et al., 2004a; Son et al., 2012). Another robust system for solvent annealing is the continuous flow setup as reported by Gotrik et al. (2012), in which inert gas flow carries solvent saturated vapors to the BCP film by bubbling through solvents (Figure 6.21). The equilibrium mole fraction of solvent in the polymer melt is determined as a function of volumetric flow rates of the inert gas and temperature, as well as operating pressure. Recently, more sophisticated devices have been developed for enhanced solvent annealing. Zhang et al. introduced microwave-assisted solvothermal annealing process to reduce the annealing time (Zhang et al., 2010). Albert et al. designed a microfluidic mixing device that enables testing different solvent compositions in a single experiment (Albert et al., 2011).

In contrast with thermal annealing where BCPs are heated to temperatures greater than T_g but lower than their ODT temperature, solvent exposure can effectively lower the T_g of the block polymers below room temperature, such that the BCPs can equilibrate at room temperature. It is proposed that the assembly of BCPs via solvent annealing follows a "quasi-equilibrium and quenching" mechanism, in which quasi-equilibrium of the assembly is achieved in a solvated state. The microdomain structures of the swollen film at equilibrium depend on the solvent selectivity and the swelling ratio.

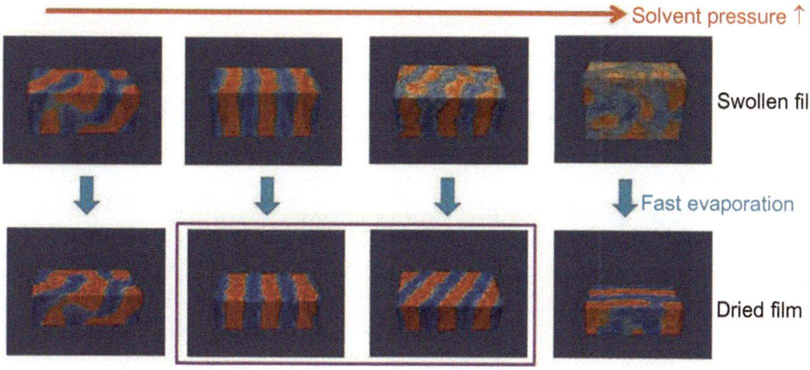

Figure 6.22 Self-assembly of symmetric BCP thin films after swelling (top) and after solvent evaporation (bottom).
Reprinted with permission from Hur et al. (2015). Copyright © 2015, ACS Publishing Group.

By quenching in a short time, the final morphology in the rapidly dried films is maintained, except for shrinking in the direction normal to the substrate (Paik et al., 2010).

Hur et al. used a TICG model of BCPs to study the evolution of microstructure and the transformations between various phases during solvent swelling and evaporation (Hur et al., 2015). As shown in Figure 6.22, when polymer volume fractions ϕ_p is large, the amount of solvent in the swollen film is not sufficient to enhance polymer chain mobility, and defects are kinetically trapped. For intermediate ϕ_p, defects in the film can be removed in the swollen state, and thus the dry film is free of defects. During the swelling process, the solvent dilutes unfavorable interactions between dissimilar blocks, and the effective Flory–Huggins parameter is $\chi_{eff}=\phi_p^a\chi_{AB}$, with a in the range of 1.0–1.6 (Lodge et al., 2002, 2003; Huang and Lodge, 1998). As ϕ_p further decreases, an order–disorder phase transition occurs, and the BCP is in the disordered swollen state. Evaporation of the solvent makes the system re-enter the ordered phase and defective dried film appears. There is a lower limit of ϕ_p corresponding to $(\chi N)_{ODT}$, and below that limit the ordered phase in a swollen film cannot be formed. Considering a higher limit of ϕ_p, polymer with low molecular weight is favored due to increased polymer chain mobility. Therefore, a short polymer with higher χ_{AB} has a wider process window for defect removal.

It has been proposed that solvent annealing provides an effective DSA strategy by lowering the kinetic barriers and impeding defect annihilation to create defect-free, self-assembled BCP films over large areas (Hur et al., 2015). The introduction of solvent annealing to graphoepitaxy where the BCP is guided by various topographic patterns has been investigated extensively. Ross et al. reported highly ordered cylindrical, lamellar and square arrays obtained by solvent annealing polystyrene-*block*-polydimethylsiloxane (PS-*b*-PDMS) films on various topographic patterns (Jung and Ross, 2009; Son et al., 2011, 2012; Yang et al., 2010; Tavakkoli et al., 2012). Bosworth et al. showed that DSA of poly(α-methylstyrene)-*block*-poly (4-hydroxystyrene) (PαMS-*b*-PHOST) in trenches could form ordered parallel cylindrical and spherical phases in a nonselective solvent (tetrahydrofuran) and a selective solvent (acetone), respectively (Bosworth et al., 2008). Kim et al. used solvent-vapor annealing followed by thermal annealing to produce a perfectly assembled perpendicular lamellar morphology of high molecular weight PS-*b*-PMMA with very low chain mobility in the trenched confinement (Kim et al., 2013). Cushen et al. reported DSA of poly(4-trimethylsilylstyrene)-*block*-poly(D,L-lactide) (PTMSS-*b*-PLA) with parallel cylinder morphology on topographical patterns with domain periodicity as small as 13.4 nm after solvent annealing (Cushen et al., 2014).

DSA on chemical patterns by solvent annealing is less studied. Bosworth et al. reported DSA of PαMS-*b*-P4VP annealing in THF formed a hexagonal dot and stripe pattern with 20-nm pitch on chemical patterns with density multiplication (Bosworth et al., 2010). Xu et al. studied the applicability of polystyrene-*block*-poly(ethylene oxide) (PS-*b*-PEO) system as well as the solvent annealing approach on a chemically patterned surface to assemble perpendicular cylinders (Xu et al., 2011).

The Nealey and de Pablo research groups have focused on DSA of P2VP-*b*-PS-*b*-P2VP triblock copolymer on chemical patterns using solvent annealing (Hur et al., 2015; Xiong et al., 2015; Khaira et al., 2014). Long-range-ordered, defect-free, registered perpendicular lamellae with a period of 16.5 nm on sparse chemical patterns for 3× and 4× density multiplication have been demonstrated (Figure 6.23) (Khaira et al., 2014). The extent of defects and the relation of defect quantity to L_S/L_0 commensurabil-

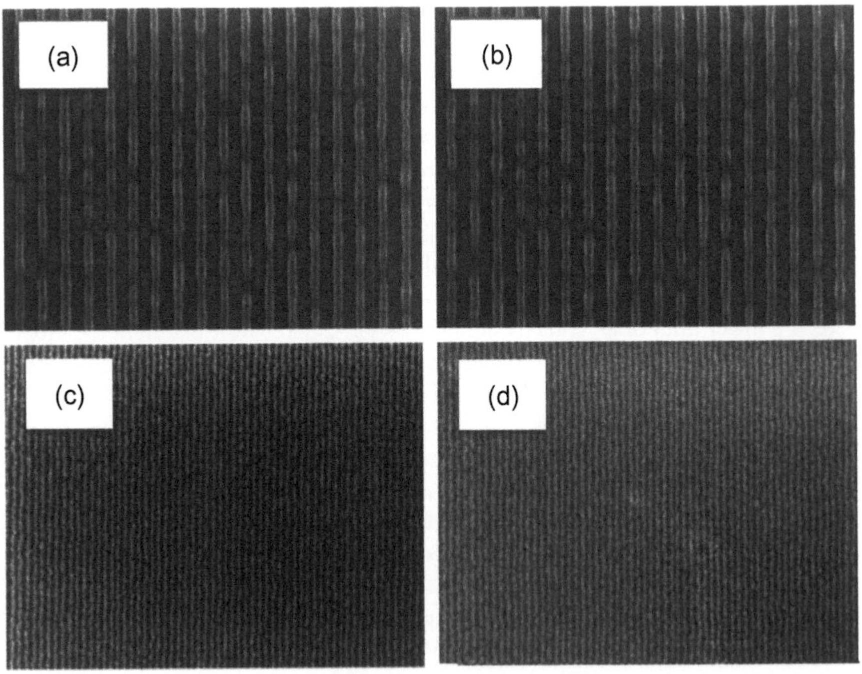

Figure 6.23 Underlying chemical patterns employed for directed self-assembly and final morphology of the P2VP-*b*-PS-*b*-P2VP triblock copolymer with bulk period 24.5 nm assembled on those patterns. (a) Chemical patterns for 3× and (b) 4× density multiplication. The triblock copolymer assembled into perpendicular lamellae with period of 16.5 nm for both (c) 3× and (d) 4× density multiplication.

ity of the guiding patterns have been studied (Khaira et al., 2014). Using a covariance matrix adaptation evolution strategy (CMA-ES), two regions of the combined parameter W/L_0, where defect-free perpendicular lamellae can form, were found. In the case of $W=0.5L_0$, only one lamellar domain anchored to the guiding stripe and a strong preferential substrate was required for a defect-free assembly. When $W=1.5L_0$, three lamellar domains sit on the guiding stripe, and a moderate anchoring was sufficient to drive the assembly. The parameter fluctuations in evolutionary computations showed that 4× multiplication is less permissive than 3× and has a much narrower range of processing conditions for optimal assembly.

6.6 Applications

6.6.1 Overview

The most important applications of self-assembly of BCPs are in the various facets of nanomanufacturing. Since 2007, the ITRS roadmap has included DSA as one among the three or four potential alternative lithography tools for next-generation microelectronic fabrication (S.I. Association, 2007). The addition of DSA, combined with other lithography tools, enables a new type of nanopatterning with lower cost and high throughput, with the resolution that can hardly be achieved by conventional photolithography. The dense pitch patterns not only enhance the resolution, but also repair any defects present in the chemical pattern.

Initial concept-proving studies included density multiplication by Ruiz et al. (2008) and DSA of device-oriented structures by Stoykovich et al. (2005). To verify the amenability of DSA for industry fabrication, defect studies and evaluation of line-edge roughness (Kim, 2014; Patrone and Gallatin, 2012, 2014) have also been frequently reported. In this section, we will briefly review some of the representative implementations of DSA for device-oriented purposes, both in hard disk drive fabrication and in semiconductor industry pilot lines.

6.6.2 Bit-patterned media

The evolution of hard disk drive relies on scaling the feature size of periodic structures to make high-density data storage possible. However, superparamagnetism imposed a limit for the smallest grain size in the traditional granular media. Instead, each bit in bit-patterned media (BPM) is lithographically patterned and the magnetic islands are comparably larger in size than the grains in this conventional case. This allows for the reduction of signal-to-noise ratio as well as provides thermal stability. The challenge shifts to the production of patterns with high area density, beyond 1 Tbits/in.^{2}, which falls in the resolution range of self-assembling BCPs.

DSA can be incorporated in the initial steps to create templates for subsequent pattern transfer or replication. The sparse chemical patterns for DSA of BCPs were prepared by a rotary E-beam tool. After DSA, the patterns with density multiplication were transferred to a substrate, serving as a master template for nanoimprinting to create a high volume of replicates and then disks with magnetic media (Figure 6.24).

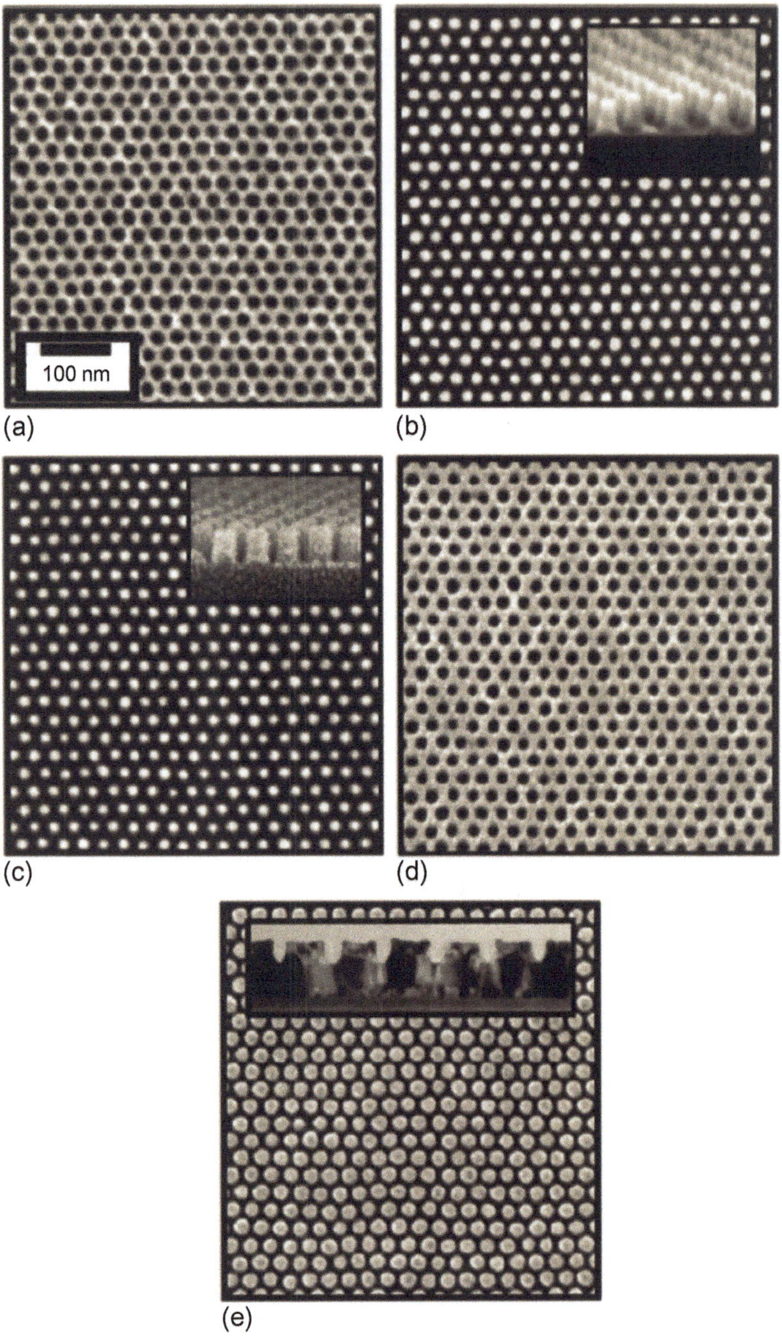

Figure 6.24 SEM micrographs at various stages of bit patterned media fabrication: (a) developed block copolymer pattern on Si master template, (b) Si master template, (c) quartz replica working template, (d) imprinted resist on disk, and (e) finished magnetic islands (inset: X-section TEM). An island placement tolerance of nm is achieved at all steps. Reprinted with permission from Albrecht et al. (2013). Copyright © 2013, IEEE.

Two geometries of BCPs have been explored to meet the density demand of BPM over 1 Tbits/in.^{2}. The first are the hexagonally close-packed cylinders (Hellwig et al., 2010) or spherical (Yang et al., 2013; Xiao et al., 2013) BCP domains, which provides the highest area density. The second geometry uses an orthogonal line cut to fabricate a rectangular unit.

Although the close-packed patterns provide the highest area density of features, the integration with recording systems has found some difficulties. It has been suggested to use rectangular bits with a bit aspect ratio (the ratio of the cross-track to the down-track pitch) greater than 1. With a line cut strategy (Ruiz et al., 2011), rectangular bit cells can be fabricated via a double imprinting process. Briefly, one submaster can be patterned with circumferential lines and the second is patterned with radial lines (Figure 6.25).

Other than the bits, servo patterns are also required to be created in BPM for the head to locate data precisely. Servo patterns are mostly irregular in shape, and thus need a specific integration strategy with the BCP assembly. For the chevron design, first an EBL sparse pattern was defined and the BCPs tended to follow the bends with different angles. Because of the free-energy penalty, defects formed at the chevron apexes. Accordingly, additional vertical lines were added at the junction of horizontal tracks and chevrons. This helps to minimize the defective area at the interfaces by half, as shown in Figure 6.26 (Liu et al., 2011c). Subsequently, a new integration method, termed "offset burst," was proposed and the position error signal for track following could be encoded (Lille et al., 2012). From the EBL step, a set of circumferential lines was intentionally shifted along the cross-track direction. One approach to integrate servo and data patterns is to run a separate patterning process after the DSA of data region. This usually is not favored because the registration of the two patterns would be extremely difficult.

6.6.3 Fin field-effect transistor

Using DSA for Fin field-effect transistor fabrication represents the first device-oriented application of BCPs in microelectronics. Figure 6.27 shows a schematic of the silicon-on-insulator (SOI) fin fabrication process. A periodic line-space pattern formed by DSA of PS-*b*-PMMA was formed, and then a cut lithography step was used to protect the functional areas. The overlay requires precise alignment.

A new strategy to customize the patterns, named Chemical Epitaxy Etch Trim using a Self-aligned Hardmask (CHEETAH), has been developed. The CHEETAH process sequence is shown in Figure 6.28. The process starts from the design of a chemical pattern that is composed of two regions: guided and unguided. The guided regions are still kept the same as before, but the unguided regions are supported by a neutral hard mask. The BCP domains can be perpendicularly oriented and aligned on an unpatterned, nonpreferential region. Only after the pattern transfer step, the hard mask in unguided region acts as an etch stop. As a test, the CHEETAH process has been validated using PS-*b*-PMMA with a natural period of 25 nm. An isolated group of lines can be fabricated using single line gaps. Vertical trim with a tip-to-tip distance of 30 nm can be generated (Doerk et al., 2014).

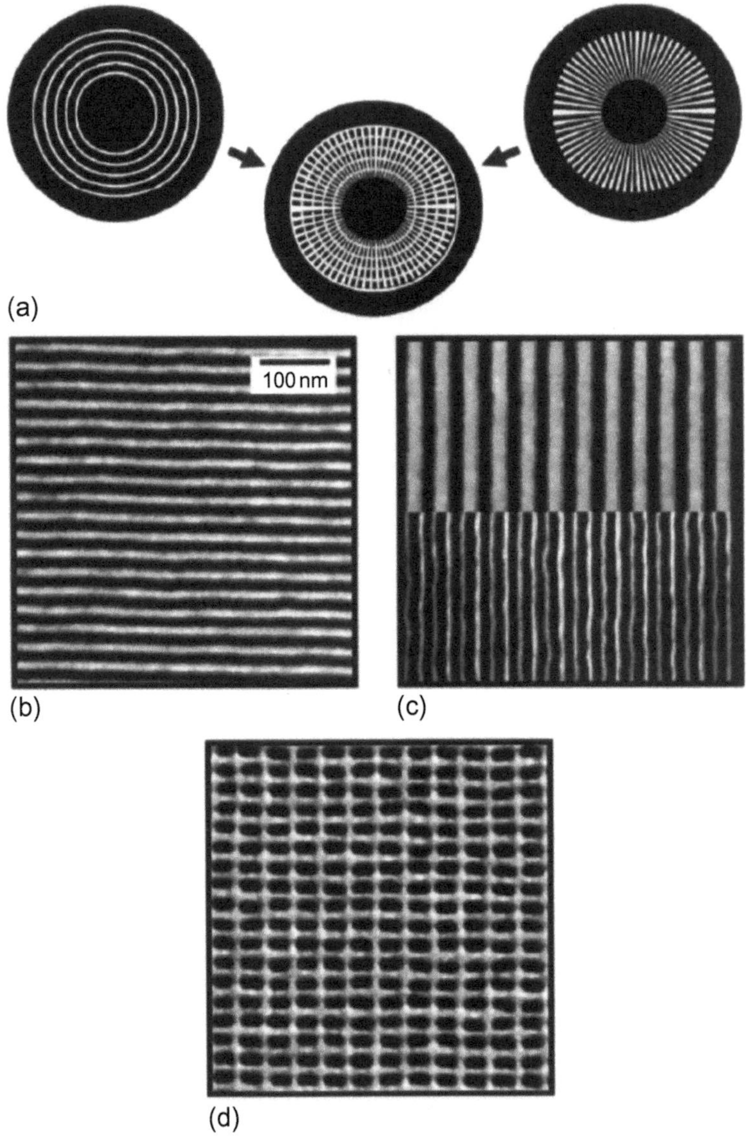

Figure 6.25 (a) Schematic representation of a double imprint process to produce rectangular features. Two "submaster" templates, one with circumferential stripes and a second one with radial lines, are imprinted onto a separate "master" template. The intersection of the two sets results in rectangular features. (b) Circumferential block copolymer lines with a 27-nm full pitch. (c) Sidewall spacer line doubling from block copolymer lines with a 41-nm full pitch (top of image) to 20.5-nm pitch lines after line doubling (bottom). (d) High bit aspect ratio rectangular features imprinted in resist by the master template.
Reprinted with permission from Albrecht et al. (2013). Copyright © IEEE.

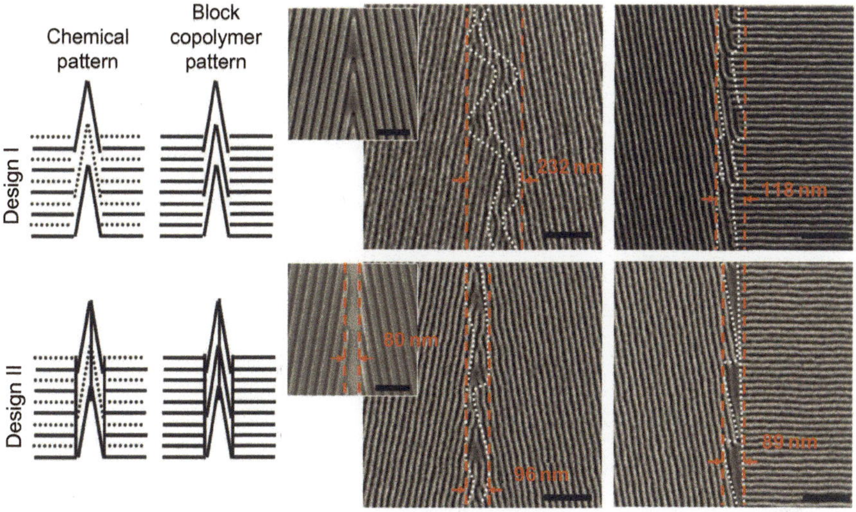

Figure 6.26 Chevron design schematics and resulting lamellar patterns. For design I and II schematics (shown at the top), the solid lines represent the e-beam patterns at a period of $2L_0 = 54$ nm, while the dotted lines represent lamellae interpolation giving 2× density multiplication. In design II extra vertical e-beam patterns were added. SEM images of the resulting lamellae ordering are shown to the right. The red dashed lines indicate the boundaries between defective and nondefective areas. The widths of the design II defective areas are smaller as a result of the additional vertical patterns.
Reprinted with permission from Liu et al. (2011c). Copyright © AVS.

6.6.4 300-mm Line integration

The Nealey group has been one of the pioneers in the coupling of DSA with 193i lithography (Liu et al., 2010c, 2011b). This work has helped pave the path for scaling up DSA on the wafer level. The LiNe flow described above has been implemented at IMEC for the study of wafer scale fabrication (Delgadillo et al., 2012; Rincon-Delgadillo et al., 2013).

Translating the DSA process from lab to fab on full 300-mm wafers allows us to closely investigate production factors such as processing window and defect origins. In the production fab environment, each processing step can be reviewed and optimized. Those steps that are incompatible with industry requirements can be reworked to make them compatible. The IMEC 300-mm DSA flow serves as an effective platform for the examination of line-space frequency multiplication, both on process conditions and materials, with the involvement of material vendors, equipment manufacturers, and circuit designers.

As reported in previous publications, there is a commensurability window in terms of guiding line width W and pitch L_S. For PS-*b*-PMMA, L_S is generally set to be an integer multiple of L_0. W was controlled both by the exposure dose in lithographic patterning of chemical pattern and the plasma trimming that followed. A broad window has been found for the lithography dose range (with depth of focus >200 nm and

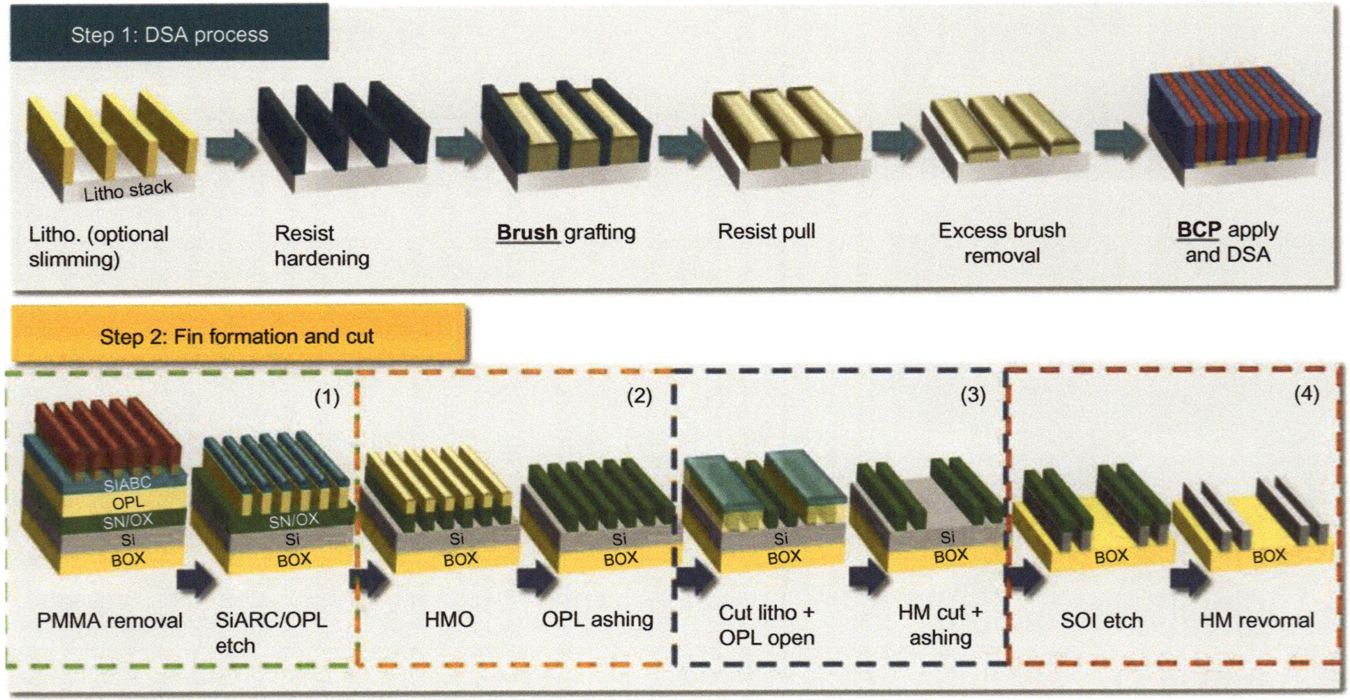

Figure 6.27 The SOI Fin formation process flow using the DSA process.

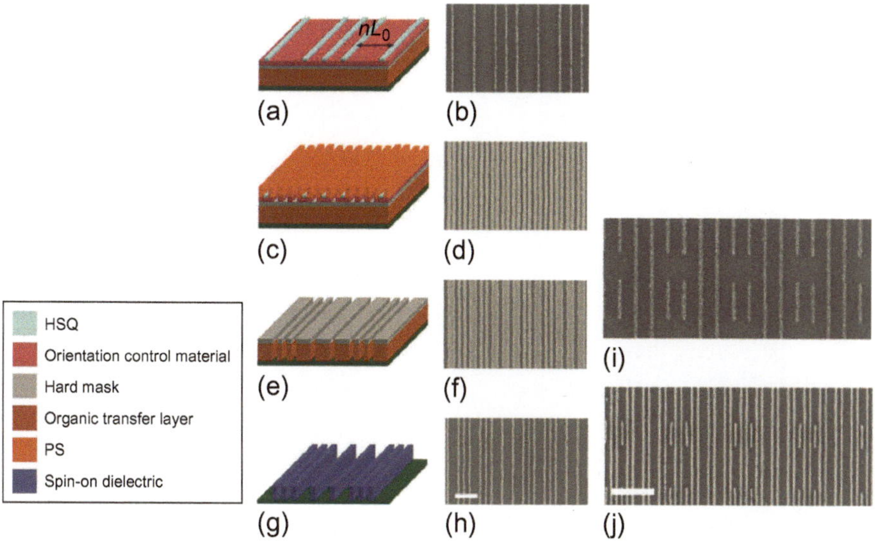

Figure 6.28 Schematic (a,c,e,g) and SEM images (b,d,f,h) of the fabrication process flow for self-aligned line cutting: (a,b) HSQ pattern. (c,d) DSA and O2 RIE to remove PMMA domains. The remaining PS domains and HSQ pattern form a composite mask for following pattern transfer. (e,f) Pattern transfer of the composite mask into the organic transfer layer using CF4 and O2 RIEs. (g,h) Pattern tone inversion. (i) SEM image of a fragmented HSQ pattern and (j) a SEM image of the resulting fragmented DSA-generated grating. Scale bars are 100 nm (b,d,f,h) and 200 nm (i,j).
Reprinted with permission from Doerk et al. (2014). Copyright © Nature.

exposure latitude >40%, CD varying from 42.6 to 32.9 nm, where CD is the critical dimension of XPS guiding stripes) to achieve good alignment of lamellae in the assembly of 25 nm PS-*b*-PMMA. A comparison of the pattern and the assembled BCP is shown in Figure 6.29. Further study of the through-film structure relies on the examination after selective removal of one block or even after pattern transfer to the underlayer (Chan et al., 2014).

Optical inspection tools were utilized to analyze the dislocation and declination defects on the 300-mm wafers. To study the defect source, the overall defect density and breakdown of major contributors are analyzed by each process step, as shown in Figure 6.30 (Delgadillo et al., 2013). Before the annealing of the BCP, most of the defects (>80%) come from the XPS layer of the chemical pattern, and 99% of them correspond to the pinholes in XPS layer, which are believed to be due to the dewetting of XPS thin film in coating and baking. The final defect density after thermal annealing of BCP was about 564/cm^2, and about 90% is from the defects in BCP assembly. Recently, the evolution of defects has been investigated batchwise by monitoring the change after each generation of material alternation. As a result, the defect density has been continuously declining with improvement on material supply and processing.

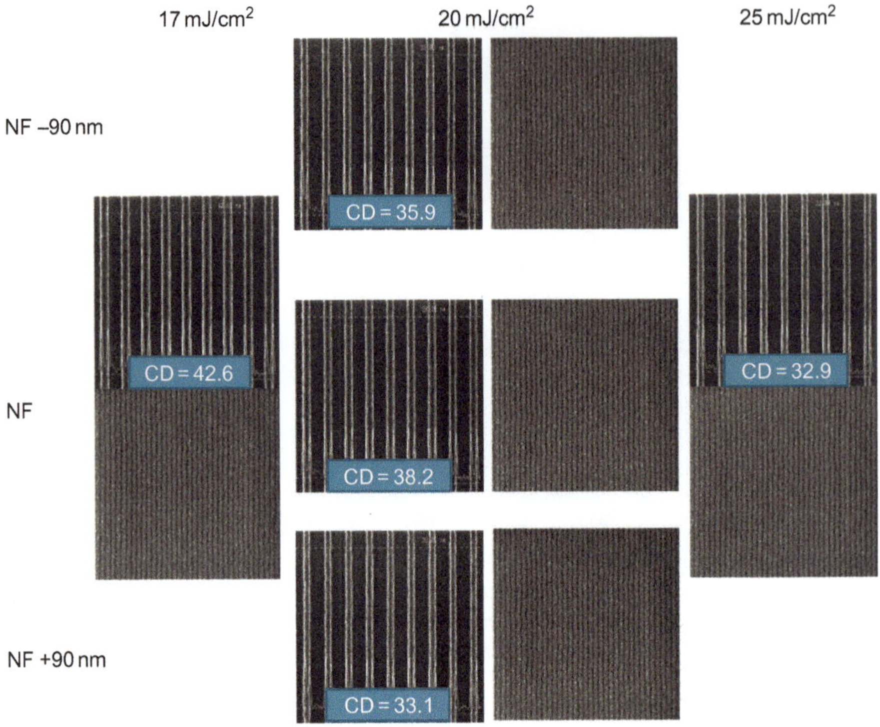

Figure 6.29 Top-down SEM images of the chemical pattern (left or top images, $L_S = 100$ nm) and assembled block copolymers ($L_0 = 25$ nm) of the DSA process as a function of pattern dose and focus settings.

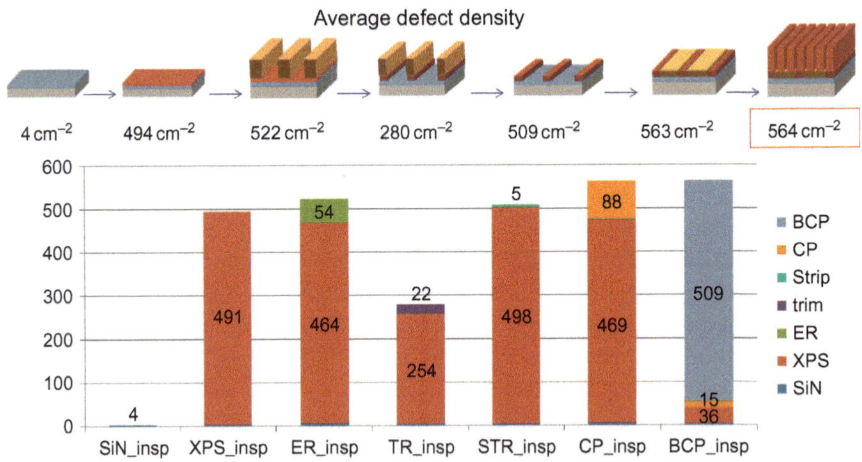

Figure 6.30 Evolution of defects in defect source analysis of DSA experiment.

6.7 Conclusion

As described above, BCPs can be directed to self-assemble to form densely packed features with highly uniform dimensions and shapes in ordered arrays at the scale of 3–50 nm. Chemical patterns can be defined using traditional lithographic materials and processes such as 193i lithography or electron beam lithography at the scale of 20–40 nm. By directing the assembly of BCP films on the chemical patterns, the overall resolution of the lithographic process may be increased by three- to four-fold or more. The understanding of the thermodynamics of the DSA system that has developed over the past decade has enabled the development of nonregular device-oriented geometries, high-resolution patterning, and assembly-defect reduction. The understanding of the dynamics of DSA is significant because process times play a critical role in the viability of the DSA process in manufacturing.

The interest and exponential growth in research activity and expenditure directed at patterning with reduced feature size is driven in the semiconductor industry by the relentless push to increase memory and processing capacity. For hard drives, DSA of BCPs is the only known technology that is feasible to fabricate nanoimprint masters to manufacture BPM at the required storage densities (at least greater than 2 Tbit/in.^{2}). Currently, DSA of PS-*b*-PMMA films on chemical patterns is a primary focus of activity, and the main research objectives revolve around demonstrations that DSA can meet manufacturing requirements related to degrees of perfection, processing latitude, integration of the technology with existing infrastructure, and device design for use with DSA patterns.

Critical research issues remain that must be addressed in order to drive DSA technology to widespread implementation in nanomanufacturing. A key roadblock is the establishment of proven pathways to realize sub-10-nm resolution and scale to 5 nm; the resolution limit of PS-*b*-PMMA is ~12 nm. Neither the semiconductor industry nor the hard drive industry will implement DSA for only a single generation of products. New materials and processes must be developed to lower the resolution limit, such that the DSA manufacturing process can serve three to four future generations of products. Whereas substantial effort is being expended by many groups to identify BCP systems capable of self-assembling into the sub-10-nm regime, technology gaps exist in fundamental and technological understanding as to how those materials may be processed and directed to assemble and continue to meet all of the constraints of manufacturing (assembly, pattern transfer, pattern quality, pattern geometry for devices). Research is in its infancy related to overall engineering of BCP systems specifically for high-resolution DSA with respect to polymer chemistry, block architecture, chain stiffness, thermophysical properties, surface and interfacial properties, and pattern transfer mechanisms. Moreover, to our knowledge, little or no activity aims to develop specialized tools and processes that will be applicable at this ultimate end-of-the-roadmap length scale. Finally, development of a comprehensive suite of metrology tools and techniques is a basic, currently unmet need to properly access and characterize DSA materials and processes at the nanoscale.

Despite these challenges, DSA of BCPs seems to be passing the test of time. Initially, DSA offered advanced lithography the ability to improve process control.

But as with any developing technology, new challenges were raised. Could device-oriented features be fabricated? Could sub-10-nm resolution be achieved? Could this technology, developed in university laboratories, actually be used in large-scale, commercial manufacturing? As the material in this chapter shows, through more than a decade of research, at a variety of academic institutions and in partnership with companies such as HGST, IBM, and IMEC, the answers to these questions are all "yes." Looking forward, we are optimistic about DSA, in large part because it involves self-assembly. Materials that self-assemble spontaneously form nanostructures in three dimensions at the molecular scale. The understanding of the underlying physics of DSA should enable these molecular scale structures to be incorporated on the macroscopic scale of devices. The integration of self-assembling materials with traditional manufacturing processes, that is, DSA, promises to extend far beyond the initial successes reviewed here.

References

Albalak, R., Capel, M., Thomas, E., 1998. Solvent swelling of roll-cast triblock copolymer films. Polymer 39 (8–9), 1647–1656.

Albert, J., Bogart, T., Lewis, R., Beers, K., Fasolka, M., Hutchison, J., Vogt, B., Epps, T., 2011. Gradient solvent vapor annealing of block copolymer thin films using a microfluidic mixing device. Nano Lett. 11 (3), 1351–1357.

Albrecht, T.R., Bedau, D., Dobisz, E., Gao, H., Grobis, M., Hellwig, O., Kercher, D., Lille, J., Marinero, E., Patel, K., Ruiz, R., Schabes, M.E., Wan, L., Weller, D., Wu, T.W., 2013. Bit patterned media at 1 tdot/in^2 and beyond. IEEE Trans. Magn. 49 (2), 773–778.

Banaszak, M., Whitmore, M., 1992. Self-consistent theory of block copolymer blends—selective solvent. Macromolecules 25 (13), 3406–3412.

Bates, F.S., Fredrickson, G.H., 1990. Block copolymer thermodynamics—theory and experiment. Annu. Rev. Phys. Chem. 41, 525–557.

Bates, F.S., Fredrickson, G.H., 1999. Block copolymers—designer soft materials. Phys. Today 52 (2), 32–38.

Bencher, C., Smith, J., Miao, L., Cai, C., Chen, Y., Cheng, J.Y., Sanders, D.P., Tjio, M., Truong, H.D., Holmes, S., Hinsberg, W.D., 2011. Self-assembly patterning for sub-15nm half-pitch: a transition from lab to fab. Proc. SPIE-Int. Soc. Opt. Eng. 7970, 79700F.

Black, C.T., 2007. Polymer self-assembly as a novel extension to optical lithography. ACS Nano 1 (3), 147–150.

Black, C.T., Ruiz, R., Breyta, G., Cheng, J.Y., Colburn, M.E., Guarini, K.W., Kim, H.C., Zhang, Y., IBM, J., 2007. Polymer self-assembly in semiconductor microelectronics. Res. Dev. 51 (5), 605–633.

Bosworth, J., Paik, M., Ruiz, R., Schwartz, E., Huang, J., Ko, A., Smilgies, D., Black, C., Ober, C., 2008. Control of self-assembly of lithographically patternable block copolymer films. ACS Nano 2 (7), 1396–1402.

Bosworth, J., Dobisz, E., Ruiz, R., 2010. 20nm pitch directed block copolymer assembly using solvent annealing for bit patterned media. J. Photopolym. Sci. Technol. 23 (2), 145–148.

Cavicchi, K., Berthiaume, K., Russell, T., 2005. Solvent annealing thin films of poly(isoprene-b-lactide). Polymer 46 (25), 11635–11639.

Chan, B.T., Tahara, S., Parnell, D., Delgadillo, P.A.R., Gronheid, R., de Marneffe, J.F., Xu, K.D., Nishimura, E., Boullart, W., 2014. 28nm pitch of line/space pattern transfer into silicon substrates with chemo-epitaxy directed self-assembly (dsa) process flow. Microelectron. Eng. 123, 180–186.

Cheng, J.Y., Rettner, C.T., Sanders, D.P., Kim, H.C., Hinsberg, W.D., 2008. Dense self-assembly on sparse chemical patterns: rectifying and multiplying lithographic patterns using block copolymers. Adv. Mater. 20 (16), 3155–3158.

Cheng, J.Y., Sanders, D.P., Truong, H.D., Harrer, S., Friz, A., Holmes, S., Colburn, M., Hinsberg, W.D., 2010. Simple and versatile methods to integrate directed self-assembly with optical lithography using a polarity-switched photoresist. ACS Nano 4 (8), 4815–4823.

Cushen, J., Wan, L., Pandav, G., Mitra, I., Stein, G., Ganesan, V., Ruiz, R., Willson, C., Ellison, C., 2014. Ordering poly(trimethylsilyl styrene-block-(d),(l)-lactide) block copolymers in thin films by solvent annealing using a mixture of domain-selective solvents. J. Polym. Sci. B Pol. Phys. 52 (1), 36–45.

Daoulas, K.C., Muller, M., Stoykovich, M.P., Kang, H., de Pablo, J.J., Nealey, P.F., 2008. Directed copolymer assembly on chemical substrate patterns: a phenomenological and single-chain-in-mean-field simulations study of the influence of roughness in the substrate pattern. Langmuir 24 (4), 1284–1295.

Delgadillo, P.A.R., 2014. Origin of Defects in Directed Self-Assembly of Diblock Copolymers Using Feature Multiplication. Ph.D. Thesis. KU Leuven, Leuven, Belgium. https://lirias.kuleuven.be/bitstream/123456789/451580/1/Thesis_Rincon_KUL.pdf.

Delgadillo, P.A.R., Gronheid, R., Thode, C.J., Wu, H.P., Cao, Y., Neisser, M., Somervell, M., Nafus, K., Nealey, P.F., 2012. Implementation of a chemo-epitaxy flow for directed self-assembly on 300-mm wafer processing equipment. J. Micro-Nanolith. Mem., 11(3), 031302.

Delgadillo, P.R., Harukawa, R., Suri, M., Durant, S., Cross, A., Nagaswami, V.R., Van Den Heuvel, D., Gronheid, R., Nealey, P., 2013. Defect source analysis of directed self-assembly process (DSA of DSA). In: SPIE Advanced Lithography. International Society for Optics and Photonics, San Jose, California, USA.

Detcheverry, F.A., Liu, G., Nealey, P.F., de Pablo, J.J., 2010. Interpolation in the directed assembly of block copolymers on nanopatterned substrates: simulation and experiments. Macromolecules 43 (7), 3446–3454.

Di, Z., Posselt, D., Smilgies, D., Papadakis, C., 2010. Structural rearrangements in a lamellar diblock copolymer thin film during treatment with saturated solvent vapor. Macromolecules 43 (1), 418–427.

Doerk, G.S., Cheng, J.Y., Singh, G., Rettner, C.T., Pitera, J.W., Balakrishnan, S., Arellano, N., Sanders, D.P., 2014. Enabling complex nanoscale pattern customization using directed self-assembly. Nat. Commun. 5, 5805.

Edwards, E.W., Montague, M.F., Solak, H.H., Hawker, C.J., Nealey, P.F., 2004. Precise control over molecular dimensions of block-copolymer domains using the interfacial energy of chemically nanopatterned substrates. Adv. Mater. 16 (15), 1315–1319.

Edwards, E.W., Stoykovich, M.P., Müller, M., Solak, H.H., De Pablo, J.J., Nealey, P.F., 2005. Mechanism and kinetics of ordering in diblock copolymer thin films on chemically nanopatterned substrates. J. Polym. Sci., Part B: Polym. Phys. 43 (23), 3444–3459.

Gotrik, K., Hannon, A., Son, J., Keller, B., Alexander-Katz, A., Ross, C., 2012. Morphology control in block copolymer films using mixed solvent vapors. ACS Nano 6 (9), 8052–8059.

Guarini, K.W., Black, C.T., Milkove, K.R., Sandstrom, R.L., 2001. Nanoscale patterning using self-assembled polymers for semiconductor applications. J. Vac. Sci. Technol. B 19, 2784–2788.

Hamley, I.W., 1998. Solid state structure of block copolymers. In: Anonymous (Eds.), The Physics of Block Copolymers. Oxford University Press, New York. http://ukcatalogue.oup.com/product/9780198502180.do.

Harant, A., Bowman, C., 2005. Solvent vapor-annealed block copolymer films on organosilane self-assembled monolayers. J. Vac. Sci. Technol. B 23 (4), 1615–1621.

Hellwig, O., Bosworth, J.K., Dobisz, E., Kercher, D., Hauet, T., Zeltzer, G., Risner-Jamtgaard, J.D., Yaney, D., Ruiz, R., 2010. Bit patterned media based on block copolymer directed assembly with narrow magnetic switching field distribution. Appl. Phys. Lett. 96 (5), 052511. http://scitation.aip.org/content/aip/journal/apl/96/5/10.1063/1.3293301.

Hinsberg, W., Cheng, J., Kim, H.-C., Sanders, D.P., 2010. Self-assembling materials for lithographic patterning: overview, status, and moving forward. Proc. SPIE-Int. Soc. Opt. Eng. 7637, 76370G.

Ho, R., Tseng, W., Fan, H., Chiang, Y., Lin, C., Ko, B., Huang, B., 2005. Solvent-induced microdomain orientation in polystyrene-b-poly (l-lactide) diblock copolymer thin films for nanopatterning. Polymer 46 (22), 9362–9377.

Hong, A.J., Liu, C.C., Wang, Y., Kim, J., Xiu, F.X., Ji, S.X., Zou, J., Nealey, P.F., Wang, K.L., 2010. Metal nanodot memory by self-assembled block copolymer lift-off. Nano Lett. 10 (1), 224–229.

Huang, C., Lodge, T., 1998. Self-consistent calculations of block copolymer solution phase behavior. Macromolecules 31 (11), 3556–3565.

Hur, S.-M., Khaira, G.S., Ramírez-Hernández, A., Müller, M., Nealey, P.F., de Pablo, J.J., 2015. Simulation of defect reduction in block copolymer thin films by solvent annealing. ACS Macro Lett. 4 (1), 11–15.

International Technology Roadmap for Semiconductors, 2011. http://www.itrs.net/ITRS%20 1999-2014%20Mtgs,%20Presentations%20&%20Links/2011ITRS/2011Chapters/ 2011Lithography.pdf.

Jeong, S.J., Moon, H.S., Kim, B.H., Kim, J.Y., Yu, J., Lee, S., Lee, M.G., Choi, H., Kim, S.O., 2010. Ultralarge-area block copolymer lithography enabled by disposable photoresist prepatterning. ACS Nano 4 (9), 5181–5186.

Jung, Y., Ross, C., 2009. Solvent-vapor-induced tunability of self-assembled block copolymer patterns. Adv. Mater. 21 (24), 2540.

Khaira, G., Qin, J., Garner, G., Xiong, S., Wan, L., Ruiz, R., Jaeger, H., Nealey, P., de Pablo, J., 2014. Evolutionary optimization of directed self-assembly of triblock copolymers on chemically patterned substrates. ACS Macro Lett. 3 (8), 747–752.

Kim, S.K., 2014. Stochastic simulation studies of line-edge roughness in block copolymer lithography. J. Nanosci. Nanotechnol. 14 (8), 6143–6145.

Kim, G., Libera, M., 1998. Morphological development in solvent-cast polystyrene-olybutadiene-polystyrene (sbs) triblock copolymer thin films. Macromolecules 31 (8), 2569–2577.

Kim, S.O., Solak, H.H., Stoykovich, M.P., Ferrier, N.J., de Pablo, J.J., Nealey, P.F., 2003. Epitaxial self-assembly of block copolymers on lithographically defined nanopatterned substrates. Nature 424 (6947), 411–414.

Kim, S., Misner, M., Xu, T., Kimura, M., Russell, T., 2004a. Highly oriented and ordered arrays from block copolymers via solvent evaporation. Adv. Mater. 16 (3), 226.

Kim, S., Misner, M., Russell, T., 2004b. Solvent-induced ordering in thin film diblock copolymer/homopolymer mixtures. Adv. Mater. 16 (23–24), 2119–2123.

Kim, S., Nealey, P.F., Bates, F.S., 2012. Decoupling bulk thermodynamics and wetting characteristics of block copolymer thin films. ACS Macro Lett. 1 (1), 11–14.

Kim, E., Ahn, H., Park, S., Lee, H., Lee, M., Lee, S., Kim, T., Kwak, E., Lee, J., Lei, X., Huh, J., Bang, J., Lee, B., Ryu, D., 2013. Directed assembly of high molecular weight block copolymers: highly ordered line patterns of perpendicularly oriented lamellae with large periods. ACS Nano 7 (3), 1952–1960.

Kim, S., Nealey, P.F., Bates, F.S., 2014. Directed assembly of lamellae forming block copolymer thin films near the order-disorder transition. Nano Lett. 14 (1), 148–152.

Lille, J., Ruiz, R., Wan, L., Gao, H., Dhanda, A., Zeltzer, G., Arnoldussen, T., Patel, K., Tang, Y., Kercher, D., Albrecht, T.R., 2012. Integration of servo and high bit aspect ratio data patterns on nanoimprint templates for patterned media. IEEE Trans. Magn. 48 (11), 2757–2760.

Liu, G.L., Stoykovich, M.P., Ji, S.X., Stuen, K.O., Craig, G.S.W., Nealey, P.F., 2009. Phase behavior and dimensional scaling of symmetric block copolymer-homopolymer ternary blends in thin films. Macromolecules 42 (8), 3063–3072.

Liu, G., Thomas, C.S., Craig, G.S.W., Nealey, P.F., 2010a. Integration of density multiplication in the formation of device-oriented structures by directed assembly of block copolymer-homopolymer blends. Adv. Funct. Mater. 20 (8), 1251–1257.

Liu, C.C., Craig, G.S.W., Kang, H.M., Ruiz, R., Nealey, P.F., Ferrier, N.J., 2010b. Practical implementation of order parameter calculation for directed assembly of block copolymer thin films. J. Polym. Sci., Part B: Polym. Phys. 48 (24), 2589–2603.

Liu, C.C., Raub, A.K., Hakeem, P.J., Brueck, S.R.J., Nealey, P.F., 2010c. The integration of block copolymer directed assembly with 193 immersion lithography. J. Vac. Sci. Technol. B 28 (6), C6B30–C6B34.

Liu, G., Delcambre, S.P., Stuen, K.O., Craig, G.S.W., De Pablo, J.J., Nealey, P.F., Nygard, K., Satapathy, D.K., Bunk, O., Solak, H.H., 2010d. Mechanism and dynamics of block copolymer directed assembly with density multiplication on chemically patterned surfaces. J. Vac. Sci. Technol. B 28 (6), C6B13–C6B19.

Liu, C.C., Han, E., Onses, M.S., Thode, C.J., Ji, S.X., Gopalan, P., Nealey, P.F., 2011a. Fabrication of lithographically defined chemically patterned polymer brushes and mats. Macromolecules 44 (7), 1876–1885.

Liu, C.C., Thode, C.J., Delgadillo, P.A.R., Craig, G.S.W., Nealey, P.F., Gronheid, R., 2011b. Towards an all-track 300 mm process for directed self-assembly. J. Vac. Sci. Technol. B 29 (6), 06F203.

Liu, G.L., Nealey, P.F., Ruiz, R., Dobisz, E., Patel, K.C., Albrecht, T.R., 2011. Fabrication of chevron patterns for patterned media with block copolymer directed assembly. J. Vac. Sci. Technol. B, 29(6), 06F204.

Liu, C.C., Ramirez-Hernandez, A., Han, E., Craig, G.S.W., Tada, Y., Yoshida, H., Kang, H.M., Ji, S.X., Gopalan, P., de Pablo, J.J., Nealey, P.F., 2013a. Chemical patterns for directed self-assembly of lamellae-forming block copolymers with density multiplication of features. Macromolecules 46 (4), 1415–1424.

Liu, C.C., Estrada-Raygoza, C., Abdallah, J., Holmes, S., Yin, Y., Schepis, A., Cicoria, M., Hetzer, D., Tsai, H., Guillorn, M., Tijo, M., Cheng, J., Somervell, M., Colburn, M., 2013b. Directed self-assembly process implementation in a 300 mm pilot line environment. Alternative Lithographic Technologies V. SPIE Proceedings, San Jose, CA.

Lodge, T., Pudil, B., Hanley, K., 2002. The full phase behavior for block copolymers in solvents of varying selectivity. Macromolecules 35 (12), 4707–4717.

Lodge, T., Hanley, K., Pudil, B., Alahapperuma, V., 2003. Phase behavior of block copolymers in a neutral solvent. Macromolecules 36 (3), 816–822.

Nagpal, U., Mueller, M., Nealey, P.F., de Pablo, J.J., 2012. Free energy of defects in ordered assemblies of block copolymer domains. ACS Macro Letters 1 (3), 418–422.

Naughton, J., Matsen, M., 2002. Limitations of the dilution approximation for concentrated block copolymer/solvent mixtures. Macromolecules 35 (14), 5688–5696.

Paik, M., Bosworth, J., Smilges, D., Schwartz, E., Andre, X., Ober, C., 2010. Reversible morphology control in block copolymer films via solvent vapor processing: an in situ gisaxs study. Macromolecules 43 (9), 4253–4260.

Park, S.M., La, Y.H., Ravindran, P., Craig, G.S.W., Ferrier, N.J., Nealey, P.F., 2007. Combinatorial generation and replicating directed assembly of complex and varied geometries with thin films of diblock copolymers. Langmuir 23 (17), 9037–9045.

Park, S.H., Shin, D.O., Kim, B.H., Yoon, D.K., Kim, K., Lee, S.Y., Oh, S.H., Choi, S.W., Jeon, S.C., Kim, S.O., 2010. Block copolymer multiple patterning integrated with conventional arf lithography. Soft Matter 6 (1), 120–125.

Patrone, P.N., Gallatin, G.M., 2012. Modeling line edge roughness in templated, lamellar block copolymer systems. Macromolecules 45 (23), 9507–9516.

Patrone, P.N., Gallatin, G.M., 2014. Response of block copolymer thin-film morphology to line-width roughness on a chemoepitaxial temp. Macromolecules 47 (14), 4824–4829.

Peng, J., Xuan, Y., Wang, H., Yang, Y., Li, B., Han, Y., 2004. Solvent-induced microphase separation in diblock copolymer thin films with reversibly switchable morphology. J. Chem. Phys. 120 (23), 11163–11170.

Peng, J., Xuan, Y., Wang, H., Li, B., Han, Y., 2005. Solvent vapor induced dewetting in diblock copolymer thin films. Polymer 46 (15), 5767–5772.

Ramirez-Hernandez, A., Liu, G., Nealey, P.F., de Pablo, J.J., 2012. Symmetric diblock copolymers confined by two nanopatterned surfaces. Macromolecules 45 (5), 2588–2596.

Ramirez-Hernandez, A., Suh, H.S., Nealey, P.F., de Pablo, J.J., 2014. Control of directed self-assembly in block polymers by polymeric topcoats. Macromolecules 47 (10), 3520–3527.

Rincon Delgadillo, P.A., Gronheid, R., Thode, C.J., Wu, H., Cao, Y., Somervell, M., Nafus, K., Nealey, P.F., 2012. All track directed self-assembly of block copolymers: process flow and origin of defects. In: Tong, W.M. (Ed.), Alternative Lithographic Technologies IV. In: Proceedings of SPIE, vol. 8323. http://proceedings.spiedigitallibrary.org/proceeding.aspx?articleid=1353161.

Rincon-Delgadillo, P., Craig, G., Gronheid, R., Nealey, P.F., 2013. Scale-up of a chemo-epitaxy flow for feature multiplication using directed self-assembly of block-copolymers. J. Photopolym. Sci. Technol. 26 (6), 831–839.

Ross, C.A., Cheng, J.Y., 2008. Patterned magnetic media made by self-assembled block-copolymer lithography. MRS Bull. 33 (9), 838–845.

Ruiz, R., Kang, H., Detcheverry, F.A., Dobisz, E., Kercher, D.S., Albrecht, T.R., de Pablo, J.J., Nealey, P.F., 2008. Density multiplication and improved lithography by directed block copolymer assembly. Science 321, 936–939.

Ruiz, R., Dobisz, E., Albrecht, T.R., 2011. Rectangular patterns using block copolymer directed assembly for high bit aspect ratio patterned media. ACS Nano 5 (1), 79–84.

S.I. Association (Ed.), 2007. International technology roadmap for semiconductors—lithography. In: International Technology Roadmap for Semiconductors. 2007 ed. International SEMATECH, Austin, TX. http://www.itrs.net/ITRS%201999-2014%20Mtgs,%20Presentations%20&%20Links/2007ITRS/2007_Chapters/2007_Lithography.pdf.

Sinturel, C., Vayer, M., Morris, M., Hillmyer, M., 2013. Macromolecules 46 (14), 5399–5415.

Somervell, M., Gronheid, R., Hooge, J., Nafus, K., Rincon Delgadillo, P., Thode, C., Younkin, T., Matsunaga, K., Rathsack, B., Scheer, S., Nealey, P., 2012. Comparison of directed self-assembly integrations. In: Somervell, M.H., Wallow, T.I. (Eds.), Advances in Resist Materials and Processing Technology XXIX. In: Proceedings of SPIE, vol. 8325. http://spie.org/Publications/Proceedings/Paper/10.1117/12.916406.

Son, J., Gwyther, J., Chang, J., Berggren, K., Manners, I., Ross, C., 2011. Highly ordered square arrays from a templated abc triblock terpolymer. Nano Lett. 11 (7), 2849–2855.

Son, J., Gotrik, K., Ross, C., 2012. High-aspect-ratio perpendicular orientation of ps-b-pdms thin films under solvent annealing. ACS Macro Lett. 1 (11), 1279–1284.

Stoykovich, M.P., Müller, M., Kim, S.O., Solak, H.H., Edwards, E.W., de Pablo, J.J., Nealey, P.F., 2005. Directed assembly of block copolymer blends into nonregular device-oriented structures. Science 308 (5727), 1442–1446.

Stoykovich, M.P., Kang, H., Daoulas, K.C., Liu, G., Liu, C.C., de Pablo, J.J., Mueller, M., Nealey, P.F., 2007. Directed self-assembly of block copolymers for nanolithography: fabrication of isolated features and essential integrated circuit geometries. ACS Nano 1 (3), 168–175.

Stoykovich, M.P., Daoulas, K.C., Muller, M., Kang, H.M., de Pablo, J.J., Nealey, P.F., 2010. Remediation of line edge roughness in chemical nanopatterns by the directed assembly of overlying block copolymer films. Macromolecules 43 (5), 2334–2342.

Stuen, K.O., Liu, C., Welander, A.M., Liu, G., de Pablo, J.J., Nealey, P.F., Satapathy, D.K., Nygard, K., Bunk, O., Solak, H.H., van der Veen, J.F., 2008. In situ characterization of block copolymer ordering on chemically nanopatterned surfaces by time-resolved small angle x-ray scattering. J. Vac. Sci. Technol. B 26 (6), 2504–2508.

Stuen, K.O., Thomas, C.S., Liu, G.L., Ferrier, N., Nealey, P.F., 2009. Dimensional scaling of cylinders in thin films of block copolymer-homopolymer ternary blends. Macromolecules 42 (14), 5139–5145.

Tada, Y., Akasaka, S., Yoshida, H., Hasegawa, H., Dobisz, E., Kercher, D., Takenaka, M., 2008. Directed self-assembly of diblock copolymer thin films on chemically-patterned substrates for defect-free nano-patterning. Macromolecules 41 (23), 9267–9276.

Tada, Y., Akasaka, S., Chen, F., Yoshida, H., Takenaka, M., Hasegawa, H., 2009a. Density multiplication by directed self-assembly of block copolymer binary blends. J. Photopolym. Sci. Technol. 22 (2), 229–233.

Tada, Y., Akasaka, S., Takenaka, M., Yoshida, H., Ruiz, R., Dobisz, E., Hasegawa, H., 2009b. Nine-fold density multiplication of hcp lattice pattern by directed self-assembly of block copolymer. Polymer 50 (17), 4250–4256.

Tavakkoli, K., Gotrik, K., Hannon, A., Alexander-Katz, A., Ross, C., Berggren, K., 2012. Templating three-dimensional self-assembled structures in bilayer block copolymer films. Science 336 (6086), 1294–1298.

Tong, Q.Q., Sibener, S.J., 2013. Visualization of individual defect mobility and annihilation within cylinder-forming diblock copolymer thin films on nanopatterned substrates. Macromolecules 46 (21), 8538–8544.

Vayer, M., Hillmyer, M.A., Dirany, M., Thevenin, G., Erre, R., Sinturel, C., 2010. Perpendicular orientation of cylindrical domains upon solvent annealing thin films of polystyrene-b-polylactide. Thin Solid Films 518 (14), 3710–3715.

Welander, A.M., Kang, H., Stuen, K.O., Solak, H.H., Mueller, M., de Pablo, J.J., Nealey, P.F., 2008. Rapid directed assembly of block copolymer films at elevated temperatures. Macromolecules 41 (8), 2759–2761.

Whitmore, M., Vavasour, J., 1992. Self-consistent mean field-theory of the microphase diagram of block copolymer neutral solvent blend. Macromolecules 25 (7), 2041–2045.

Wilmes, G.M., Durkee, D.A., Balsara, N.P., Liddle, J.A., 2006. Bending soft block copolymer nanostructures by lithographically directed assembly. Macromolecules 39 (7), 2435–2437.

Xiao, S., Yang, X., Park, S., Weller, D., Russell, T.P., 2009. A novel approach to addressable 4 teradot/in.^{2} patterned media. Adv. Mater. 21 (24), 2516–2519.

Xiao, S.G., Yang, X.M., Lee, K.Y., Hwu, J.J., Wago, K., Kuo, D., 2013. Directed self-assembly for high-density bit-patterned media fabrication using spherical block copolymers. J. Micro-Nanolith. Membr. 12 (3), 031110.

Xiong, S., Wan, L., Chapuis, Y.-A.C., Ruiz, R., Nealey, P.F., 2015. Directed self-assembly of ABA triblock copolymer on chemical contrast pattern via solvent annealing: molecular architecture, solvent selection, and pattern transfer. In: SPIE Advanced Lithography, pp. 9423–9424.

Xu, J., Hong, S., Gu, W., Lee, K., Kuo, D., Xiao, S., Russell, T., 2011. Fabrication of silicon oxide nanodots with an area density beyond 1 teradots inch-2. Adv. Mater. 23 (48), 5755.

Xuan, Y., Peng, J., Cui, L., Wang, H., Li, B., Han, Y., 2004. Morphology development of ultrathin symmetric diblock copolymer film via solvent vapor treatment. Macromolecules 37 (19), 7301–7307.

Yang, J., Jung, Y., Chang, J., Mickiewicz, R., Alexander-Katz, A., Ross, C., Berggren, K., 2010. Complex self-assembled patterns using sparse commensurate templates with locally varying motifs. Nat. Nanotechnol. 5 (4), 256–260.

Yang, X.M., Xiao, S.G., Hsu, Y., Feldbaum, M., Lee, K., Kuo, D., 2013. Directed self-assembly of block copolymer for bit patterned media with areal density of 1.5 teradot/inch^2 and beyond. J Nanomater, 615896.

Yoshida, H., Suh, H.S., Ramirez-Herunandez, A., Lee, J.I., Aida, K., Wan, L., Ishida, Y., Tada, Y., Ruiz, R., de Pablo, J., Nealey, P.F., 2013. Topcoat approaches for directed self-assembly of strongly segregating block copolymer thin films. J. Photopolym. Sci. Technol. 26 (1), 55–58.

Zhang, X., Harris, K., Wu, N., Murphy, J., Buriak, J., 2010. Fast assembly of ordered block copolymer nanostructures through microwave annealing. ACS Nano 4 (11), 7021–7029.

X-ray characterization of directed self-assembly block copolymers

7

D.F. Sunday, R.J. Kline
NIST, Gaithersburg, MD, USA

7.1 Introduction

The development of new lithographic strategies, materials, and equipment has enabled the semiconductor industry to match or exceed the targets set by Moore's law, resulting in an incredible rate of improvement in the cost of processing power and memory. A recently developed addition to the lithographic toolbox is the emergence of directed self-assembly (DSA) block copolymer (BCP) lithography, where a chemical or topographical template is patterned via traditional lithographic techniques and a BCP is deposited on the pattern (Yang et al., 2000; Kim et al., 2003; Stoykovich, 2005). This technology has the potential to provide a relatively low cost solution to extend conventional lithography to sub-25 nm pitches. Details regarding the processing and materials used in this technique are covered elsewhere in this book, and extensively in the external literature (Edwards et al., 2004, 2005; Stoykovich and Nealey, 2006; Stoykovich et al., 2007, 2010; Saavedra et al., 2010; Herr, 2011; Wan et al., 2012; Ruiz et al., 2011; Cheng et al., 2010; Liu et al., 2011a,b, 2013; Doerk et al., 2013). The focus of this chapter is the characterization of BCPs with X-ray techniques and how these measurements can help solve outstanding challenges in BCP DSA.

BCP DSA has seen significant progress throughout the last decade and is nearing implementation in some sectors of the semiconductor industry. Challenges remaining include the development of new materials capable of achieving smaller pitches (Kennemur et al., 2012; Cushen et al., 2012; Cheng et al., 2012; Chang et al., 2012), developing improved measures of the surface energy (Maher et al., 2014; Bates et al., 2012), and most importantly reducing the concentration of defects (Bencher et al., 2012; Delgadillo et al., 2013; Ito et al., 2014; Simão et al., 2014; Cao et al., 2013; Rincon Delgadillo et al., 2012), which are currently at an unacceptably high level. Reducing defect concentrations will require a better understanding of the factors that lead to changes in internal structure and the parameter space where the probability of defect formation is reduced. Improving the understanding of the polymer physics, which control the self-assembly process, will require a cooperative approach between experimental measurements and modeling. Measurements for BCP DSA systems must be capable of providing detailed information about the internal structure of the film, including subsurface changes in the morphology and interfacial roughness. X-ray scattering measurements are sensitive to compositional variations

Directed Self-assembly of Block Copolymers for Nano-manufacturing. http://dx.doi.org/10.1016/B978-0-08-100250-6.00007-9

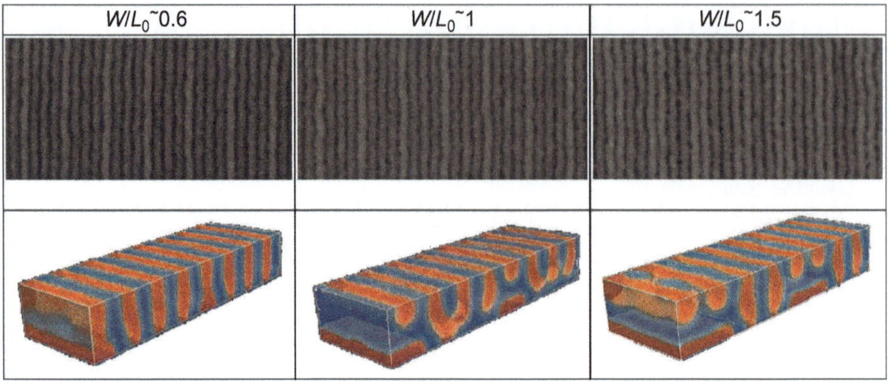

Figure 7.1 Top-down SEM and three-dimensional simulations of PS-*b*-PMMA assembled on a chemical template as a function of the ratio of the guiding line width (W) to the BCP period (L_0). These simulations demonstrate the potential morphologies that may occur as W/L_0 is varied beyond the ideal ratio of 0.5.
Reprinted with permission from Liu et al. (2013). Copyright 2013 American Chemical Society.

on subnanometer length scales, making them excellent candidates to determine the internal structure of a DSA BCP film. This is particularly important as the screening techniques typically used to determine the long-range orientation of the BCP film are often poorly suited to observe internal morphological changes. This is illustrated in Figure 7.1, which shows how a set of three separate samples, which each appear to be well ordered from top-down scanning electron microscopy (SEM) images, can have different internal morphologies (Liu et al., 2013). Liu and coworkers conducted both measurements and simulations on samples consisting of PS-*b*-PMMA patterned on a chemical template, where the width of the guiding stripe was varied. Close to the ideal ratio of $W/L_0=0.5$ (where W is the guiding stripe width and L_0 is the BCP pitch) the simulations show well-defined lamella oriented perpendicular to the substrate. At higher multiples of this ratio ($W/L_0=1$, 1.5) deviations from the desired structure are observed where bridging between PMMA blocks breaks the continuity of the PS lamella. In all cases, the SEM images show no hint of the buried structure. X-ray measurements offer the ability to differentiate between different buried morphologies. Historically, X-ray measurements have been enormously valuable to the BCP community, providing ways to measure morphology, kinetics, interfacial behavior, and thermodynamics. X-ray measurements are primarily conducted in three different geometries: transmission, specular reflectivity, and grazing incidence (GISAXS). We will start by briefly reviewing how X-rays interact with matter, including resonant soft X-rays, which have the potential to be enormously useful in the study of thin film BCPs. We will then provide an overview on the use of each X-ray configuration, including the application to patterned DSA films and in evaluating the relevant thermodynamic parameters.

7.2 Interactions of X-rays with matter

In this section, we will briefly review the relevant theory on the interaction of X-rays with materials. More detailed information on the scattering of electromagnetic waves can be found in external references (Daillant and Gibaud, 2009; Attwood, 2000; Guiner, 1994; Warren, 1969). For traditional high-energy X-rays, the scattering power of a material is proportional to the electron density, and the contrast between two different components in a sample is solely a function of the relative electron density of those components. This is not the case near an atomic absorption edge, where the contrast can change significantly. X-rays interact with the electron cloud surrounding an atom and are, therefore, sensitive to the electronic structure of materials. This property has been taken advantage of extensively with near edge X-ray absorption fine structure (NEXAFS) spectroscopy (Ade and Hitchcock, 2008; Krishnan et al., 2010), and more recently for resonant soft X-ray scattering (RSoXS) (Ade and Hitchcock, 2008). The strength of an X-ray's interaction with a material is a function of the complex refractive index (n), which has both a real and imaginary component (dispersive and absorptive components, respectively) according to Equation (7.1). Where n_a is the atomic density, δ is the real (dispersive) component, β is the imaginary (absorptive) component, λ is the wavelength, r_e is the classic electron radius, and f_1^0 and f_2^0 are the real and imaginary components of the atomic scattering factor. Away from atomic absorption edges f_1^0 and f_2^0 are slowly varying functions of the electron density and wavelength of the radiation. As a result, there is little difference between the scattering strength for most polymeric materials. For example, the ubiquitous PS-*b*-PMMA system has a scattering contrast of about 10% at 17 keV. Near an atomic absorption edge though, f_1^0 and f_2^0 can vary sharply. The carbon K edge (the K edge corresponds to absorption of the X-rays by electrons in the 1s shell) is generally listed as 285 eV, but the fine structure is determined by the type of chemical bonds. This means that the scattering contrast between materials can be made chemical bond specific. Araki and coworkers demonstrated this effect through NEXAFS measurements on PS and PMMA, as shown in Figure 7.2a (Araki et al., 2006). A strong absorbance is observed for PS at 285 eV corresponding to the C $1s \rightarrow \pi^*$ transition for the carbon in the aromatic rings, while the absorbance for β(PMMA) at 289 eV results from the C $1s \rightarrow \pi^*$ transition for the carbonyl carbon. As a result of this shift in the location of the absorbance, there is a strong enhancement in the contrast between PS and PMMA in the energy region near the absorbance peaks as shown in Figure 7.2c. The enhanced absorbance is a doubled-edged sword. While it leads to improved contrast for the scattering it also limits the thickness of a film that may be measured and increases rate of beam damage. As a result, the use of soft X-rays are limited to thin films, generally in the 100 nm range, and care must be taken to monitor beam damage during the experiment. Additionally, the samples have to be mounted on thin windows similar to those used for transmission electron microscopy (TEM) in order to minimize absorption.

$$n = 1 - \delta - i\beta = 1 - \frac{n_a r_e \lambda^2}{2\pi}\left(f_1^0 - if_2^0\right) \quad (7.1)$$

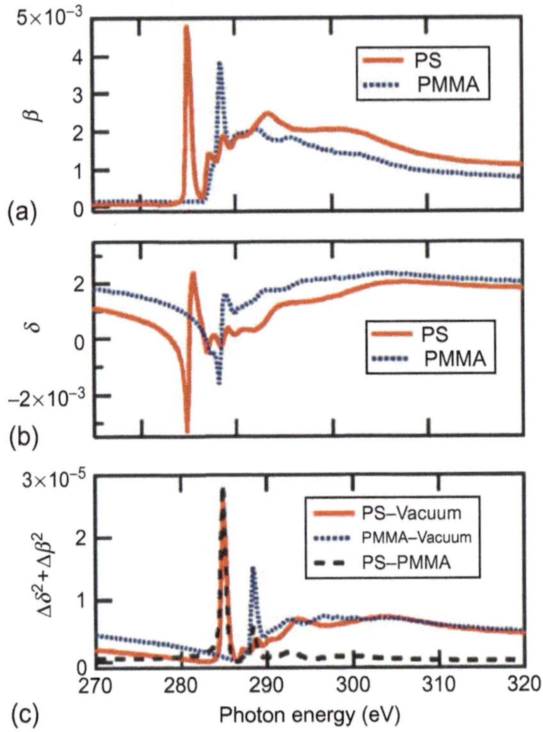

Figure 7.2 (a) Absorptive term (β) of the complex index of refraction for PS and PMMA, measured by NEXAFS. (b) Dispersive term (δ) of the complex index of refraction for PS and PMMA calculated from the absorptive term through the Kramers–Kronig integral relation. (c) Calculated scattering contrast between PS, PMMA, and vacuum.
Reprinted with permission from Araki et al. (2006). Copyright 2006 AIP Publishing LLC.

In a multiatom sample, scattered waves from individual atoms interfere, giving rise to characteristic intensities as a function of the scattering angle. This is most often represented as intensity (I) as a function of the scattering vector q, which is defined in Equation (7.2). In the case where the sample has uniform spacing between atoms (such as in a silicon lattice), these interferences give rise to sharp peaks in intensity known as Bragg peaks. The spacing between the atoms can then be reconstructed according to Bragg's law (Equation 7.3) (n in this equation will represent a given number of lattice spacings and should not be confused with the use of n to represent the complex index of refraction throughout the rest of the text, d represents a lattice spacing). This concept can be extended to a more macroscopic scale. The scattering from uniformly spaced regions of constant composition interferes in the same way that scattering from uniformly spaced atoms does, leading to the appearance of sharp scattering peaks. Nonuniform fluctuations in composition will still result in scattering, although the resulting pattern will lack the sharp, well-defined intensity peaks characteristic of scattering off of a uniform lattice. The most general equation for calculating the scattering profile from either an amorphous or crystalline sample is given in Equation (7.4), where ρ represents the scattering length density (SLD), ϕ_i and V_i are the volume fraction and volume of component *i*, *P*(*q*) is the form factor, and *S*(*q*) is the structure factor. The form factor describes the scattering from the most basic structure within the sample, that is, an arrangement of atoms within a lattice, a sphere in a matrix, or a lamella. The structure factor describes how the scattering from those individual shapes interferes with one another and is related to the distance between those structures, such as the periodicity in the lattice or the distribution of spheres in a matrix. In order to calculate the microstructure that gives rise to a given scattering profile, an inverse approach must be taken (the exception being the small number of cases where coherent diffraction allows the phase change in the sample can be measured in addition to the magnitude of the scattered intensity) (Robinson et al., 2001; Nishino et al., 2009). An educated guess is made as to the nature of the microstructure in the sample, and the scattering profile is calculated from the assumed structure. This is then compared to the experimental profile, and the original structure is iterated until a good match to the measured profile is obtained. The challenge is that there may be more than one structure that can give a simulated scattering profile that matches the experimental data. As a result, as much *a priori* information, or complementary experimental data about the sample as possible should be obtained in order to determine which solution is correct.

$$q = \frac{4\pi}{\lambda}\sin(\theta) \tag{7.2}$$

$$n\lambda = 2d\sin(\theta) \tag{7.3}$$

$$I(q) = (\rho_i - p_j)^2 \phi_i V_i P(q) S(q) \tag{7.4}$$

It should be noted that the majority of the discussion in this section also applies to characterization through neutron scattering, the sole difference being the contrast mechanism. The neutron scattering cross-section interacts with the atomic nucleus and can vary significantly between elements and isotopes. Neutron scattering can take advantage of this by selectively replacing hydrogen atoms with deuterium to control the scattering contrast, a technique known as contrast variation.

7.3 SAXS and RSoXS

The history of BCP research is intertwined with X-ray measurements, which have enabled a better understanding of morphology, kinetics, and thermodynamics. BCP morphology is typically determined in one of two ways: TEM on thin films stained with heavy metals for contrast or through small-angle scattering (SAS). In comparison to real space methods such as TEM, scattering measurements are able to sample the structure over a large, three-dimensional area and obtain a much larger statistical sampling of the microstructure. Small-angle X-ray scattering (SAXS) is one of the mainstays in polymer characterization, typically used to investigate length scales between 1 and 100 nm. BCP morphologies such as lamella, cylinders, spheres, and gyroids all have distinct scattering profiles. As a result, SAS techniques are frequently utilized to distinguish between morphologies and to study thermodynamic behavior such as order–order and order–disorder transitions (ODT).

Measuring the structure of thin films in a transmission geometry is difficult for standard scattering techniques due to the small contrast between components and the minimal amount of material to scatter from (Equation 7.4 shows how the scattering intensity is proportional to both the contrast between the components and the volume of material). This has created an opportunity for resonant scattering to find its niche as a tool to investigate the structure of thin films. For example, Virgili and coworkers used a combination of SAXS and RSoXS to examine the difference in order of asymmetric polystyrene-*b*-polyisoprene (SI) when the sample is prepared in the bulk compared to a thin film (Virgili et al., 2007). Two different samples were examined, one that formed spheres on a body-centered cubic lattice labeled SI(7-79), and the other that formed cylinders on a hexagonally closed packed lattice and is labeled SI(4-35). The scattering profiles for both the bulk SAXS and thin film RSoXS are presented in Figure 7.3. The RSoXS results on the sphere forming SI(7-79) show a transition to a

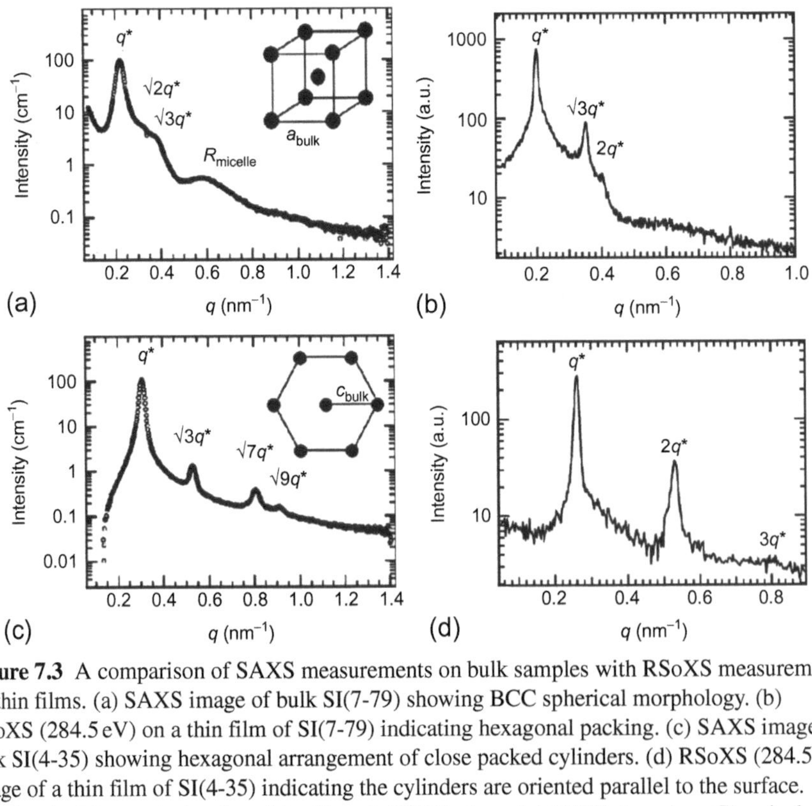

Figure 7.3 A comparison of SAXS measurements on bulk samples with RSoXS measurements on thin films. (a) SAXS image of bulk SI(7-79) showing BCC spherical morphology. (b) RSoXS (284.5 eV) on a thin film of SI(7-79) indicating hexagonal packing. (c) SAXS image of bulk SI(4-35) showing hexagonal arrangement of close packed cylinders. (d) RSoXS (284.5 eV) image of a thin film of SI(4-35) indicating the cylinders are oriented parallel to the surface. Reprinted with permission from Virgili et al. (2007). Copyright 2007 American Chemical Society.

hexagonal arrangement when the thickness of the film is reduced. The cylinder forming SI(4-35) shows peaks only at unit multiples of q^*, indicating that the cylinders arrange themselves parallel to the surface in a fingerprint pattern. This illustrates the utility of combining the two measurements to probe the different behaviors in bulk compared to thin films.

RSoXS can serve not only to enhance contrast in a thin film, but also to vary that contrast sharply as a function of energy. Wang and coworkers took advantage of this property to evaluate the morphology of a triblock copolymer. The real and imaginary refractive index components for PS, PI, and poly(2-vinyl pyridine) (P2VP) are shown in Figure 7.4 (Wang et al., 2011). A sharp change in both components can be seen around 284 eV, although the magnitude of that change varies for each component. The polymer film consists of IS2VP triblock copolymer, which formed hexagonally packed cylinders of both PI and P2VP within a PS matrix. The contrast of each component was varied by changing the incident energy. At 284 eV, the scattering contrast between PI and P2VP was identical, allowing determination of the hexagonal cylinder morphology, while transitioning the energy to 280 eV resulted in contrast between all three of the components. As a result, the lattice structure of each component in the blend could be uniquely determined. This ability to "tune" the contrast between different components by adjusting the incident beam energies is one of the powerful advantages of resonant X-ray scattering over nonresonant versions where the contrast is fixed by the atomic compositions.

7.4 Thermodynamics

The majority of the computational methods for predicting or modeling the phase behavior of BCPs typically utilize the Flory–Huggins interaction parameter (χ) to characterize the strength of interactions between each block. This parameter accounts for both enthalpic and entropic contributions and has been remarkably successful when used to calculate the phase behavior in the system (Matsen, 1995; Leibler, 1980; Bates and Fredrickson, 1990; Bates et al., 1990). A BCP can exist in either a disordered or ordered state, the disordered state is characterized by weak concentration fluctuations where the components are at least partially mixed. In the ordered state the concentration fluctuations become much stronger, and the two components segregate into regions made up exclusively of one component segregated by a mixed interface region of varying width. The state of the system is typically characterized by the product of χN, where for a symmetric BCP a value of χN below 10.5 indicates that the material will be in the disordered state, while systems with larger values of χN will be ordered (Bates and Fredrickson, 1990). The difference in the distribution of the component concentration in the ordered versus disordered state allows the ODT to be determined by scattering methods, which are very sensitive to changes in the component distribution. The scattering behavior of a polymer at the ODT is show in Figure 7.5. For most systems, χ is inversely dependent on temperature, therefore, the sample can be heated and the scattering profile is measured at a series of temperatures in order to determine the ODT temperature. Kennemur and coworkers developed

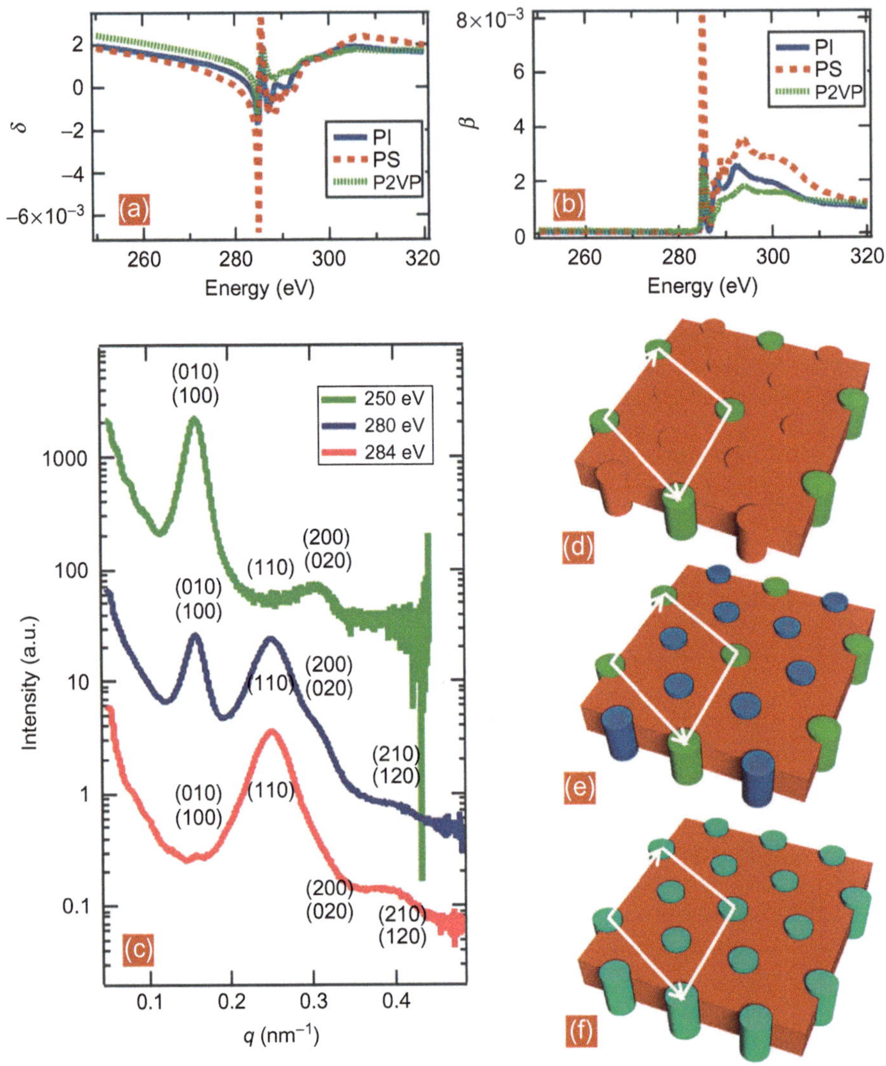

Figure 7.4 (a,b) Real and imaginary part of the complex index of refraction of poly(1,4-isoprene) (blue solid line), polystyrene (red dashed line), and poly(2-vinyl pyridine) (green dotted line), respectively. (c) 1D radial averaged scattering intensity versus q at 250, 280, and 284 eV, from top to bottom respectively. (d–f) schematic illustration of hexagonally packed cylindrical morphology of IS2VP triblock copolymer seen by X-rays at 250, 280, and 284 eV, respectively.

Reprinted with permission from Wang et al. (2011). Copyright 2011 American Chemical Society.

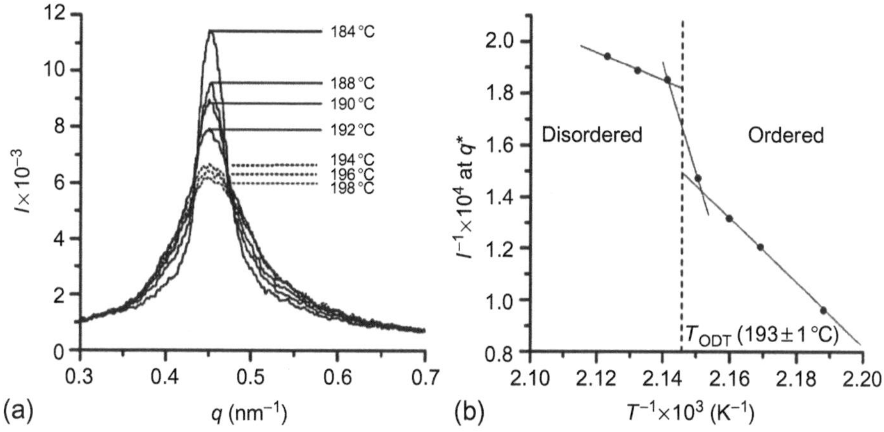

Figure 7.5 (a) SAXS pattern for PtBS–PMMA as a function of temperature. Solid lines indicate the sample is ordered, while the disordered samples are shown by the dotted line. (b) Inverse intensity at q^* as a function of inverse temperature showing the discontinuity in the maximum intensity at the ODT.

poly(tert-butylstyrene)-*b*-PMMA (PtBS-*b*-PMMA), a new material being explored for potential nanopatterning applications due to its large χ value (Kennemur et al., 2012). A symmetric sample of PtBS-*b*-PMMA with a $M_n = 17.6$ kg/mol was examined with temperature-dependent SAXS measurements. The radially averaged SAXS curves are shown in Figure 7.5a as a function of temperature. The ordered sample at 184 °C shows a sharp peak at $q = 0.45\,\text{nm}^{-1}$, and as the temperature is increased the maximum peak intensity decreases and the peak width broadens. This occurs due to the increased interpenetration between components at the interface, dampening the scattering intensity. The actual transition point is determined by the discontinuity of the plot of inverse intensity against inverse temperature, as shown in Figure 7.5b. The plot of the full width half max as a function of temperature will show a similar discontinuity and is often used in place of maximum intensity to determine the ODT.

This result provides value of χ at a given molecular mass and temperature, and by doing a series of measurements with samples at different molecular mass, the value of χ as a function of temperature can be determined. This curve can be fit to find the enthalpic (α) and entropic (β) components of χ according to Equation (7.5). This same type of analysis can be performed using dynamic mechanical analysis by observing the change in the storage modulus in the low frequency region, although this approach typically requires much larger sample volumes compared to scattering.

$$\chi = \frac{\alpha}{T} + \beta \tag{7.5}$$

The alternative to measurements on a series of molecular mass is to conduct measurements on a single sample, which is in the disordered state. Similar to determining the ODT, the scattering profile is measured as a function of temperature, and the scattering intensity is fit using the random phase approximation (RPA). The RPA framework, initially developed by Ludwig Leibler, connects the scattered intensity in the disordered phase to χ (Leibler, 1980; Feldman et al., 2010; Sakurai et al., 1992). The initial theory has been expanded to include corrections for polydispersity and asymmetry in order to more adequately reflect complexities of real polymeric materials. These examples illustrate how X-ray scattering can be a powerful tool in determining some of the fundamental materials parameters that govern the phase behavior and morphology development (self-assembly) in BCP systems. Later, we will extend this utility to include BCP-DSA lithography.

7.5 Critical dimension small-angle X-ray scattering

Critical dimension small-angle X-ray scattering (CDSAXS) was developed to meet the growing challenges in semiconductor metrology of characterizing periodic structures as the pitch continues to shrink. CDSAXS is a transmission scattering measurement where the sample is rotated to change the projection of the scattering vector, enabling reconstruction of a two- or three-dimensional nanostructure. For a grating sample, the lines are oriented parallel to the axis of rotation, and the q vector is defined such that q_y is parallel to the lines, q_x is perpendicular to the lines, and q_z is normal to the sample plane, as shown in Figure 7.6. The rotation shifts the q vector projection on the detec-

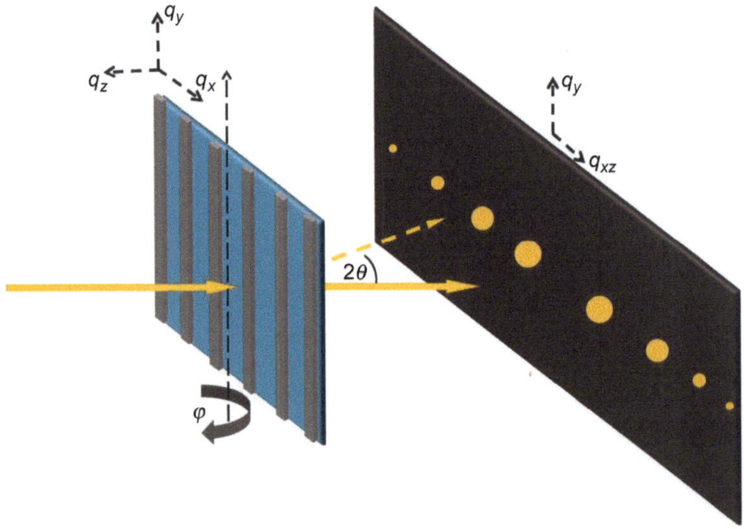

Figure 7.6 Geometry and q vectors for CDSAXS. q_z is defined as perpendicular to the substrate, q_x is in the plane of the substrate and perpendicular to the direction of the line grating, and q_y is in plane of the substrate and parallel to the direction of the line grating. Reprinted with permission from Sunday et al. (2013).

tor from being a function of q_y and q_x to a function of q_y and q_{xz}. The q_{xz} vector can then be deconvoluted according to the rotational angle and positioned on the detector to obtain q_x and q_z, according to Equations (7.6) and (7.7) (2θ is the scattering angle and φ is the rotational angle, where normal incidence is 0°). Obtaining the scattering intensity along the q_z axis in addition to q_x enables the reconstruction of the two-dimensional SLD profile of the film. This measurement was originally used to determine the shape of etched line gratings or polymer resists (Hu et al., 2004; Jones et al., 2003; Jones, 2006; Wang et al., 2007a,b, 2009a; Lemaillet et al., 2013), but has since been adapted to measure DSA BCP samples (Perera et al., 2012; Sunday et al., 2013, 2014). For a sample with an asymmetric periodic structure in both the x and y planes, rotation in both directions would be required to fully reconstruct the three-dimensional profile.

$$q_x = q_{xz} \times \cos\left(\varphi - \frac{2\theta}{2}\right) \tag{7.6}$$

$$q_z = q_{xz} \times \sin\left(\varphi - \frac{2\theta}{2}\right) \tag{7.7}$$

In order for CDSAXS to be capable of evaluating the internal structure of a BCP film, a significant fraction of the sample must have long-range orientational order. This can be determined from normal incidence (where the beam is perpendicular to sample plane) scattering. For a sample that is completely aligned in one direction, the scattering pattern will consist of diffraction spots extending perpendicular from the axis of the lines, whereas a sample with no orientational order (fingerprint) will show a ring with uniform intensity. This is demonstrated in Figure 7.7, where samples with varying degrees of alignment are measured. These samples consist of PS-*b*-PMMA patterned on a chemical template at 2× density multiplication, after which the PMMA was removed with UV ozone. The sample in Figure 7.7a is almost completely oriented in one direction and shows a well-defined diffraction spot. The spot in Figure 7.7b is still well defined, but now shows evidence of some arcing, indicating that a fraction of the lines are misoriented. The presence of the arc near the diffraction peak suggests that the portion of the BCP that is misoriented does not deviate significantly from the

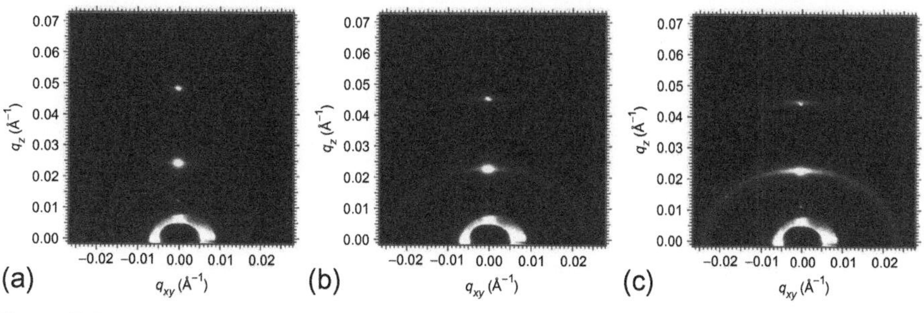

Figure 7.7 Normal incidence images of PS-*b*-PMMA after the PMMA was removed assembled on a chemical template with varying degrees of order, with (a) being the most ordered and the misalignment of the polymer to the template increasing through (b) and (c).

preferred orientation. Figure 7.7c shows a sample that has a large fraction of fingerprint region to accompany the aligned section.

In order to obtain sufficient scattering intensity to accurately reconstruct the internal structure of an unetched DSA BCP, it was necessary to use resonant scattering to enhance the contrast. To demonstrate this improvement in scattered intensity a 1:1 DSA sample was measured with both hard and soft X-rays, where the template was patterned with an e-beam at a pitch of 46 nm and PS-*b*-PMMA was self-assembled on top of the pattern (Sunday et al., 2013). Figure 7.8 shows the $I(q_x,q_z)$ maps obtained from the sample at 17 and 282 eV (Figure 7.8a and b, respectively). The measurement at 17 keV is far enough from an absorption edge that the contrast originates solely from the difference in electron density between the two components. As a result, only five peak orders are visible before becoming undifferentiated from the background scattering. This scattering pattern contains insufficient detail to determine a well-constrained fit to the structure. This is shown in Figure 7.9a, where the best fit obtained to the experimental scattering at 17 keV, as well as the 95% confidence interval (as determined by a Monte Carlo Markov chain algorithm) are shown. Figure 7.8b shows $I(q_x,q_z)$ obtained at 282 eV, 10 peak orders are now visible with a greatly improved separation from the background scattering. Additionally, satellite peaks, which bracket the primary peaks on both sides, are now visible. Those correlate to oscillatory variations in the template pattern due to the e-beam lithography tool (Perera et al., 2010). The line pitch can be obtained from the spacing between the peaks in the q_x direction. What can be seen more clearly in this image are the fringes in intensity along the scattering peaks in the q_z direction. An approximate measure of the height of the sample can be obtained from the q_z value of the first minima in the 1st order peak. This is not as accurate a measure as the pitch as variations in SLD as a function of depth can cause

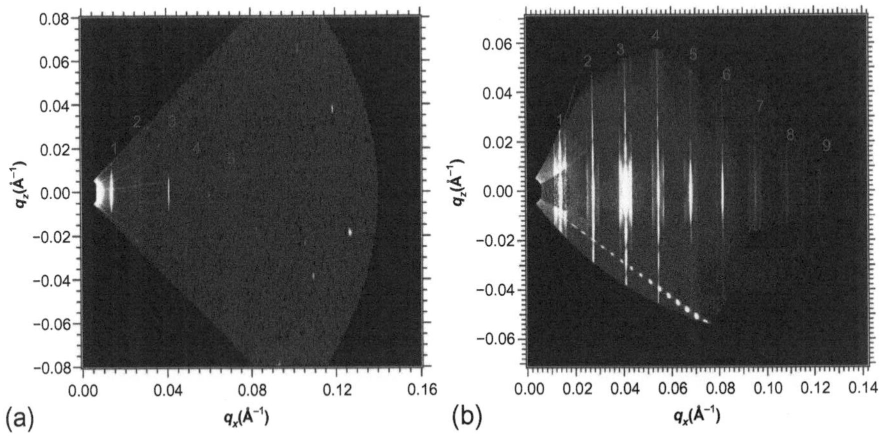

Figure 7.8 Scattering maps for 46 nm pitch PS-*b*-PMMA assembled on a 46 nm template. (a) $I(q_x,q_z)$ reconstructed from measurements at 17 keV using 121 images. (b) $I(q_x,q_z)$ reconstructed from soft X-rays at 282 eV using over 400 images. The missing region is due to a gap in the data collection.
Reprinted with permission from Sunday et al. (2013).

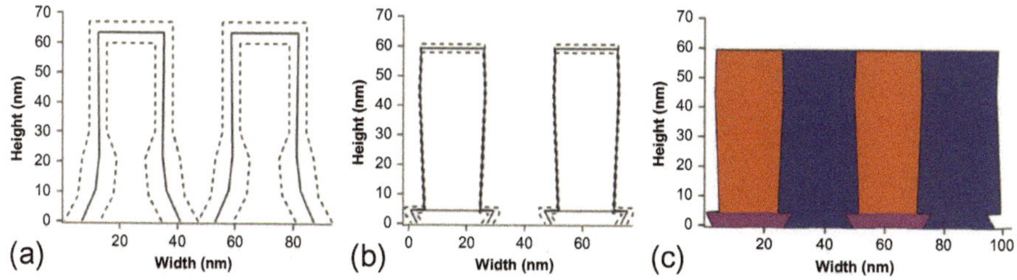

Figure 7.9 (a) Best fit (solid line) and 95% confidence intervals (dashed lines) for 17 keV scattering map using a three trapezoid model. (b) Best fit (solid line) and 95% confidence intervals (dashed lines) for 282 eV scattering map using a three trapezoid model. (c) Filled representation of the fit shown in (b) with the PMMA shown in red, PS in Blue and the oxidized PS lines in purple.
Reprinted with permission from Sunday et al. (2013).

deviations in the position of the minima, but it is an excellent first-order approximation. The structure of the sample determined from the 282 eV scattering is shown in Figure 7.9b, and a color version is shown in Figure 7.9c. The uncertainty around the modeled structure is drastically reduced, and the portion of the brush that was oxidized to prepare the guiding stripe can be distinguished from the PS and PMMA blocks.

This sample is an excellent example of the difference between CDSAXS and res-CDAXS, along with the significant improvement in scattering intensity that can be realized using RSoXS. More complex samples having buried structure were also examined with res-CDSAXS. When patterning samples with a 2× or greater density multiplication, the shape of the lamella on the guiding stripe will likely deviate from the shape of the corresponding block on the neutral stripe. It is important to be able to evaluate the difference between these two structures. To examine this, PS-*b*-PMMA was patterned on a chemical template with both topographical and chemical heterogeneity designed for 4× density multiplication. The chemical template had a pitch of 100 nm and the BCP had a pitch of 25 nm (Cheng et al., 2010). The template consisted of a neutral brush patterned on top of an antireflective coating, which the PMMA block was attracted to and which acted as the guiding stripe for alignment. Figure 7.10 shows the $I(q_x,q_z)$ scattering profile from the first sample in a series patterned with different widths of the neutral stripe measured at 282 eV (Sunday et al., 2014). Five primary peak orders are observed along with a number of satellite peaks. The spacing of the primary peaks corresponds to the BCP lamellar repeat, while the q_x spacing between the more closely spaced satellite peaks corresponds to the larger pitch of the chemical template. The information regarding the shape of the structures is convolved throughout the peaks. The intensity of the primary peaks is significantly larger than the satellites and is dominated by the scattering from the BCP, and the satellite peaks contain information from both the chemical template and from shifts in spacing of the BCP lamella.

Figure 7.11 shows the fit to the $I(q_x,q_z)$ map in Figure 7.10 with two different models, the first shows the primary peak orders being fit by a model that assumes all four of the lamella are identical and the chemical template is indistinguishable from the lamella. While this provided a decent fit to the data, significant improvements were made by including the chemical template and differentiating the shape of the lamella on and off of

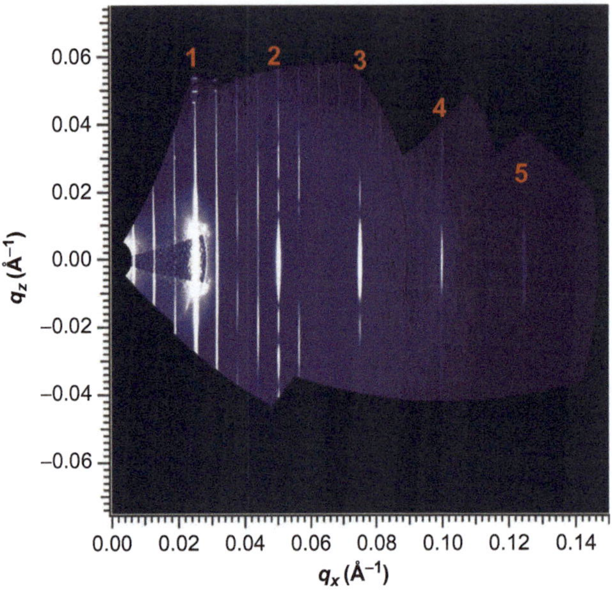

Figure 7.10 $I(q_x,q_z)$ for a 25 nm pitch PS-*b*-PMMA polymer patterned on top of a chemical template with an approximately 100 nm pitch. The primary peak orders are labeled with the red numerals, which correspond to the pitch of the BCP. The unlabeled satellite peaks correspond to the chemical template.
Reprinted with permission from Sunday et al. (2014). Copyright 2014 American Chemical Society.

the chemical template. When examining the difference in the quality of the fit between the two models it is worth noting that higher order peaks typically contain smaller scale information. Most of the changes between the scattered intensity in the two models occurs for the 3rd and 5th order peaks, with some improvement to the 2nd order peak occurring as well. The average shape for each model is similar, with the differences primarily occurring around the base of the structure on the antireflective coating and on the neutral brush. It was these smaller scale differences which gave rise to the change in the scattering around the higher order peaks. There are several noticeable features to the shape. The first feature is that there is a significant enrichment of PMMA on the antireflective coating, which is hydrophilic and strongly attracts PMMA. This results in a necking of the PMMA layer about 10 nm above the surface of the substrate. The lamella, which are located on the chemical template, are nominally unchanged along the vertical profile, and only small variations in the width are observed.

The fits to the remainder of the series are shown in Figure 7.12. The sample discussed above had the smallest template width of the series and increasing it resulted in considerable changes to the film structure. The enrichment and necking of PMMA near the substrate was observed in all cases, and, additionally, there was an enrichment of PS at the surface due to the lower surface of energy of PS compared to PMMA. This resulted in a concentration of 70% by volume PS at the surface, a value that closely

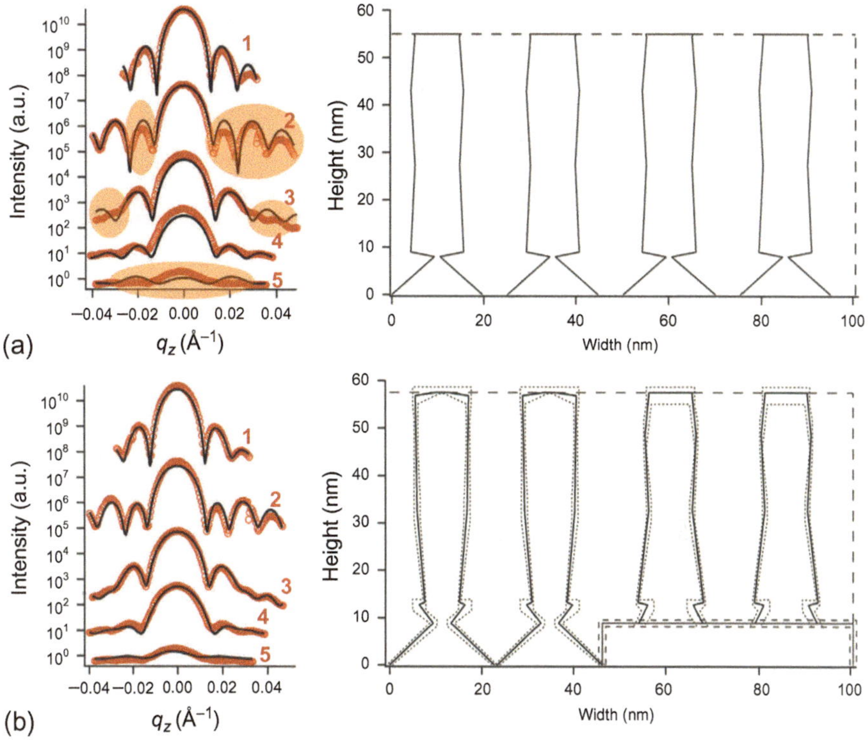

Figure 7.11 q_z cuts and model fits to the data from the $I(q_x,q_z)$ map of DSA1 (282 eV) (○ experimental data, – simulated fit). (a) Model fits from a single lamella structure, showing four repeat units encompassing the template defined superlattice pitch. Solid black lines outline the PMMA lamella; PS lamella are bordered by long dashed lines. The simulated scattering from the single grating provides inadequate fit to the experimental data at the 2nd, 3rd, and 5th order peaks as shown in the highlighted regions. (b) Model fits using a multiple lamellar profiles. Solid black lines indicate the outline of the PMMA lamella; PS lamella are bordered by the long dashes. 95% confidence intervals for PMMA lamella are shown by the dotted lines, and the purple solid and dashed lines represent best fits and 95% confidence intervals for the neutral brush, respectively.

matched the results of surface NEXAFS. For the wider templates (between 59 and 60 nm), a narrowing of the PMMA lamella is observed part in the upper half of the line. There is no immediately obvious reason why any given line would result in that shape, but scattering techniques produce an ensemble average. The image shown represents the average along the length of the lamella, and this result suggests that there are variations in the average structure, possibly including a mixture of compete and broken lamella that combine to provide the concentration profile. This effect is even more prominent when the template width is increased to an average of 61.5 nm. This sample shows a narrowing of the PMMA lamella to below 1 nm at two separate spots

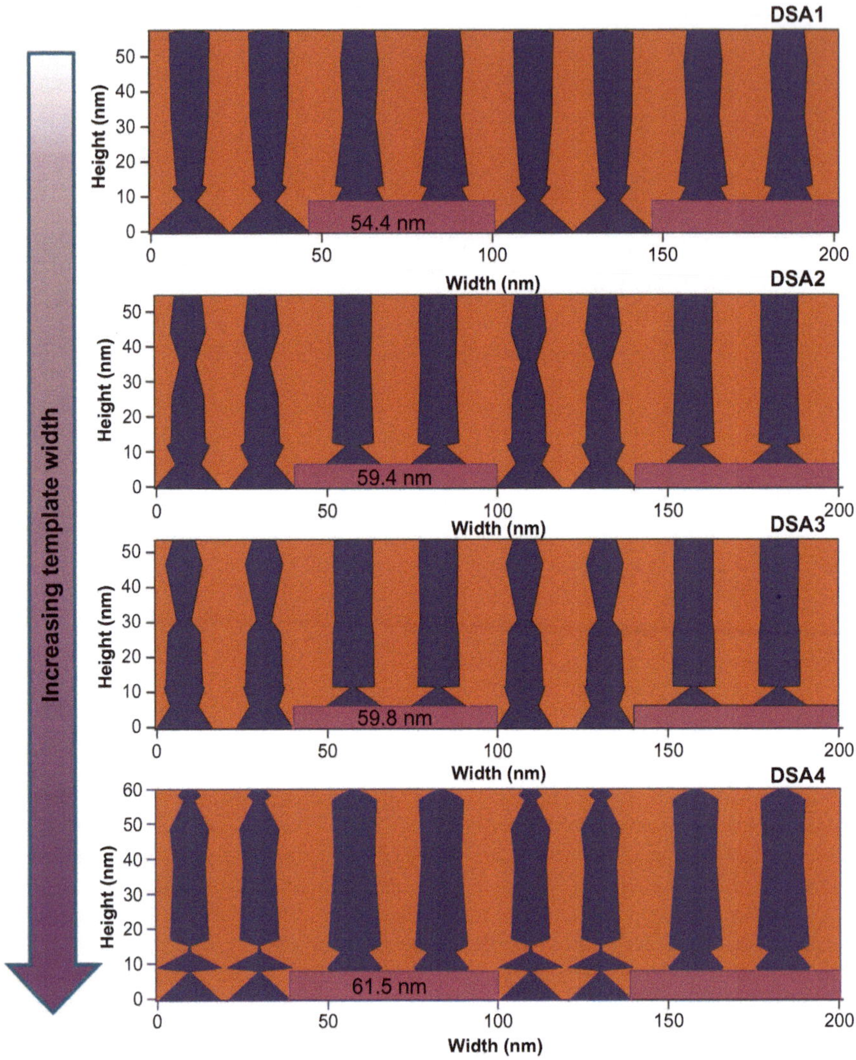

Figure 7.12 Internal film structures of DSA1-4 in order of increasing template width. Purple areas represent the neutral PS-*r*-PMMA brush, the blue area is PMMA, and the red are is PS. Two-unit cells of the overall superlattice are shown for each sample to provide a better visual representation of the structure.
Reprinted with permission from Sunday et al. (2014). Copyright 2014 American Chemical Society.

in the lower half of the film. This likely occurs due to a mixture of perpendicular and parallel orientations, similar to those presented by the models in Figure 7.1. A key point to notice is that at the surface all of these samples presented well aligned lamella, and the variations in morphology that could disrupt the etch process only appear in the internal structure of the film.

Going forward, this technique has the potential to be used to screen new materials and processes for the parameter space, which minimizes the stresses on the BCP structure by finding which conditions lead to an average structure with the smallest deviation from the native shape of the BCP. There are limitations, including the need to conduct the measurement on an extremely thin substrate (<100 nm) to minimize the absorption and loss of signal and that the measurement can only be done at highly specialized synchrotron beam lines.

GISAXS. Grazing incidence small-angle X-ray scattering (GISAXS) offers a complementary alternative to transmission scattering measurements. GISAXS measurements are conducted by probing the sample with an incident beam at an angle around the critical angle of the film. The scattering is then recorded on a two-dimensional detector to record both the in-plane and out-of-plane scattering. The critical angle (θ_c) is proportional to the real component of the complex refractive index according to Equation (7.8) (assuming the real component is much larger than the imaginary component, which is typically the case for hard X-ray energies). For soft materials, the critical angle is generally on the order of 0.1°, and, as a result, the footprint of the beam on the sample is many times larger than the size of the incident beam.

$$\theta_c \propto \sqrt{2\delta} \tag{7.8}$$

There are both advantages and disadvantages to this geometry. The large beam size means that a much larger number of structures are sampled compared to a transmission measurement. This can enable much faster measurement times than in transmission geometries, and, as a result, GISAXS is an excellent tool for *in situ* kinetic measurements on thin films (Leroy et al., 2004; Revenant et al., 2004; Richard et al., 2006; Paik et al., 2010). However, the increased path length of the X-rays through the material at a grazing incidence angle also means that the assumption of a single scattering event used to justify the application of the Born approximation for the analysis of transmission data is no longer valid (Renaud et al., 2009). This means that multiple scattering events must be accounted for, which is a significant challenge for the analysis of the data (there are a number of ongoing attempts to develop software to deal with this analysis (Lazzari, 2002; Chourou et al., 2013). Additionally, by controlling the incident angle, it is possible to gain some degree of control over the penetration depth into the film. When operating at an angle below the critical edge, the X-rays will only probe the surface of the film, while angles above the critical edge will penetrate the entire depth of the film. This provides an avenue to investigate difference between the surface of the film and the buried structure. The geometry of the GISAXS measurement is detailed in Figure 7.13 where the vertical axis is labeled q_z and the horizontal axis q_y (the q_z labeling convention is constant throughout the literature, but the horizontal axis is often labeled q_{xy} or $q_{||}$). An attenuator is placed along $q_y=0$ in order to prevent damage to the detector by intense specular reflections of the beam. Typically, this attenuator is narrow enough that the edges of the specular spots are visible. This provides information about the vertical composition of the film and the film thickness. At an exit angle equal to the critical angle an intense band is visible. This is referred to as the Yoneda band and originates due to interference within the film.

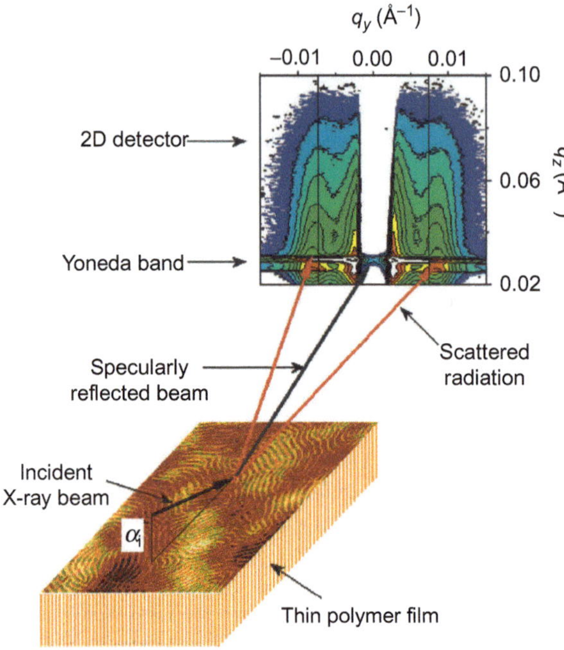

Figure 7.13 Experimental description of the GISAXS measurement. The sample is probed by X-rays around the critical angle of the film. The scattering is recorded by a two-dimensional detector allowing for simultaneous recording of in-plane and out-of-plane scattering. A beamstop is present to protect the detector from damage due to the specular reflection, which typically is several orders of magnitude larger than the off-specular scattering.
Reprinted with permission from Busch et al. (2007). Copyright 2007 American Chemical Society.

The wide range of two-dimensional information makes GISAXS an excellent measurement for determining orientation in thin films. This is particularly useful for polymers that are being explored as DSA candidates, where lamellar BCP must be oriented perpendicular to the substrate. Controlling orientation is often a complex balance between the surface energies of the polymer at the substrate and surface and can even also be influenced by factors such as the film thickness (Maher et al., 2014; Bates et al., 2012; Vu et al., 2011; Coulon et al., 1989; Choi et al., 2012; Son et al., 2008; Zucchi et al., 2010).

The ability of GISAXS to readily differentiate between perpendicular and parallel orientations is demonstrated in Figure 7.14 (Choi et al., 2012). Choi and coworkers prepared substrates by grafting PS-*r*-PMMA with a range of compositions in order to control the surface energy. PS-*b*-PMMA was then spin coated on top of the sample, and a combination of TEM and GISAXS was used to evaluate the morphology. Representative GISAXS patterns are shown in Figure 7.14a and b. For substrates where the grafted brush had a PS fraction of 0.20, two peaks are observable along the q_z axis. This indicates that the BCP formed lamella parallel to the surface, and the spacing between q_z^* and $2q_z^*$ is reflective of the thickness of the BCP lamella. The

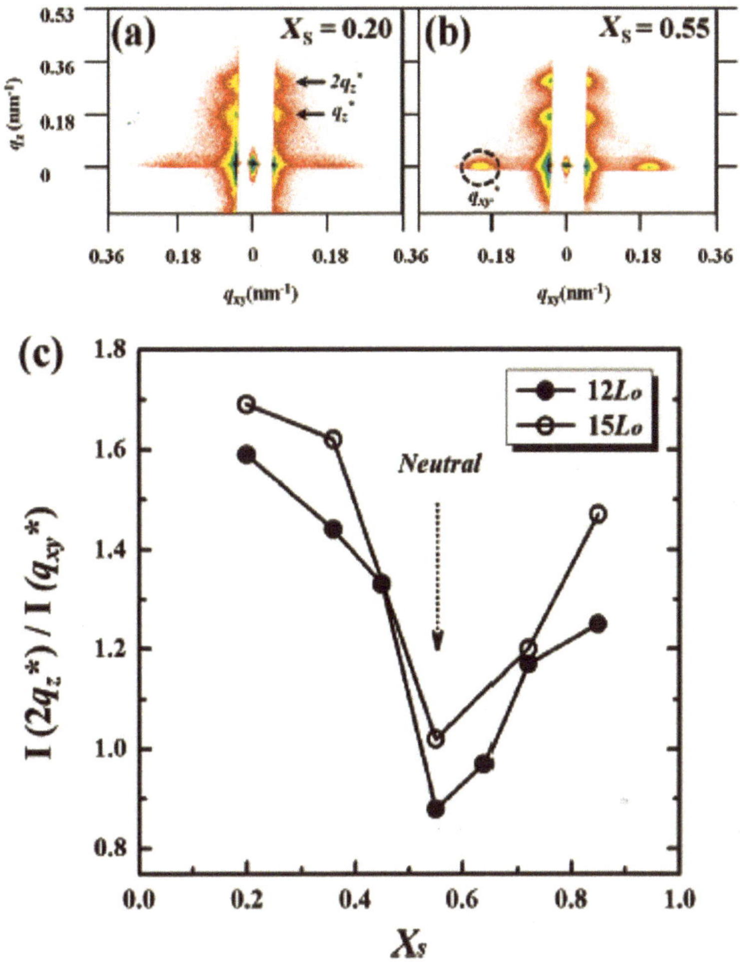

Figure 7.14 GISAXS patterns for PS-*b*-PMMA on substrates grafted with a random copolymer having polystyrene fractions of 20 (a) and 55 (b). Ratio of intensity at the second maxima for the out-of-plane scattering ($I(2q_z^*)$) to the first maxima for the in-plane scattering ($I(q_{xy}^*)$) as a function of PS fraction in the brush (X_s).
Reproduced from Choi et al. (2012) with permission of The Royal Society of Chemistry.

out-of-plane peaks are still present when the BCP was prepared on a substrate where the PS fraction in the brush was 0.55, but they now are accompanied by in-plane peaks (labeled as q_{xy}^*). The presence of the in-plane peaks indicates that some of the lamella are oriented perpendicular to the surface, while the continued presence of the out-of-plane peaks shows that there is a mixed morphology, where some of the lamella are parallel and some are perpendicular. This was confirmed via the TEM results, showing the lower half of the lamella oriented perpendicular, while the upper portion had mostly parallel orientation. Figure 7.14c plots the ratio of the intensity of the second order out of plane peak versus the first order in plane peak ($I(2q_z^*)/I(q_{xy}^*)$) as a function of the fraction of polystyrene in the brush. The minima at a PS fraction of 0.55 indicates that there is a relatively small window of brush composition, which will induce a perpendicular orientation in the BCP.

In addition to differentiating between perpendicular and parallel domain orientations, recent results from Vu and coworkers have shown that GISAXS measurements are also sensitive to tilted lamella (Vu et al., 2011). Diblock and triblock copolymers of PS and PMMA were assembled on a neutral substrate where the majority of the lamella were perpendicularly oriented. Two-dimensional GISAXS patterns were collected at varying incident angles. As the incident angle was increased, the presence of curved features in the scattering pattern become increasingly visible. Simulating data for lamella with varying tilt angles allowed the average tilt of the lamella to be determined. It was found that there was a greater average tilt for the diblock compared to the triblock structures, and these tilted lamella were likely kinetically trapped. GISAXS experiments are not limited to symmetric morphologies. They have also proven useful when examining the behavior of thin films of asymmetric polymers. Polystyrene-*b*-poly(2-vinylpyridine) (PS-*b*-P2VP), which formed spheres of P2VP, was spin coated in layers of varying thickness in order to understand the impact of confinement on the orientation of the spheres (Stein et al., 2007). When there were fewer than four layers, the spheres relieved confinement frustration by packing in a hexagon pattern. As soon as five layers were present, the in-plane distribution of the spheres began shifting to a body-centered cubic arrangement. This was determined by evaluating the in-plane scattering in the GISAXS pattern.

Reflectivity. Specular reflectivity is a special case of the reflection geometry where the incident and scattered angles are equal. This allows measurement of only the out-of-plane vector, which contains information about the SLD as a function of film depth. In this case, the intensity profile as a function of angle can be solved for exactly using the Parratt formalism (Parratt, 1954). This technique has historically been used to examine the behavior at interfaces, in particular, to study the interface between polymer layers, a problem that it is uniquely suited to solve (Anastasiadis et al., 1990, 1998; Russell et al., 1991a,b; Harton et al., 2006; Russell, 1996; Lin, 2002; Mayes et al., 1994; Scherble et al., 1999; Wang et al., 2005, 2009b; Sferrazza et al., 1997; Ma et al., 2013; Ade et al., 2009; Yan et al., 2011; Lee et al., 2005). The width of polymer interfaces can theoretically be determined from transmission scattering measurements (Ruland, 1987; Stribeck and Ruland, 1978), but in practice this information is convolved with the morphological information, the resolution function, and the incoherent background, making accurate determination difficult (Stamm, 1991).

Neutron reflectivity was the early workhorse in this field. By synthesizing BCPs with one of the blocks being selectively deuterated, the contrast could be dramatically improved. One of the early series of experiments in this field probed the structure and interface of dPS-*b*-PMMA. Anastasiadis and coworkers assembled thin films of dPS-*b*-PMMA and probed them with neutron reflectivity. Representative results of those experiments are shown in Figure 7.15 (Anastasiadis et al., 1990). The dPS-*b*-PMMA was assembled in a lamellar morphology where the PMMA block was found to coat the substrate and the PS block was present at the surface. This configuration minimized the enthalpic interactions at both surfaces. Several different molecular masses

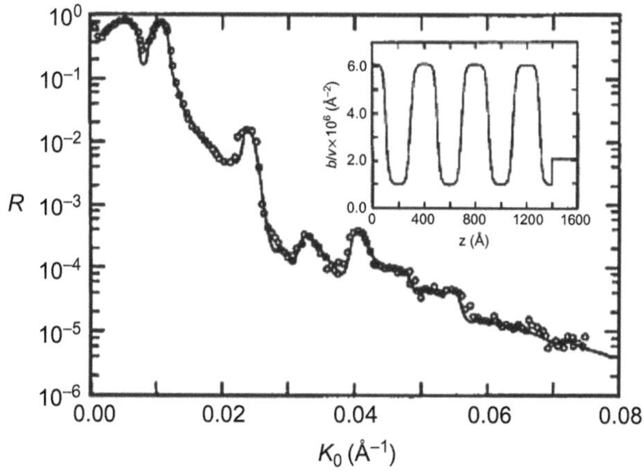

Figure 7.15 Neutron reflectivity curve of dPS-*b*-PMMA (Mw = 100 K). The circles represent experimental data, and the line represents the calculated model fit using the Parratt formalism. The inset shows the SLD profile used to calculate the simulated reflectivity curve.
Reprinted with permission from Anastasiadis et al. (1990). Copyright 1990 AIP Publishing LLC.

were measured and they were able to determine the length of each block along with the widths of the interface, which they observed to be constant throughout the molecular masses series. They also examined the behavior of homopolymer bilayers upon the addition of a symmetric diblock copolymer. By selectively deuterating one of the components in the system, they were able to determine the distribution of all three components throughout the interfacial region (Russell et al., 1991a). They also observed that the interface broadened upon the addition of the diblock (Russell et al., 1991b).

Recently, soft X-ray reflectivity (RSoXR) has become a popular technique for probing polymer films. The ability to vary the contrast by changing the X-ray energy near the absorption edge enables the study of polymeric materials without isotopic labeling. Experiments by Ade and coworkers also show how varying the energy can be used to effectively screen out interactions between layers by contrast matching (Ade et al., 2009). Figure 7.16 shows reflectivity curves from a stack of three different materials. By varying the contrast, they were able to screen out the impact of certain interfaces by contrast matching the scattering between adjacent layers. This is demonstrated when comparing the curves measured at 282.6–284.6 eV. At these two energies two very different length scales dominate the scattering. The uniqueness problem discussed earlier is particularly problematic when attempting to deconvolute the contributions of multiple interfaces in a stack. Unlike the layer thickness, which will result in fringing at distinct spacings, the interfacial width results in decay of the fringes and the effects of multiple widths will be convolved. Simultaneous solution of a sufficient number of reflectivity curves with different contrast functions allow for differentiation of the individual widths. The same type of approach has also been applied toward deconvoluting the distribution of a homopolymer throughout a diblock copolymer. Polyvinylphenol (PVPH) was blended with PS-*b*-PMMA and spin coated

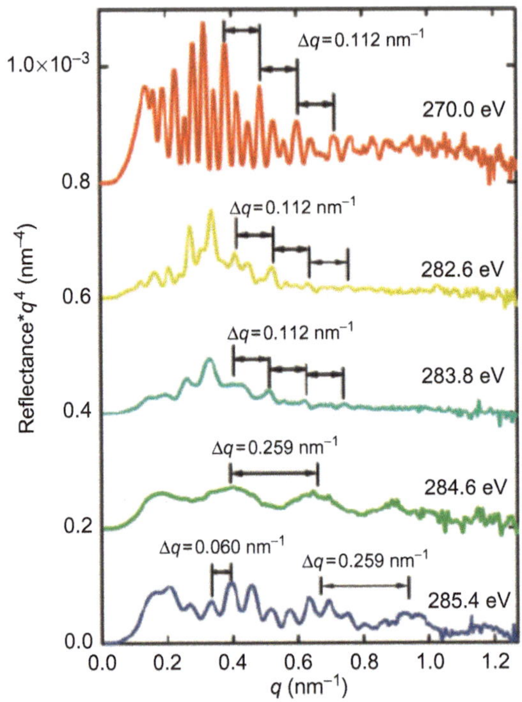

Figure 7.16 RSoXR curves of a trilayer polymer stack at different energies, where the energy variation resulted in screening out the contributions from specific interfaces.
Reprinted with permission from Ade et al. (2009). Copyright 2009 John Wiley and Sons.

into a multilayer structure. Using RSoXR at both the carbon and oxygen edge, it was determined that the PVPH was uniformly distributed throughout the PMMA layer (Sunday and Kline, 2015).

7.6 Conclusions and future direction

X-rays play a crucial role in understanding the behavior of materials with nanoscale structure and have already provided a wealth of information on BCPs. Overcoming the final technical challenges that prevent the integration of DSA as a lithography extension technique will require a combination of measurements, and X-ray techniques have the potential to play a large role. The ability of BCP DSA films to form structures with complex three-dimensional morphologies means that the uniqueness problem is likely to be a serious challenge in converting scattering profiles to the correct real space solution. One option for alleviating this issue is integrating simulations into the solution of scattering profiles. This may also present a route for connecting the underlying polymer physics to the scattering data. In addition to providing morphological data, X-rays will continue to contribute crucial thermodynamic and kinetic information. Soft X-rays are also likely to play a major role, due to their ability to control contrast between materials and characterize thin films.

References

Ade, H., Hitchcock, A.P., 2008. NEXAFS microscopy and resonant scattering: composition and orientation probed in real and reciprocal space. Polymer 49, 643–675.

Ade, H., Wang, C., Garcia, A., Yan, H., Sohn, K.E., Hexemer, A., Bazan, G.C., Nguyen, T.-Q., Kramer, E.J., 2009. Characterization of multicomponent polymer trilayers with resonant soft X-ray reflectivity. J. Polym. Sci. B: Polym. Phys. 47, 1291–1299.

Anastasiadis, S.H., Russell, T.P., Satija, S.K., Majkrzak, C.F., 1990. The morphology of symmetric diblock copolymers as revealed by neutron reflectivity. J. Chem. Phys. 92, 5677.

Anastasiadis, S.H., Retsos, H., Toprakcioglu, C., Menelle, A., Hadziioannou, G., 1998. On the interfacial width in triblock versus diblock copolymers: a neutron reflectivity investigation. Macromolecules 31, 6600–6604.

Araki, T., Ade, H., Stubbs, J.M., Sundberg, D.C., Mitchell, G.E., Kortright, J.B., Kilcoyne, A.L.D., 2006. Resonant soft X-ray scattering from structured polymer nanoparticles. Appl. Phys. Lett. 89, 124106.

Attwood, D., 2000. Soft X-Rays and Extreme Ultraviolet Radiation Principles and Applications. Cambridge University Press, Cambridge, United Kingdom.

Bates, F.S., Fredrickson, G.H., 1990. Block copolymer thermodynamics: theory and experiment. Annu. Rev. Phys. Chem. 41, 525–557.

Bates, F.S., Rosedale, J.H., Fredrickson, G.H., 1990. Fluctuation effects in a symmetric diblock copolymer near the order–disorder transition. J. Chem. Phys. 92, 6255.

Bates, C.M., Seshimo, T., Maher, M.J., Durand, W.J., Cushen, J.D., Dean, L.M., Blachut, G., Ellison, C.J., Willson, C.G., 2012. Polarity-switching top coats enable orientation of sub-10-nm block copolymer domains. Science 338, 775–779.

Bencher, C., Yi, H., Zhou, J., Cai, M., Smith, J., Miao, L., Montal, O., Blitshtein, S., Lavi, A., Dotan, K., et al., 2012. Directed self-assembly defectivity assessment. Part II. In: Proc. SPIE 8323, 83230N.

Busch, P., Posselt, D., Smilgies, D.-M., Rauscher, M., Papadakis, C.M., 2007. Inner structure of thin films of lamellar poly(styrene-*b*-butadiene) diblock copolymers as revealed by grazing-incidence small-angle scattering. Macromolecules 40, 630–640.

Cao, Y., Her, Y., Delgadillo, P.R., Vandenbroeck, N., Gronheid, R., Chan, T., Hashimoto, Y., Romo, A., Somervell, M., Nafus, K., et al., 2013. Using process monitor wafers to understand directed self-assembly defects. In: Proc. SPIE 8680, 86801S.

Chang, S.-W., Vogel, E.E., Ginzburg, V.V., Murray, D.J., Kramer, J.W., Weinhold, J.D., Chuang, V.P.W., Sharma, R., Evans, J.P., Landes, B., et al., 2012. Designing new materials and processes for directed self-assembly applications. In: Proc. SPIE 8323, 83231M.

Cheng, J.Y., Sanders, D.P., Truong, H.D., Harrer, S., Friz, A., Holmes, S., Colburn, M., Hinsberg, W.D., 2010. Simple and versatile methods to integrate directed self-assembly with optical lithography using a polarity-switched photoresist. ACS Nano 4, 4815–4823.

Cheng, J., Lawson, R.A., Yeh, W.-M., Jarnagin, N.D., Peters, A., Tolbert, L.M., Henderson, C.L., 2012. Directed self-assembly of poly(styrene)-block-poly(acrylic acid) copolymers for sub-20 nm pitch patterning. In: Proc. SPIE 8323, 83232R.

Choi, S., Kim, E., Ahn, H., Naidu, S., Lee, Y., Ryu, D.Y., Hawker, C.J., Russell, T.P., 2012. Lamellar microdomain orientation and phase transition of polystyrene-*b*-poly(methyl methacrylate) films by controlled interfacial interactions. Soft Matter 8, 3463.

Chourou, S.T., Sarje, A., Li, X.S., Chan, E.R., Hexemer, A., 2013. HipGISAXS: a high-performance computing code for simulating grazing-incidence X-ray scattering data. J. Appl. Crystallogr. 46, 1781–1795.

Coulon, G., Russell, T.P., Deline, V.R., Green, P.F., 1989. Surface-induced orientation of symmetric, diblock copolymers: a secondary ion mass-spectrometry study. Macromolecules 22, 2581–2589.

Cushen, J.D., Otsuka, I., Bates, C.M., Halila, S., Fort, S., Rochas, C., Easley, J.A., Rausch, E.L., Thio, A., Borsali, R., et al., 2012. Oligosaccharide/silicon-containing block copolymers with 5 nm features for lithographic applications. ACS Nano 6, 3424–3433.

Daillant, J., Gibaud, A., 2009. X-Ray and Neutron Reflectivity Principles and Applications. Springer, Berlin, Heidelberg.

Delgadillo, P.R., Harukawa, R., Suri, M., Durant, S., Cross, A., Nagaswami, V.R., Heuvel, D.V.D., Gronheid, R., Nealey, P., 2013. Defect source analysis of directed self-assembly process (DSA of DSA). Proc. SPIE 8680, 86800L.

Doerk, G.S., Liu, C.-C., Cheng, J.Y., Rettner, C.T., Pitera, J.W., Krupp, L.E., Topuria, T., Arellano, N., Sanders, D.P., 2013. Pattern placement accuracy in block copolymer directed self-assembly based on chemical epitaxy. ACS Nano 7, 276–285.

Edwards, E.W., Montague, M.F., Solak, H.H., Hawker, C.J., Nealey, P.F., 2004. Precise control over molecular dimensions of block-copolymer domains using the interfacial energy of chemically nanopatterned substrates. Adv. Mater. 16, 1315–1319.

Edwards, E.W., Stoykovich, M.P., Müller, M., Solak, H.H., de Pablo, J.J., Nealey, P.F., 2005. Mechanism and kinetics of ordering in diblock copolymer thin films on chemically nanopatterned substrates. J. Polym. Sci. B: Polym. Phys. 43, 3444–3459.

Feldman, K.E., Kade, M.J., Meijer, E.W., Hawker, C.J., Kramer, E.J., 2010. Phase behavior of complementary multiply hydrogen bonded end-functional polymer blends. Macromolecules 43, 5121–5127.

Guiner, A., 1994. X-Ray Diffraction in Crystals, Imperfect Crystals, and Amorphous Bodies. Dover Publications, New York, NY.

Harton, S.E., Stevie, F.A., Zhu, Z., Ade, H., 2006. Changes in thermodynamic interactions at highly immiscible polymer/polymer interfaces due to deuterium labeling. J. Phys. Chem. B 110, 10602–10605.

Herr, D.J.C., 2011. Directed block copolymer self-assembly for nanoelectronics fabrication. J. Mater. Res. 26, 122–139.

Hu, T., Jones, R.L., Wu, W., Lin, E.K., Lin, Q., Keane, D., Weigand, S., Quintana, J., 2004. Small angle X-ray scattering metrology for sidewall angle and cross section of nanometer scale line gratings. J. Appl. Phys. 96, 1983.

Ito, C., Durant, S., Lange, S., Harukawa, R., Miyagi, T., Nagaswami, V., Rincon Delgadillo, P., Gronheid, R., Nealey, P., 2014. Inspection of directed self-assembly defects. In: Proc. SPIE 9049, 90492D.

Jones, R.L., 2006. Pattern fidelity in nanoimprinted films using critical dimension small angle X-ray scattering. J. Micro Nanolithogr. MEMS MOEMS 5, 013001.

Jones, R.L., Hu, T., Lin, E.K., Wu, W.-L., Kolb, R., Casa, D.M., Bolton, P.J., Barclay, G.G., 2003. Small angle X-ray scattering for sub-100 nm pattern characterization. Appl. Phys. Lett. 83, 4059.

Kennemur, J.G., Hillmyer, M.A., Bates, F.S., 2012. Synthesis, thermodynamics, and dynamics of poly(4-*tert*-butylstyrene-*b*-methyl methacrylate). Macromolecules 45, 7228–7236.

Kim, S.O., Solak, H.H., Stoykovich, M.P., Ferrier, N.J., de Pablo, J.J., Nealey, P.F., 2003. Epitaxial self-assembly of block copolymers on lithographically defined nanopatterned substrates. Nature 424, 411–414.

Krishnan, S., Paik, M.Y., Ober, C.K., Martinelli, E., Galli, G., Sohn, K.E., Kramer, E.J., Fischer, D.A., 2010. NEXAFS depth profiling of surface segregation in block copolymer thin films. Macromolecules 43, 4733–4743.

Lazzari, R., 2002. IsGISAXS: a program for grazing-incidence small-angle X-ray scattering analysis of supported islands. J. Appl. Crystallogr. 35, 406–421.

Lee, H.-J., Soles, C.L., Ro, H.W., Jones, R.L., Lin, E.K., Wu, W., Hines, D.R., 2005. Nanoimprint pattern transfer quality from specular X-ray reflectivity. Appl. Phys. Lett. 87, 263111.

Leibler, L., 1980. Theory of microphase separation in block copolymers. Macromolecules 13, 1602–1617.

Lemaillet, P., Germer, T.A., Kline, R.J., Sunday, D.F., Wang, C., Wu, W., 2013. Intercomparison between optical and X-ray scatterometry measurements of FinFET structures. In: Proc. SPIE 8681, 86810Q.

Leroy, F., Revenant, C., Renaud, G., Lazzari, R., 2004. Situ GISAXS study of the growth of Pd on MgO(001). Appl. Surf. Sci. 238, 233–237.

Lin, E.K., 2002. Direct measurement of the reaction front in chemically amplified photoresists. Science 297, 372–375.

Liu, C.-C., Thode, C.J., Rincon Delgadillo, P.A., Craig, G.S.W., Nealey, P.F., Gronheid, R., 2011a. Towards an all-track 300 mm process for directed self-assembly. J. Vac. Sci. Technol. B: Microelectron. Nanometer Struct. Process. Meas. Phenom. 29, 06F203.

Liu, C.-C., Han, E., Onses, M.S., Thode, C.J., Ji, S., Gopalan, P., Nealey, P.F., 2011b. Fabrication of lithographically defined chemically patterned polymer brushes and mats. Macromolecules 44, 1876–1885.

Liu, C.-C., Ramírez-Hernández, A., Han, E., Craig, G.S.W., Tada, Y., Yoshida, H., Kang, H., Ji, S., Gopalan, P., de Pablo, J.J., et al., 2013. Chemical patterns for directed self-assembly of lamellae-forming block copolymers with density multiplication of features. Macromolecules 46, 1415–1424.

Ma, W., Vodungbo, B., Nilles, K., Liu, Y., Theato, P., Luning, J., 2013. Precise structural investigation of symmetric diblock copolymer thin films with resonant soft X-ray reflectivity. Soft Matter 9, 8820.

Maher, M.J., Bates, C.M., Blachut, G., Sirard, S., Self, J.L., Carlson, M.C., Dean, L.M., Cushen, J.D., Durand, W.J., Hayes, C.O., et al., 2014. Interfacial design for block copolymer thin films. Chem. Mater. 26, 1471–1479.

Matsen, M.W., 1995. Phase behavior of block copolymer/homopolymer blends. Macromolecules 28, 5765–5773.

Mayes, A.M., Russell, T.P., Deline, V.R., Satija, S.K., Majkrzak, C.F., 1994. Block copolymer mixtures as revealed by neutron reflectivity. Macromolecules 27, 7447–7453.

Nishino, Y., Takahashi, Y., Imamoto, N., Ishikawa, T., Maeshima, K., 2009. Three-dimensional visualization of a human chromosome using coherent X-ray diffraction. Phys. Rev. Lett. 102, 018101.

Paik, M.Y., Bosworth, J.K., Smilges, D.-M., Schwartz, E.L., Andre, X., Ober, C.K., 2010. Reversible morphology control in block copolymer films via solvent vapor processing: an in situ GISAXS study. Macromolecules 43, 4253–4260.

Parratt, L.G., 1954. Surface studies of solids by total reflection of X-rays. Phys. Rev. 95, 359–369.

Perera, G.M., Stein, G.E., Liddle, J.A., 2010. Spatial coherence in electron-beam patterning. J. Vac. Sci. Technol. B: Microelectron. Nanometer Struct. Process. Meas. Phenom. 28, 1048.

Perera, G.M., Wang, C., Doxastakis, M., Kline, R.J., Wu, W., Bosse, A.W., Stein, G.E., 2012. Directed self-assembly of lamellar copolymers: effects of interfacial interactions on domain shape. ACS Macro Lett. 1, 1244–1248.

Renaud, G., Lazzari, R., Leroy, F., 2009. Probing surface and interface morphology with grazing incidence small angle X-ray scattering. Surf. Sci. Rep. 64, 255–380.

Revenant, C., Leroy, F., Lazzari, R., Renaud, G., Henry, C., 2004. Quantitative analysis of grazing incidence small-angle X-ray scattering: Pd/MgO(001) growth. Phys. Rev. B 69, 035411:1–17.

Richard, M.-I., Schülli, T.-U., Wintersberger, E., Renaud, G., Bauer, G., 2006. Growth mode, strain state and shape of Ge islands during their growth at different temperatures: a combined in situ GISAXS and GIXD study. Thin Solid Films 508, 213–217.

Rincon Delgadillo, P.A., Gronheid, R., Thode, C.J., Wu, H., Cao, Y., Somervell, M., Nafus, K., Nealey, P.F., 2012. All track directed self-assembly of block copolymers: process flow and origin of defects. In: Proc. SPIE 8323, 83230D.

Robinson, I., Vartanyants, I., Williams, G., Pfeifer, M., Pitney, J., 2001. Reconstruction of the shapes of gold nanocrystals using coherent X-ray diffraction. Phys. Rev. Lett. 87 195505:1–4.

Ruiz, R., Dobisz, E., Albrecht, T.R., 2011. Rectangular patterns using block copolymer directed assembly for high bit aspect ratio patterned media. ACS Nano 5, 79–84.

Ruland, W., 1987. Small-angle scattering of the statistical structure of domain boundaries. Macromolecules 20, 87–93.

Russell, T.P., 1996. On the reflectivity of polymers: neutrons and X-rays. Phys. B Condens. Matter 221, 267–283.

Russell, T.P., Anastasiadis, S.H., Menelle, A., Felcher, G.P., Satija, S.K., 1991a. Segment density distribution of symmetric diblock copolymers at the interface between two homopolymers as revealed by neutron reflectivity. Macromolecules 24, 1575–1582.

Russell, T.P., Menelle, A., Hamilton, W.A., Smith, G.S., Satija, S.K., Majkrzak, C.F., 1991b. Width of homopolymer interfaces in the presence of symmetric diblock copolymers. Macromolecules 24, 5721–5726.

Saavedra, H.M., Mullen, T.J., Zhang, P., Dewey, D.C., Claridge, S.A., Weiss, P.S., 2010. Hybrid strategies in nanolithography. Rep. Prog. Phys. 73, 036501.

Sakurai, S., Mori, K., Okawara, A., Kimishima, K., Hashimoto, T., 1992. Evaluation of segmental interaction by small-angle X-ray scattering based on the random-phase approximation for asymmetric polydisperse triblock copolymers. Macromolecules 25, 2679–2691.

Scherble, J., Stark, B., Stühn, B., Kressler, J., Budde, H., Höring, S., Schubert, D.W., Simon, P., Stamm, M., 1999. Comparison of interfacial width of block copolymers of d_8-poly(methyl methacrylate) with various poly(*n*-alkyl methacrylate)s and the respective homopolymer pairs as measured by neutron reflection. Macromolecules 32, 1859–1864.

Sferrazza, M., Xiao, C., Jones, R.A.L., Bucknall, D.G., Webster, J., Penfold, J., 1997. Evidence for capillary waves at immiscible polymer/polymer interfaces. Phys. Rev. Lett. 78, 3693.

Simão, C., Tuchapsky, D., Khunsin, W., Amann, A., Morris, M.A., Sotomayor Torres, C.M., 2014. Defect analysis and alignment quantification of line arrays prepared by directed self-assembly of a block copolymer. In: Proc. SPIE 9050, 905028.

Son, J.G., Bulliard, X., Kang, H., Nealey, P.F., Char, K., 2008. Surfactant-assisted orientation of thin diblock copolymer films. Adv. Mater. 20, 3643–3648.

Stamm, M., 1991. Investigation of the interface between polymers: a comparison of scattering and reflectivity techniques. J. Appl. Crystallogr. 24, 651–658.

Stein, G.E., Kramer, E.J., Li, X., Wang, J., 2007. Layering transitions in thin films of spherical-domain block copolymers. Macromolecules 40, 2453–2460.

Stoykovich, M.P., 2005. Directed assembly of block copolymer blends into nonregular device-oriented structures. Science 308, 1442–1446.

Stoykovich, M.P., Nealey, P.F., 2006. Block copolymers and conventional lithography. Mater. Today 9, 20–29.

Stoykovich, M.P., Kang, H., Daoulas, K.C., Liu, G., Liu, C.-C., de Pablo, J.J., Müller, M., Nealey, P.F., 2007. Directed self-assembly of block copolymers for nanolithography: fabrication of isolated features and essential integrated circuit geometries. ACS Nano 1, 168–175.

Stoykovich, M.P., Daoulas, K.C., Müller, M., Kang, H., de Pablo, J.J., Nealey, P.F., 2010. Remediation of line edge roughness in chemical nanopatterns by the directed assembly of overlying block copolymer films. Macromolecules 43, 2334–2342.

Stribeck, N., Ruland, W., 1978. Determination of the interface distribution function of lamellar two-phase systems. J. Appl. Crystallogr. 11, 535–539.

Sunday, D.F., Kline, R.J., 2015. Reducing block copolymer interfacial widths through polymer additives. Macromolecules 48, 679–686. http://dx.doi.org/10.1021/ma502015u.

Sunday, D.F., Hammond, M.R., Wang, C., Wu, W., Kline, R.J., Stein, G.E., 2013. Three-dimensional X-ray metrology for block copolymer lithography line-space patterns. J. Micro/Nanolithogr. MEMS MOEMS 12, 031103.

Sunday, D.F., Hammond, M.R., Wang, C., Wu, W., Delongchamp, D.M., Tjio, M., Cheng, J.Y., Kline, R.J., Pitera, J.W., 2014. Determination of the internal morphology of nanostructures patterned by directed self assembly. ACS Nano 8, 8426–8437.

Virgili, J.M., Tao, Y., Kortright, J.B., Balsara, N.P., Segalman, R.A., 2007. Analysis of order formation in block copolymer thin films using resonant soft X-ray scattering. Macromolecules 40, 2092–2099.

Vu, T., Mahadevapuram, N., Perera, G.M., Stein, G.E., 2011. Controlling domain orientations in thin films of AB and ABA block copolymers. Macromolecules 44, 6121–6127.

Wan, L., Ruiz, R., Gao, H., Patel, K.C., Lille, J., Zeltzer, G., Dobisz, E.A., Bogdanov, A., Nealey, P.F., Albrecht, T.R., 2012. Fabrication of templates with rectangular bits on circular tracks by combining block copolymer directed self-assembly and nanoimprint lithography. J. Micro/Nanolithogr. MEMS MOEMS 11, 031405.

Wang, C., Araki, T., Ade, H., 2005. Soft X-ray resonant reflectivity of low-Z material thin films. Appl. Phys. Lett. 87, 214109.

Wang, C., Jones, R.L., Lin, E.K., Wu, W.-L., Rice, B.J., Choi, K.-W., Thompson, G., Weigand, S.J., Keane, D.T., 2007a. Characterization of correlated line edge roughness of nanoscale line gratings using small angle X-ray scattering. J. Appl. Phys. 102, 024901.

Wang, C., Jones, R.L., Lin, E.K., Wu, W.-L., Leu, J., 2007b. Small angle X-ray scattering measurements of lithographic patterns with sidewall roughness from vertical standing waves. Appl. Phys. Lett. 90, 193122.

Wang, C., Choi, K.-W., Chen, Y.-C., Price, J., Ho, D.L., Jones, R.L., Soles, C., Lin, E.K., Wu, W.-L., Bunday, B.D., 2009a. Nonplanar high-k dielectric thickness measurements using CD-SAXS. In: Proc. SPIE 7272, 72722M.

Wang, C., Garcia, A., Yan, H., Sohn, K.E., Hexemer, A., Nguyen, T.-Q., Bazan, G.C., Kramer, E.J., Ade, H., 2009b. Interfacial widths of conjugated polymer bilayers. J. Am. Chem. Soc. 131, 12538–12539.

Wang, C., Lee, D.H., Hexemer, A., Kim, M.I., Zhao, W., Hasegawa, H., Ade, H., Russell, T.P., 2011. Defining the nanostructured morphology of triblock copolymers using resonant soft X-ray scattering. Nano Lett. 11, 3906–3911.

Warren, B.E., 1969. X-Ray Diffraction. Dover Publications, New York, NY.

Yan, H., Wang, C., Garcia, A., Swaraj, S., Gu, Z., McNeill, C.R., Schuettfort, T., Sohn, K.E., Kramer, E.J., Bazan, G.C., et al., 2011. Interfaces in organic devices studied with resonant soft X-ray reflectivity. J. Appl. Phys. 110, 102220.

Yang, X.M., Peters, R.D., Nealey, P.F., Solak, H.H., Cerrina, F., 2000. Guided self-assembly of symmetric diblock copolymer films on chemically nanopatterned substrates. Macromolecules 33, 9575–9582.

Zucchi, I.A., Poliani, E., Perego, M., 2010. Microdomain orientation dependence on thickness in thin films of cylinder-forming PS-*b*-PMMA. Nanotechnology 21, 185304.

Self-assembly of block copolymers by graphoepitaxy

8

Samuel M. Nicaise, K.G. Amir Tavakkoli, Karl K. Berggren
Massachusetts Institute of Technology, Cambridge, MA, USA

8.1 Introduction

Graphoepitaxy for block copolymer (BCP) self-assembly is used to control the arrangement of BCP microdomains both locally and over large areas. This approach was first introduced by Segalman et al. (2001), and has been developed and improved significantly since. BCP graphoepitaxy takes advantage of topographical patterns on a substrate to template the long- and short-range direction of the otherwise uncontrolled BCP microdomains (Bang et al., 2009). Typically, this approach aims to incorporate the benefits of top-down fabrication with bottom-up self-assembly. While top-down fabrication can be expensive, and possibly include complicated mechanisms or machines, it can also produce near-arbitrary features with low defectivity. Bottom-up self-assembly naturally forms a limited set of geometries, such as spheres, cylinders, lamellae, and gyroids; although, it is inexpensive, simple, and can produce features smaller than other top-down patterning processes. The benefits of topographical templating of BCP self-assembly include strict control over the microdomains—namely, bends, junctions, and morphology—as well as templating for multiple layers of patterns. In comparison to chemoepitaxy of BCPs (Ruiz et al., 2008; Ouk Kim et al., 2003; Stoykovich et al., 2005, 2007; Cheng et al., 2008; Rockford et al., 1999), graphoepitaxy is beneficial because it also can use the confinement effect for generation of complicated patterns (Tavakkoli et al., 2012a,b, 2014; Yu et al., 2006; Wu et al., 2004; Jeong et al., 2009; Xiang et al., 2004). Also, it typically leaves behind the templates, which is not always desirable, though these can be removed via sacrificial templating (Ilievski and Ross, 2010; Moon et al., 2012; Park et al., 2011; Voet et al., 2011; Jeong et al., 2010).

Research has shown that BCP graphoepitaxy mainly operates by using commensuration matching/mismatching, lateral alignment, and the confinement effect, as will be explained in later sections. First, topographical templates can impose order by matching/mismatching with the natural geometric period of BCPs in a commensurate/incommensurate manner (Bita et al., 2008; Yang et al., 2010; Chang et al., 2012). If the spacing between templating features is an integer multiple of the BCP natural period, the microdomains will typically order in the direction(s) perpendicular to the commensuration, as in chemoepitaxy. This commensuration effect does not require strict period matching, and the following sections will discuss that the BCP can self-heal up to a 10% mismatch between the template period and the BCP period (Stoykovich et al., 2005; Yang et al., 2010; Cheng et al., 2003). Furthermore, templates have been shown to direct various morphologies or orientations when the template period is

Directed Self-assembly of Block Copolymers for Nano-manufacturing. http://dx.doi.org/10.1016/B978-0-08-100250-6.00008-0

otherwise commensurate with the nonequilibrium geometry (Tavakkoli et al., 2012a,b, 2014; Xiang et al., 2004; Park et al., 2009; Hong et al., 2012a,b; Moon and McCarthy, 2003; Chu et al., 2014; Dobriyal et al., 2009; Shin, 2004). Moreover, graphoepitaxy can induce long-range order by imposing lateral ordering on the BCP. Lateral ordering typically occurs when a linear templating feature induces proximal ordering of a similar shape on the BCP domains, such as a trench edge inducing parallel lines of BCP in the direction parallel to the trench direction (Jeong et al., 2009, 2010; Hammond and Kramer, 2006; Han et al., 2010; Cheng et al., 2004a; Jung et al., 2010). The confinement effect can also control BCP self-assembly by restricting BCPs in a specific area (Yu et al., 2006; Wu et al., 2004; Jeong et al., 2009; Xiang et al., 2004; Tavakkoli et al., 2012b, 2014).

Though this chapter does not go into detail on the applications of BCP graphoepitaxy, recent research has been primarily applied to integrated circuit and memory technologies, but also could be useful for sensors, computation, energy devices, biological substrates, filtration, optical metamaterials, and much more (Milliron et al., 2007; Cheng et al., 2004b; Black, 2005; Thurn-Albrecht et al., 2000; Naito et al., 2002; Jung et al., 2008a; Phillip et al., 2010; Jackson and Hillmyer, 2010). This chapter opens by describing the predominant directed self-assembly (DSA) mechanisms, which can be harnessed for graphoepitaxy. Next, the focus shifts to the template fabrication methods for the DSA of uniform BCP patterns such as dots and lines. Many different fabrication techniques for graphoepitaxial templating have been employed in order to direct a wide range of final nanopatterns. Fabrication for templates has been achieved by photo-, interference, electron-beam, and nanoimprint lithographies, along with self-forming templates of BCP, surface-reconstructed ridges, and porous materials. The chapter then moves away from uniform patterns to focus on complex patterns. Aperiodic DSA structures are initially reviewed, followed by a look at multilayer DSA structures. Each section includes a discussion of the observed research phenomenon as well as projection of future research.

8.2 DSA for uniform 1D and 2D patterned arrays

8.2.1 *Mechanisms of DSA for uniform 1D and 2D array templates*

There are many mechanisms of DSA that are common for templating uniform patterns of BCPs, and they are briefly discussed herein. The later sections refer back to this section, as each share these general mechanisms, but also detail specific examples and exceptions to the generalities in this section.

8.2.1.1 *Commensuration*

The matching, or commensuration, of the template dimensions to the natural pitch of the BCP, L_0, allows for significant control over microdomain ordering. This mechanism is diagramed in Figure 8.1. In a 1D, trench/mesa template, the width of the trench can

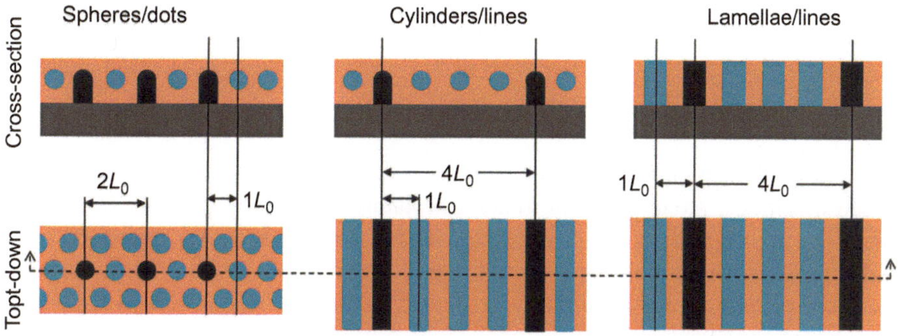

Figure 8.1 Schematic diagrams of the natural period, L_0, for different morphologies. Templating features are placed at integer number of L_0 spacings to be commensurate to the block copolymer spacing. In these images, pink is the majority block of the BCP, blue is the minority block, light gray is the substrate, and black is the templating features.

be an integer, N, multiple of L_0. In a 2D, post array template, the distance between two chosen posts—in either a Cartesian direction or angled direction—could be an integer, N, multiple of L_0 as described by Bita et al. (2008). Commensuration is typically important for templating periods of around $10L_0$ or fewer (Ross et al., 2008). While the BCP microdomains could accommodate templating features that were not precisely $N \times L_0$ (where N is an integer), control over the feature distance could reduce incommensuration-induced energy (Cheng et al., 2003).

8.2.1.2 Lateral ordering

Aside from commensuration, microdomains can be locally ordered based on the impact of an adjacent templating feature. Typically in a trench/mesa template, the lateral sidewall of the trench directed the nearby microdomains to assemble parallel to the sidewall. This effect diminished for microdomains further away from the templating feature.

8.2.1.3 Kinetic flow

In some trench/mesa templating cases, the impact of the kinetic polymer flow from the mesa into the trench was dominant over other templating mechanisms. In these cases, the orientation of the microdomains was typically perpendicular to the trench direction, as opposed to parallel ordering as a function of commensuration or lateral ordering. The impact was larger for nanoimprint lithography (NIL), and is discussed in that section.

8.2.1.4 Confinement

When the spacing of templating features is small, typically less than $2L_0$, the confinement of the BCP significantly controls self-assembly. In many cases, this confinement can change the morphology of the BCP from the typical geometry to a metastable morphology (Yu et al., 2006; Wu et al., 2004; Jeong et al., 2009; Xiang et al., 2004;

Tavakkoli et al., 2012, 2014). Confinement can control BCP assembly in one, two, or three dimensions.

8.2.1.5 Surface chemistry

For most graphoepitaxy DSA, the template is functionalized with a polymer brush or self-assembled monolayer. For trench templating, trench-bottom functionalization controlled the orientation of the microdomains, either in- or out-of-plane, and side-wall functionalization helped to induce lateral ordering. For NIL, functionalization increased pattern fidelity during mold removal. In the case of post array templating, all reports have shown use of functionalization brushes, as the posts act as surrogates of the appropriate block.

8.2.1.6 Template size

Last, the size of the template, including trench height or width and post height or diameter, affects the pattern. This mechanism is widely varied, and is therefore discussed in the following sections. Commensuration and lateral ordering are not the only important properties of the templates; other geometries must be appropriate in order for the discussed phenomena to occur.

8.2.2 Photolithography

Photolithography (PL) is typically employed to fabricate micro and nanostructures in a photoresist thin film through light exposure of a chosen pattern. For the purposes of graphoepitaxy, the PL pattern is produced by exposing light, generally 193 nm or longer wavelength, through a mask and, typically, spatially demagnifying the image at the photoresist film, though the final structure is not necessarily the same as the mask pattern (e.g., in double-patterning or phase-shifting masks). The photoresist structure can be used directly for graphoepitaxy, or transferred into the substrate, in which case the substrate pattern is used for graphoepitaxy. The photoresists are specifically designed and formulated for a particular light source and typically target high contrast and resolution.

PL has many benefits for graphoepitaxy, including rapid research prototyping, and therefore has been widely employed for templated self-assembly (Segalman et al., 2001; Moon et al., 2012; Jeong et al., 2010; Hammond and Kramer, 2006; Black and Bezencenet, 2004; Gu et al., 2012; Segalman et al., 2003; Sanders et al., 2010; Yi et al., 2012). Many of the benefits are similar to interference lithography (IL) and NIL, and we therefore refer to those later sections in this chapter. PL for BCP graphoepitaxy was employed by early investigators (Segalman et al., 2001) in order to understand BCP templating mechanisms, and these mechanisms have since been applied to other templating processes. In laboratory setups, the Abbé diffraction limits the smallest patternable pitch to greater than half the light wavelength, which typically is not smaller than 200 nanometers. This pitch limitation is a drawback though the parallel nature of PL has recently allowed for industrially relevant large-area, highly ordered

BCP patterns (for example, 193i, 193 nm, and extreme UV lithography) (Moon et al., 2012; Jeong et al., 2010; Cheng et al., 2010, 2012).

Figure 8.2 shows examples of oriented BCP features by PL trench templating. In all cases, the features were oriented parallel to the trenches, which had widths from a few hundred nanometers to 1 μm. Figure 8.2a is one of the earliest examples of PL-based BCP graphoepitaxy by Black et al. (Black and Bezencenet, 2004). Figure 8.2b shows the PL template and dot BCP patterns that Chang et al. fabricated by assembling out-of-plane cylinders (Chang and Wong, 2006). This early work was expanded upon to show more potential for large-area fabrication where, in Figure 8.2c, Moon et al. produced BCP ordered cylinder patterns over the majority of the sample and then grafted lamellae BCP on top of the primary pattern (Moon et al., 2012).

Generally, the templating mechanism induced by PL is similar to the introductory discussion. The lateral ordering and commensuration in the trench arrays were important for effective direction of line and dot patterns (Black and Bezencenet, 2004; Borah et al., 2013; Chang et al., 2006, 2009). As array pitches grew larger, above 700 nm in one case (Jeong et al., 2010) or 5 μm in another case (Segalman et al., 2001), defect density increased. At a higher trench width, lateral ordering reduced defects near the trench sidewall compared to the middle of the trench (Segalman et al., 2001; Hammond and Kramer, 2006; Black and Bezencenet, 2004; Borah et al., 2013a). In contrast to the typical importance to trench width, Hammond et al. suggested that thermal annealing temperatures slightly below the nematic-isotropic transition temperature had a larger impact on reducing defects than controlling trench width (Hammond and Kramer, 2006). As a separate contrast to the typical importance of lateral ordering,

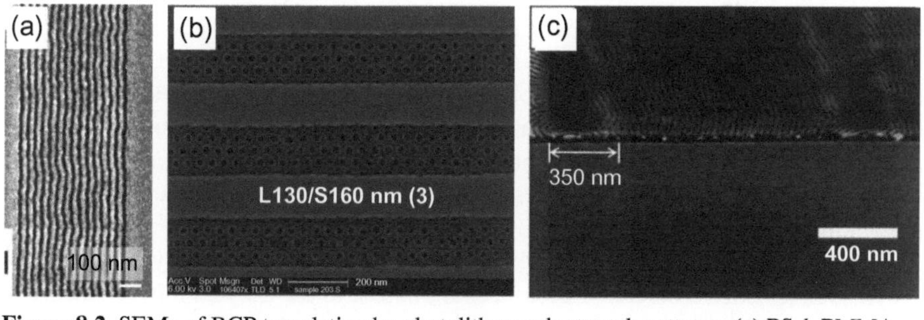

Figure 8.2 SEMs of BCP templating by photolithography trench patterns. (a) PS-*b*-PMMA cylinders are aligned along the trench axis for commensurate trench widths (Black and Bezencenet, 2004). (b) PS-*b*-PMMA out-of-plane cylinders are aligned in different widths of trench templates with half-cylinders at the trench sidewall (Chang and Wong, 2006). (c) PS-*b*-PMMA cylinders were first templated by 1-μm wide trenches on the bottom and then grafted with thicker lamellae BCP on the top (Moon et al., 2012). One print or electronic copy may be made for personal use only. Systematic electronic or print reproduction and distribution, duplication of any material in this paper for a fee or for commercial purposes, or modification of the content of the paper are prohibited.
Reproduced ("Adapted" or "in part") from Moon et al. (2012) with permission of The Royal Society of Chemistry. Panel c: © 2002 IEEE. Reprinted, with permission, from Black and Bezencenet (2004). Copyright 2006 Society of Photo Optical Instrumentation Engineers.

the perpendicular ordering of in-plane cylinders via short templating trenches was observed as a result of chain distortion (Gu et al., 2012). Last, the choice of trench sidewall chemistry increases the ordering of many different morphologies and orientations (Jeong et al., 2010; Borah et al., 2013a; Chang and Wong, 2006).

PL graphoepitaxy has recently targeted applications in which large-area and high-throughput BCP patterns are desirable. The parallel and industrial-standard nature of PL makes it appropriate for these applications, even in the absence of high resolution. Moon et al. made use of sparse trenches to template a first layer of BCP cylinders, on top of which high-aspect ratio BCP lamellae features were grafted (Moon et al., 2012). Jeong et al. developed an ultra-large-area two-step patterning process by first templating the initial layer of PS-*b*-PMMA lamella with disposable PL resist trenches and second patterning an underlying polymer brush with the initial lamellae for the final chemoepitaxy (Jeong et al., 2010). Gu et al. innovated on templated in- and out-of-plane PS-*b*-P2VP and PS-*b*-PEO with increased etch-selectivity by over-swelling the minority blocks (Gu et al., 2012).

In conclusion, while PL templating was the first lithography method for BCP physically templated self-assembly, and has since given way to other lithography methods, it has the potential for rapid industrial adoption. Work by Jeong et al. is a prime example of the promise for high-quality large-area patterning by pairing PL with BCP lithography (Jeong et al., 2010). Furthermore, the plentitude of lessons that have since been learned from interference and NIL templating (discussed below) may be applied to industry-directed PL templating in the future.

8.2.3 Interference lithography

Graphoepitaxial templates for BCP self-assembly can be fabricated with IL, typically, using UV-light to expose a thin film of photoresist (Ahlhorn et al., 1989). In the IL setup, a single temporally coherent beam is split into multiple beams. The beams are recombined at the substrate resulting in a standing wave, or a linear pattern of peaks and nodes, due to the interference of the coherent beams. The energy at the peaks is absorbed by the photoresist, and the resulting nanostructure is typically a periodic pattern.

There are several setups and specifications for IL. A simple, widely used setup is Lloyd's mirror interferometer (Fucetola et al., 2009). In the Lloyd's mirror, the original beam is reflected off an adjacent mirror and recombined with the original primary beam. Experiments use either a HeCd laser at 325 nm, or a diode laser at 405 nm (Cheng et al., 2006a). In a more optically complex setup, achromatic IL has been used to produce high-quality, large-area, sub-100 nm pitch gratings. A 193-nm beam from an ArF laser is diffracted from two separate transmission gratings and recombined. By using high-quality diffraction gratings, the optical path and wavelength of the original beam are not as important, thus allowing for higher quality patterns (Savas, 1996).

IL is beneficial for BCP DSA because it is able to quickly and controllably fabricate uniform large area, over many square centimeters, 1D or 2D gratings with engineerable geometry. The pitch of the linear array can be controlled based on the light wavelength and the angle of recombination. IL can achieve sub-diffraction-limit pitches at half the light wavelength. This sub-50 nm resolution of IL is on the length scale of the natural pitch of the nanofabrication-applicable BCPs, allowing for commensuration with the BCP natural pitch (Cheng et al., 2003; Jung and Ross, 2007; Switkes et al., 2000). Furthermore, the duty cycle of the pattern, or the width ratio of the trenches and mesas, can be controlled by changing the delivered dose, and gratings have been shown to have a near-zero-degree sidewall angle (Savas, 1996).

IL can be a simple fabrication process, allowing for rapid prototyping in a research and development environment. Benefits include the lack of a photomask, potentially low pattern defectivity, and applicability with both a thick or thin photoresist. On the other hand, IL has limitations compared to other lithography methods; for example, the patterning is limited to periodic arrays and therefore likely precludes complex template designs.

IL has been used in many reports to fabricate trench/mesa patterns for graphoepitaxial templating of BCP self-assembly (Black, 2005; Jung and Ross, 2007; Black and Bezencenet, 2004; Ilievski and Ross, 2010; Jung et al., 2010; Gu et al., 2012; Segalman et al., 2001; Hammond and Kramer, 2006; Sundrani et al., 2004; Cheng et al., 2002, 2003, 2006a; Chuang et al., 2009; Jung et al., 2008b). In Figure 8.3a and b, examples from Jung et al. are given of IL-patterned trenches that templated parallel alignment of cylinder-forming PS-*b*-PDMS BCP over 1 cm^2 areas. Figure 8.3c is an example of IL trench templating of sphere-forming PS-*b*-PFS BCP in an ordered hexagonal array. Trench/mesa templating from IL has been used to template all morphologies of cylindrical, lamellae, and spherical BCP (Segalman et al., 2001; Ilievski and Ross, 2010; Hammond and Kramer, 2006; Black, 2005; Black and Bezencenet, 2004; Gu et al., 2012; Jung and Ross, 2007; Sundrani et al., 2004; Cheng et al., 2002, 2003, 2006a; Jung et al., 2008b, 2010). The templating mechanisms are similar to the initial general discussion, though specifics to IL research are discussed below.

Figure 8.3 Scanning electron micrographs (SEM) of IL-templated self-assembly of different block copolymers by graphoepitaxy. (a) In-plane cylinders of PS-*b*-PDMS are templated by a sparse trench array, and the inset (b) shows the oriented lines close up (Jung and Ross, 2007). (c) Ordered spherical dots of PS-*b*-PFS are templated by an IL array (Cheng et al., 2002). Reprinted (adapted) with permission from Jung and Ross (2007). Copyright 2007, American Chemical Society. Reprinted with permission from Cheng et al. (2002). Copyright 2002, AIP Publishing LLC.

The trench can have a chemical preference to one of the blocks, or be neutral, and hence modify the degree of ordering within the trench (Jung and Ross, 2007). For the bottom of the trench, preferential wetting promotes in-plane orientation of BCP domains while a neutral surface leads to out-of-plane orientation of BCP domains (Black and Bezencenet, 2004). On the other hand, reports have shown well-ordered DSA without specific surface chemistry modification (Ilievski and Ross, 2010).

The lateral ordering of the BCP initiates at the trench sidewall, and then this order propagates into the middle of the trench (Segalman et al., 2001). Alternatively, if trench edge roughness is high, BCP cylinders have exhibited fluctuation in domain placement, although this effect is dampened for domains closer to the middle of the trench (Cheng et al., 2006a). Hammond et al. showed that when the trench width is >1 μm, ordering and defect density cease to be a function of trench width, and assembly is ordered at the trench edge and disordered in the middle (Hammond and Kramer, 2006). If the trench width is not perfectly commensurate with the BCP period due to width fluctuation, the elastic nature of the BCP will adapt (Ilievski and Ross, 2010). This is useful when the patterning approach is less precise or there exists defects in the trench sidewall. When the BCP elastically adjusts, the distance between the microdomains does not adjust uniformly across the trench. For example, Cheng et al. found that microdomain spacing is similar to L_0 in the middle of the trench but less than L_0 near the trench edge due to a minority-block wetting layer at the trench sidewall (Cheng et al., 2003).

Other mechanisms also affect the templating and ordering of the BCP within the trench. Many reports discuss the kinetic flow of BCP from the mesa into the trench, and therefore the metastable ordering of the BCP perpendicular to the trench direction (Gu et al., 2012; Sundrani et al., 2004). The orientation, either parallel or perpendicular, of cylinder-forming BCP can be controlled specifically by tuning the solvent-annealing conditions and trench width (Jung and Ross, 2007). Further annealing after flow into the trench is required for rotation away from perpendicular to equilibrium parallel orientation. Beyond the flow kinetics, the height of the mesa is important. If the mesa height is less than the BCP thin film thickness, the influence of the template is reduced (Segalman et al., 2001). On the other hand, mesas that were taller than a few L_0 had a consistently high impact.

In conclusion, IL will continue to be useful for fabrication of graphoepitaxial templates for BCP DSA. IL can produce templates with sub-50 nm resolution (Switkes et al., 2000) that can keep up with the ever-shrinking natural pitches of BCPs. These templates can be fabricated and engineered for research and prototyping environments, allowing for rapid experimentation and innovation. Certain techniques have been employed to fabricate IL patterns over cm^2 areas (Bruccoleri et al., 2012; Mukherjee et al., 2010; Ahn et al., 2008b). Beyond the research lab, IL is currently not used by industry for high-volume and resolution semiconductor manufacturing, but the combination of top-down lithography via IL and bottom-up BCP self-assembly may translate into industrial fabrication in the future. For this adoption to take place, many limitations with IL must be solved, such as wafer-scale patterning and the impact of pattern roughness (Cheng et al., 2003).

8.2.4 Electron-beam lithography

Another method for fabrication of graphoepitaxial templates is electron-beam lithography (EBL). EBL is a direct-write, maskless fabrication method with a higher resolution than other conventional lithography methods. In EBL, templates are fabricated by scanning a focused beam of electrons on a sensitive resist, with the exposing and blanking of the beam controlled by a computer. When an electron beam exposes a positive-tone or negative-tone resist via primary, secondary, and backscattered electrons, exposed areas can be removed or left behind, respectively. Beam size and the resist system contribute to the feature size and resolution of the final pattern. EBL can fabricate almost all desired 2D patterns, as long as the desired template is within the resolution limit of the EBL instrument (Yang and Berggren, 2007; Tavakkoli et al., 2011) and can fabricate a limited number of 3D patterns (Do et al., 2014). Although it is a highly flexible method, it suffers from several issues, such as high cost and low-volume production and it is better suited for prototyping or low-volume production.

Polymethyl methacrylate (PMMA) (Zailer et al., 1996; Vieu et al., 2000; Broers et al., 1996) and hydrogen silsesquioxane (HSQ) (Grigorescu and Hagen, 2009; Henschel et al., 2003; Yang et al., 2009) are the two common resists that are used for fabrication of graphoepitaxial templates due to their high resolution. Positive-tone PMMA is commonly used for the fabrication of trench arrays (Park et al., 2007) and negative-tone HSQ and PMMA are used for fabrication of trench arrays and topographical posts (Tavakkoli et al., 2014; Bita et al., 2008; Yang et al., 2010; Kim et al., 2008a; Cheng et al., 2006b).

EBL can fabricate two different physical templates: sparse posts and trench arrays for uniform, ordered BCP DSA. EBL trenches can be employed for graphoepitaxy with or without transfer of the resist pattern into the substrate (Cheng et al., 2004a, 2006b; Xiao et al., 2005; Hong et al., 2011). All of our previous discussions in the IL and PL sections are also applicable for trench templating fabricated by EBL. The only EBL-specific difference is that the trenches fabricated using EBL can have comparably less line-edge roughness, facilitating precision templating for investigation of BCP accommodation of incommensurability and smaller length BCPs (Cheng et al., 2004a; Hosaka et al., 2014).

In recent graphoepitaxial templating studies, EBL sparse post arrays were very effective for domain ordering (Tavakkoli et al., 2012a,b; Bita et al., 2008; Yang et al., 2010; Chang et al., 2012, 2014). Sparse posts can speed up the final nanopatterning throughput by a factor of 30 or more because only the posts need to be exposed, letting the BCP fill in the rest of the pattern. Also, sparse posts can locally control the orientation and morphologies of the self-assembled BCP because they are smaller than the BCP domain (Tavakkoli et al., 2012b; Yang et al., 2010; Chang et al., 2012, 2014). This sub-domain control is difficult or impossible by other lithographic patterns such as periodic structures.

Figure 8.4 shows resulting ordered spherical and cylindrical BCP patterns using sparse post templates. In an early development, Bita et al. achieved long-range order of spherical microdomains (Figure 8.4a) with 10-nm diameter HSQ posts functionalized

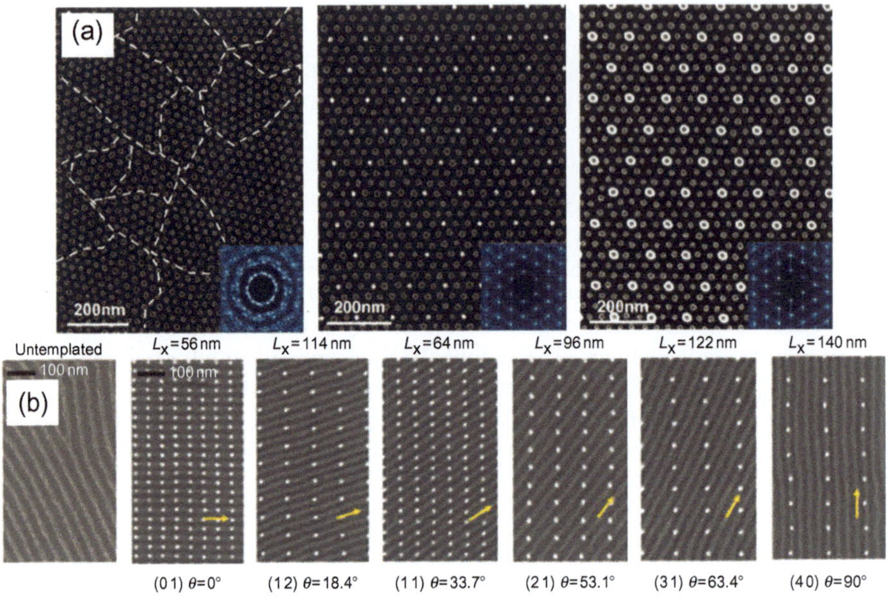

Figure 8.4 SEMs of ordered microdomains using EBL-fabricated sparse post templates. (a) Spherical (Bita et al., 2008) and (b) cylindrical morphologies (Yang et al., 2010) ordered using functionalized posts. The first image in both rows is without posts.
From Bita et al. (2008). Reprinted with permission from AAAS. Reprinted by permission from Macmillan Publishers Ltd: Yang et al. (2010), copyright 2010.

with a minority-block brush (Bita et al., 2008). The period of the posts was commensurate with the BCP period, allowing for high spatial frequency BCP arrays using a lower spatial frequency template. They supported their results with free-energy modeling, and later modeling suggested improvements for more sparse templating (Tang and Ma, 2010). Also, Yang et al. used a similar method to order cylindrical microdomains of PS-*b*-PDMS BCP (Figure 8.4b) (Yang et al., 2010). They showed that the commensurability of the post spacing and BCP period controls the orientation of lines. In later sections, we will discuss more about using EBL templating for the fabrication of complex and three-dimensional structures.

For sparse post templating, size (Chang et al., 2012), height (Chang et al., 2012), shape (dot or dash) (Yang et al., 2010), periodicity, and distribution (Chang et al., 2014; Hannon et al., 2014) of the topographic features are important. Control over order is optimized when the diameter of the posts is the same or smaller than the diameter of microdomains, minimizing microdomain distortion (Chang et al., 2012). Ordering is also optimized when the post height is near the height of the microdomains; short posts are ineffective for post templating (Chang et al., 2012). Sub-10 nm diameter and 18 nm pitch is the smallest sizes of EBL features that have been employed for BCP templating (Chang et al., 2012; Yang and Berggren, 2007).

The mechanisms for post templating are typically based on the confinement effect and commensurability between the periods of the posts and BCP natural period, rather than lateral ordering. Furthermore, post templating mechanisms were strongly dependent on the use of polymeric brushes, which were preferential to one of the blocks. One of the advantages of using HSQ and PMMA is that hydroxyl-terminated molecules or polymers can be grafted to the fabricated posts (Tavakkoli et al., 2012a, 2014). When posts are brushed with a block of polymer, the posts act as surrogates of that block of the polymer domain (Tavakkoli et al., 2012a,b, 2014; Bita et al., 2008; Yang et al., 2010; Son et al., 2011a).

EBL is a precise method for the fabrication of template for the BCP self-assembly in comparison to other methods discussed in this chapter. In order to decrease the template fabrication time, parallel beams can be used, but investigation of high throughput and more economical templating methods is needed. Nonetheless, EBL will be a good method for studying DSA processes and rapid prototyping of templates.

8.2.5 Nanoimprint lithography

NIL is a high-throughput, parallel, top-down nanofabrication method that has been used for graphoepitaxy of BCP DSA. While NIL is a more recent form of fabrication compared to other traditional lithography methods, the method's rapid adoption into the self-assembly community has allowed it to be applied widely.

NIL makes use of a master mold that is imprinted into a thin layer of a resist on a receiving substrate. Typically, the mold is pressed into the resist with physical pressure, and thereby produces an inverse of the mold pattern on the substrate. In many cases, elevated temperatures or UV exposure are combined with the pressure treatment to enhance pattern fidelity or retention (Schift, 2008; Guo, 2007; Wagner and Jayatissa, 2005; Chou et al., 1996; Zankovych et al., 2001). In the case of BCP graphoepitaxy, a BCP film, instead of a resist, can be imprinted. The BCP self-assembles in an ordered fashion as a result of the proximal mold features, which act as a directing template during annealing. After the imprint process, the mold is removed (demolding) to leave behind the assembled BCP nanopattern.

NIL templating of BCP self-assembly has been used in many different ways since it was first adopted in 2004 (Li and Huck, 2004). Use of a hard mold has been most popular, following in the footsteps of traditional NIL research. A master mold is fabricated from a crystalline material such as Si (Voet et al., 2011; Man et al., 2011; Simão et al., 2014), InP (Li and Huck, 2004), or silica (Jung et al., 2008a; Yang et al., 2014). Alternatively, soft molding, in which the mold is typically polymeric (Li and Yokoyama, 2005; Jeong et al., 2012), and other imprinting methods have also been shown (Yang et al., 2014). Soft molding for NIL uses either a polymeric daughter mold (Li and Yokoyama, 2005; Jeong et al., 2012) or a prepatterned substrate before BCP self-assembly (Yang et al., 2014; Borah et al., 2013b; Salaun et al., 2013; Kim et al., 2012). Templating via prepatterning is similar to graphoepitaxy, which was discussed in previous sections. The benefits of prepatterning include relatively fast NIL and decoupling of the imprint factors from the BCP annealing.

NIL, as a fabrication technique and also as a templating method, has several benefits and limitations. Generally, NIL is most appropriate for BCP templating because it is rapid, able to produce nanofeatures smaller than the diffraction limit, high-throughput, and large-area in nature (Park et al., 2011; Simão et al., 2014). NIL templates the BCP directly without needing any additional patterning process and has been shown to be highly repeatable with durable molds. As a result of these benefits, NIL may be appropriate for directing the self-assembly of BCPs at an industrial scale. In spite of the benefits, NIL has presented some significant limitations, especially for BCP graphoepitaxy. Many recent publications have shown innovative ways to overcome these challenges, as discussed below.

The chemistry on the NIL mold surface must be considered for two reasons: (1) postannealing removal of the mold and (2) promotion of chain-mobility while annealing. Stiction and high adhesion energy can pull up the BCP during mold release and have been the most prevalent contributors to poor final pattern results (Park et al., 2011; Voet et al., 2011). Therefore, attempts to modify the mold-BCP interface with a release chemistry have shown significant success. Voet et al. and Salaün et al. were able to incorporate a fluroalkyl-homopolymer into the BCP film to reduce the adhesion energy (Voet et al., 2011; Salaun et al., 2013). Park et al. improved on this approach by coating the mold with a PDMS brush that reduced both adhesion energy and sticking friction (Park et al., 2011).

Another implicit challenge to NIL is that portions of the imprinted area are left unpatterned, typically in the mesa areas of the mold. As a result, the substrate is not completely covered with a BCP nanopattern. This effect can be mitigated with a second BCP self-assembly, as has been reported in other research (Moon et al., 2012), though this has yet to be shown for NIL.

The reflow of the BCP has been an issue when the templating features—either the NIL mold or prepatterned substrate—do not inhibit polymer flow during thermal annealing. Salaün et al. first imprinted the BCP film into a mesa/trench form, followed by immediate mold release, and then performed a mold-free thermal anneal (Salaun et al., 2013). In this approach, the imprinted pattern reflowed into an undulating-thickness BCP film, and therefore the graphoepitaxy templating effect was lost.

Last, because many NIL processes require high temperatures or high pressures, the BCP can be damaged during NIL or thermal annealing (Kim et al., 2012). To prevent damage, low-pressure processing or a prepatterned templating can be employed. Furthermore, soft molding masks must be able to withstand heating if thermal annealing is employed (Yang et al., 2014). Solvent annealing of the BCP (Simão et al., 2014; Yang et al., 2014; Salaun et al., 2013) was used. In this case, in order to avoid template swelling, the solubility of the solvent must not match that of the NIL mold or substrate prepattern (Borah et al., 2013b; Salaun et al., 2013).

Graphoepitaxial templating of BCP self-assembly via NIL has adopted a wide variety of approaches, as discussed above, and, therefore, the mechanisms and issues of BCP direction are similarly diverse. Figure 8.5 surveys reports of NIL-directed BCP patterning. As shown in Figure 8.5a, Li et al. provided the first simple and foundational NIL-based BCP graphoepitaxy (Li and Huck, 2004). In this research, the PS-*b*-PMMA BCP was annealed during NIL to form an ordered two-row array of

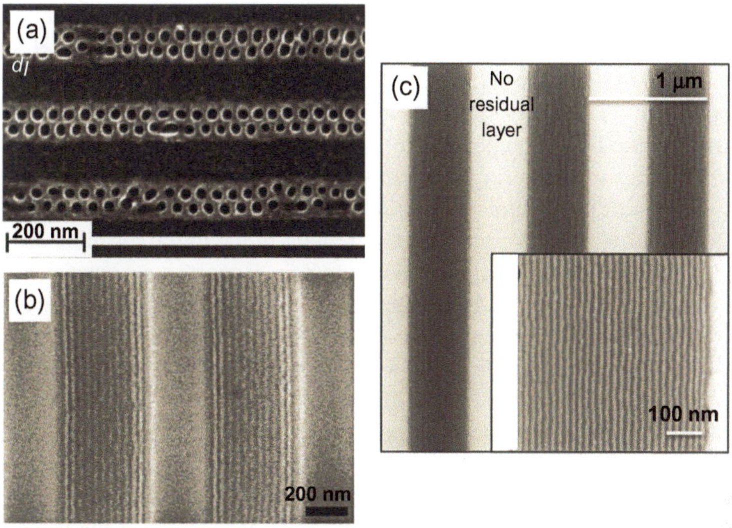

Figure 8.5 SEMs of directed BCP self-assembly with nanoimprint-lithography-defined templates. (a) Out-of-plane cylinders were templated into a two-row line defined by a NIL trenches (Li and Huck, 2004). (b) In-plane cylinders of PS-*b*-PMMA were templated by sol–gel ZnO-based trenches fabricated by NIL (Kim et al., 2012). (c, inset) In-plane cylinders of PS-*b*-PDMS were templated by NIL and successfully released with the aid of PDMS brush on the templating mold (Park et al., 2011).
Reprinted (adapted) with permission from Li and Huck (2004). Copyright 2004, American Chemical Society. Reprinted (adapted) with permission from Kim et al. (2012). Copyright 2012 Wiley Materials. Reprinted (adapted) with permission from Park et al. (2011). Copyright 2011 American Chemical Society.

out-of-plane cylinders. Figure 8.5b shows annealed in-plane PS-*b*-PMMA cylinders in ZnO trenches fabricated by soft molding of a ZnO sol-gel (Kim et al., 2012). Figure 8.5c and inset show a prime example of the impact of incorporating a release layer into the BCP layer, wherein the mold was released without any damage to the patterned BCP film (Park et al., 2011). The resulting ordered cylinders were some of the most successful examples of NIL-based BCP graphoepitaxy. Different morphologies of BCP have been used in NIL research: both in- and out-of-plane cylinders (Park et al., 2011; Voet et al., 2011; Yang et al., 2014; Li and Yokoyama, 2005; Borah et al., 2013b; Kim et al., 2012; Salaün et al., 2011) and out-of-plane lamellae (Man et al., 2011; Salaün et al., 2011; Thébault et al., 2012; Kim et al., 2008b).

Control of the contours of the mold and imprinted polymer film were important for obtaining quality self-assembly. NIL imprint patterns have a particular depth and pitch, and the thickness of the polymer film conforms to fill the mold profile. Protruding mold features, instead of recessed features, resulted in better BCP layer conformation to the mold profile with a thinner residual layer (Park et al., 2011). One of the ways to circumvent dependence on mold profile is using reverse NIL (Tavakkoli et al., 2012c). A BCP residual layer was generally found to adversely affect templating, and in some cases the natural order of the residual layer propagated into the templated region (Park et al., 2011; Li and Huck, 2004; Man et al., 2011; Simão et al., 2014). On the other hand, as discussed later, reports showed specific templating of the residual layer, further highlighting the importance of understanding and controlling the BCP film geometries.

Similar to other methods for lithographic templating, commensuration of the NIL template width with the natural pitch, L_0, of the BCP was necessary for highly ordered graphoepitaxy. Li et al. suggested this requirement in the initial example of NIL templating (Li and Huck, 2004) as shown in Figure 8.5a. Reports have consistently shown that defectivity increases with incommensurate trench widths. Furthermore, similar to the discussion with IL trenches, if the trench width is too large, commensuration becomes less important and ordering in the middle of the trench is reduced (Man et al., 2011). While most reports show ordering of the nanodomains parallel to the axis of the trench patterns, if the trench wall is not sharply orthogonal to the substrate surface, the ordering changes. Li et al. showed that an NIL mold of sinusoidal undulations templated in-plane cylinder domains to orient perpendicular to the undulations (Li and Yokoyama, 2005). Beyond trench templating, Yang et al. showed the use of an NIL prepatterned hexagonal array of posts or holes to template the ordered alignment of PS-*b*-PMMA BCP. The BCP was solvent annealed for low-defect arrays of hexagonally packed out-of-plane cylinders over square-mm areas (Yang et al., 2014).

While commensuration is important for domain alignment, many reports of flow-based ordering showed stronger, though possibly less controlled, ordering. Here, we highlight some of the research that investigated the kinetic nature of BCP alignment. While the flow phenomenon is not unique to NIL templating, the kinetic nature of the molding process seems to increase the impact of flow for these systems. Many papers reported that a short molding step resulted in cylinder domains oriented perpendicular to NIL trench axis (Park et al., 2011; Voet et al., 2011; Kim et al., 2012; Thébault et al., 2012), in contrast to most other graphoepitaxy reports that showed orientation parallel to the trench axis. Perpendicular orientation was due to the flow of the BCP from the mesa regions into the trench regions, though parallel orientation was nonetheless achieved with further annealing as the final, lowest energy, equilibrium orientation. The shear flow of polymer into the trenches was controlled by NIL pressure, BCP thickness, BCP viscosity, proximity to an open end of a trench pattern, and width of the template mesa (Park et al., 2011; Thébault et al., 2012). Furthermore, other interesting kinetic-induced orientations have been reported. Kim et al. discovered a transient effect where a preannealed in-plane alignment of lamellae layers at the top surface of a thick BCP layer was maintained during an NIL molding step (Kim et al., 2008b). In this case, the surface lamellae layers conformed to the trench/mesa template pattern, instead of reannealing in the presence of the confining trench. In a different approach, Simão et al. showed the preferential out-of-plane orientation of cylinder microdomains in the residual layer underneath the mesa template features (Simão et al., 2014). They suggested that the BCP film flowed out from underneath the mesa template features, therefore inducing in-plane alignment of polymer chains and thus out-of-plane cylinders. In general, an alignment effect was achieved over large areas for all the flow-based ordering approaches, but ordering was not well controlled.

The chemistry of the substrate and mold/template surfaces is important for adequate ordering and direction. Due to the intimate contact of the BCP with the mold/template, the surface chemistry was required to promote polymer mobility and self-assembly (Borah et al., 2013b). Salaün et al. reported that a simple surface energy explanation was not enough to describe the impact of the surface chemistry on self assembly, suggesting consideration of the polymer and surface molecule interaction. Inadequate polymer mobility at the surfaces can impede ordering around the template features. Furthermore, the chemical preferentiality of the surface can determine the orientation of the BCP microdomains. For in-plane cylinders, preferentiality of the mold surfaces was beneficial because it allowed for polymer mobility (Salaun et al., 2013). Alternately, neutral preferentiality induced out-of-plane orientation of cylinders if the film was thick enough (Yang et al., 2014; Kim et al., 2012), or allowed in-plane orientation if the film was $N \times L_0$ thick (where N is an integer). For out-of-plane lamellae, the trench top and substrate needed neutral preference, though the trench sidewalls could be preferential to one of the blocks (Man et al., 2011).

NIL DSA of BCPs is a promising, high-throughput, rapid approach and has seen significant improvements since the first implementation in 2004. In order to facilitate high-quality orientation and reuse of master molds, more understanding of the mold treatment, design, and material must be obtained (Salaün et al., 2011). As a result of some limitations, the BCP order and nanopattern quality is not as high as other forms of graphoepitaxy, though quantitative analysis of key pattern parameters may be the topic of future research. Many reports show orientation results based on polymeric flow, though this approach requires more research for higher reproducibility and understanding. To mitigate some of the challenges with polymer flow, we suggest that future work may include engineering polymer drop placement (Ahn et al., 2008; Onses et al., 2013), roll-to-roll patterning (Ahn and Guo, 2008; Guo and Ahn, 2011; Stuart and Chen, 2009), and variable depth mask profiles (Salaün et al., 2011).

8.2.6 Self-forming

Another method for fabrication of graphoepitaxial templates for BCP self-assembly is the "self-forming" method. There are several self-forming processes, but three of them are more commonly used for fabrication of physical templates. In the next sections we discuss these three self-forming methods: BCP, surface reconstruction, and porous materials.

8.2.6.1 Self-forming: BCP

One of the self-forming methods for directed BCP self-assembly is the use of a BCP-nanopatterned template (Jung et al., 2010; Son et al., 2011b). Although this method is simple and inexpensive, ordered BCP templates cannot be achieved without other ordering techniques. It requires primary ordering for the first BCP layer. In this method, a single microdomain layer is first assembled and then selectively etched to leave only one of the blocks, therefore providing a topographical template for the directed self-assembly of the second BCP.

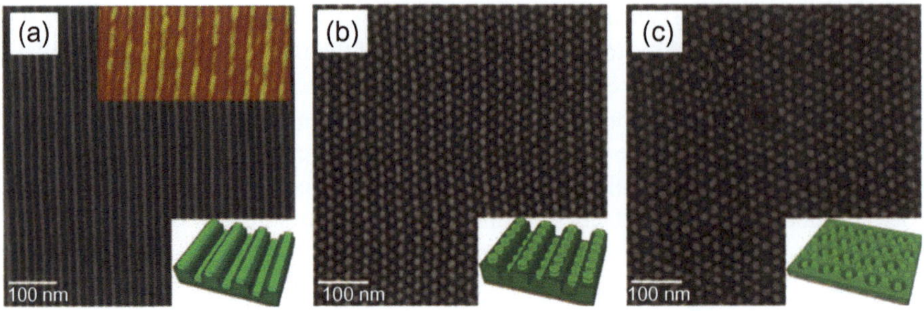

Figure 8.6 SEMs and cartoons of BCP as a template for another BCP layer with lower molecular weight. (a) Line-on-line, (b) dot-on-line, (c) dot-in-perforated lamellar patterns (Son et al., 2011b).

Self-assembled BCP patterns as self-forming templates can be used to increase pattern density Jung et al. demonstrated 45 kg/mol PS-*b*-PDMS BCP patterns as templates to direct another BCP layer with smaller, 16 kg/mol molecular weight, cylindrical PS-*b*-PDMS BCP (Jung et al., 2010). Son et al. reported the same results and also templated cylindrical 12 kg/mol PS-*b*-PDMS BCP on top of a cylindrical 45 kg/mol PS-*b*-PDMS BCP template (Son et al., 2011b).

The other advantage of the BCP self-forming method is fabrication of hierarchical, multilevel structures such as line-on-line patterns (as mentioned above), dot-on-line, and dot-on-perforated lamellar patterns (Figure 8.6). Son et al. showed the templating of 16 kg/mol spheres by 45 kg/mol in-plane cylindrical templates, in which one row of spheres assembled in the trench between the high-molecular-weight cylinders and one row of spheres formed on the templating cylinders (Son et al., 2011b). They also templated 12 kg/mol spheres by 45 kg/mol cylindrical templates. They achieved one row of spheres in the trench and alternating one or two rows of spheres on the templating cylinders. For the case of dot-on-perforated lamella structures, they showed 16 kg/mol spheres in a 45 kg/mol perforated lamellar pattern, with one sphere inside the perforation and six spheres on the adjacent lamellar mesa.

Commensurability and duty cycle are the two important parameters controlling this hierarchical method of templating. About 10% fluctuations in commensurability between the first and second BCP periods can be accommodated. Also, by increasing the duty cycle of the high-molecular-weight line template to closer to 50%, defect density of line-on-line patterns can be decreased (Kim et al., 2008a; Son et al., 2011b).

8.2.6.2 Self-forming: surface reconstruction

Another self-forming method for the fabrication of a physical template is surface reconstruction. In this method, faceted surfaces were generated by reconstruction of a single crystalline wafer such as sapphire or silicon. Cutting a sapphire substrate along most crystallographic planes produces an unstable surface. If this surface is annealed at high temperature for 24h, unidirectional ridges are formed on the entire surface of the substrate. Such sapphire templates have two kinds of natural nonuniformities: variation in periodicity and ridge dislocations (Park et al., 2009). Also faceted substrates can be fabricated by etching of patterned and off-cut silicon ⟨111⟩ using KOH. This method, in comparison to the sapphire method, generated defect-free ridge arrays, and the resulting facets showed stronger geometric asymmetry (Hong et al., 2012a). Alternatively, faceted substrates that were fabricated via alternate methods have also shown direct self-assembly (Rockford et al., 1999; Fasolka et al., 1997).

The surface reconstruction method is a simple low-cost technique. Also, it was demonstrated on a polymeric replica, poly(butylene terephthalate) (PBT), and therefore can be inexpensive (Hong et al., 2012b). One of its limitations is that it can only generate simple unidirectionally aligned ridge patterns and has not been used for fabrication of complex templates.

There are a few reports of using the surface reconstruction method to make a template for BCP self-assembly (Park et al., 2009; Hong et al., 2012a,b). So far, faceted substrates have ordered out-of-plane and in-plane cylindrical microdomains (Park et al., 2009; Hong et al., 2012a,b). The surface reconstruction method was only used for achieving single grains of BCP microdomain on a substrate and has not been used for the fabrication of complicated structures. The important parameters for this method are: ridge direction, commensurability between the period of the facets and BCP, and topography of the faceted substrate. DSA using surface reconstruction was used successfully by Park et al. (2009) to fabricate highly ordered, ultradense, out-of-plane cylinders confirmed by grazing-incidence small-angle X-ray spectroscopy over large areas ($4\,cm^2$). They oriented 3-nm-diameter out-of-plane cylinders with a 6.9-nm periodicity. For the application of bit-patterned media (BPM), where magnetic bits of information are stored in small two-dimensional circular features, the resulting areal density was over 10 terabit/in^2 (See Figure 8.7a). Although they used a faceted sapphire substrate with defects, it successfully directed and guided BCP self-assembly with minimal influence from the substrate defects. In their work, the substrate facets did not directly determine the position of the out-of-plane cylinders, and instead simply influenced overall BCP alignment. Overall, four different molecular weights of polystyrene-*block*-poly(ethylene oxide) (PS-*b*-PEO) were templated on faceted substrates. Similar approaches have been taken advantage of and are shown in Figure 8.7b and c, as detailed below.

In-plane cylinders can be perpendicular (Hong et al., 2012b) or parallel (Hong et al., 2012a) to the ridges of the substrate, depending on the BCP film thickness and topography of the faceted substrate. Hong et al. showed that by using the faceted sapphire, in-plane cylinders were perpendicular to the facets for the 38- and 47.5-nm thick BCP films (Hong et al., 2012b). Underlying cylinders spanned the facets and were important for perpendicular direction of the top-layer of cylinders. Modeling suggested that perpendicular orientation to the ridges could be achieved when the periodicity of the ridges was $>1.2L_0$ or $<0.8L_0$ (Hong et al., 2012b). In another report, they showed perpendicular orientation was only achieved in thick BCP films, where otherwise cylinders were parallel to the facets for thin BCP films (Hong et al., 2012a).

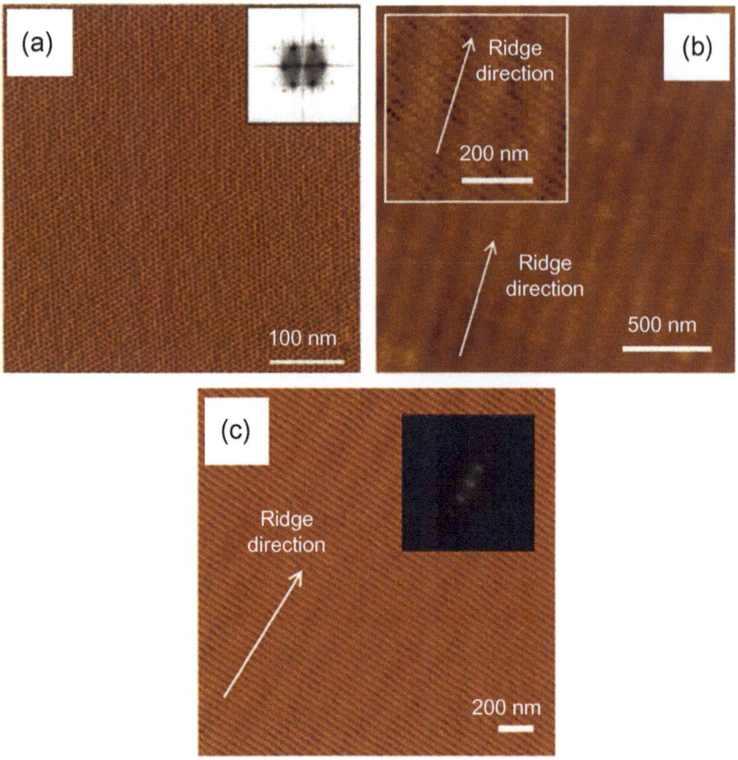

Figure 8.7 AFM results of arrays of out-of-plane and in-plane cylinders of PS-*b*-PEO on faceted sapphire and silicon substrates. Multiple different molecular weights and thin film thicknesses were used in order to template out-of-plane cylinders and parallel and perpendicular in-plane cylinders. (a) Ordered out-of-plane cylinders were obtained on a faceted sapphire substrate with 7.0 kg/mol BCP (Park et al., 2009). (b) Ordered in-plane cylinders were obtained on a faceted silicon substrate with a 26.5 kg/mol BCP, oriented both perpendicular (as shown) and parallel to the ridge direction (Hong et al., 2012a). (c) Ordered in-plane cylinders were obtained on a faceted sapphire substrate with a 26.5 kg/mol BCP, oriented primarily perpendicular to the ridge direction (Hong et al., 2012b).
From Park et al. (2009). Reprinted with permission from AAAS. Hong et al. (2012a), Copyright (c) 2012 John Wiley and Sons.

Alternatively to ridges, the use of wrinkled substrates has directed out-of-plane orientation of lamellae to be perpendicular to the wrinkle direction (Kim et al., 2014).

8.2.6.3 Self-forming: porous materials

The other self-forming method is the use of porous templates of alumina for the BCP self-assembly. Anodic aluminum oxide (AAO) can be fabricated using chemical anodization of an aluminum thin film or foil. Pore size, pore period, and length can be controlled by varying anodization conditions. Pore diameter has been varied from

18 to 400 nm, and thickness of the membrane has been varied from 100 nm to 10 μm. This method has fabricated pores over large areas but the pores were only well ordered over a few micrometers. Aside from self-assembly in AAO pores, other techniques for DSA in pores exist (Stewart-Sloan & Thomas, 2014; Shi & Li, 2013), such as within colloidal crystal voids (Arsenault et al., 2005), thin resist pits (Jung et al., 2008b), and nanobowls (Yu et al., 2011).

This method has facilitated study of BCP self-assembly in two-dimensional confinement. In this method, BCP self-assembly is unconstrained in the direction of the pore axis while it is confined radially by the surface of the pores. Some of the BCP morphologies, such as helices and toroids, have been only achieved using this method of confinement (Moon and McCarthy., 2003; Chu et al., 2014; Dobriyal et al., 2009; Shin, 2004; Wu et al., 2004; Xiang et al., 2005, Yan et al., 2015).

The first step in AAO-based BCP self-assembly is wetting the pore walls. Several methods can be used to wet the surface of the pores with BCP such as the melt method, solution method, or solvent-annealing-induced nanowetting in templates method (Moon and McCarthy, 2003; Xiang et al., 2004; Chu et al., 2014).

Morphologies of the BCP in this confinement system can be controlled using several parameters such as (Chu et al., 2014; Dobriyal et al., 2009; Shin, 2004):

- Commensurability between pore diameter and BCP period. Incommensurability resulted in compression and stretching of BCP chains and, therefore, variations in the morphology and period of the BCP microdomains.
- The interaction between the BCP and pore surface. Depending on which block of a BCP wets the pore surface—majority, minority, or neither—the morphology changed. In the case of lamellae-forming BCP, two morphologies can be observed: stacked disks with a neutral pore wall or concentric cylinders with a preferential pore wall (Xiang et al., 2004).
- The diameter of the cylindrical pores. If the diameter of cylindrical pore is large compared to the BCP period, morphologies similar to those achieved on a planar substrate are observed. If the pore diameter is small and incommensurate compared to the BCP period, unusual morphologies were observed because of the pore curvature (Shin, 2004). For the example case of lamellae-forming BCP, small-diameter pore confinement results in toroids, stacked disks, and wires. For the case of cylinders, small-diameter pore confinement results in spheres and peapods (Xiang et al., 2004; Chu et al., 2014; Dobriyal et al., 2009; Wu et al., 2004; Xiang et al., 2005).

A challenge for this templating method is imaging the DSA results since the template masks the BCP pattern from electron beams. Transmission electron microscopy (TEM) is most popular because it allows observation of the internal structure of the microdomains. In some cases, one of the blocks was selectively removed to enhance imaging contrast (Chu et al., 2014).

Figure 8.8 shows TEM results of poly(ethylene oxide)-*b*-poly(propylene oxide)-*b*-poly(ethylene oxide) BCP structures after confined self-assembly in cylindrical pores. Depending on the pore diameter, different morphologies were achieved such as single helices, single helices with a straight core channel, double helices, core–shell double helices, core–shell triple-helices, and three-layer stacked doughnuts.

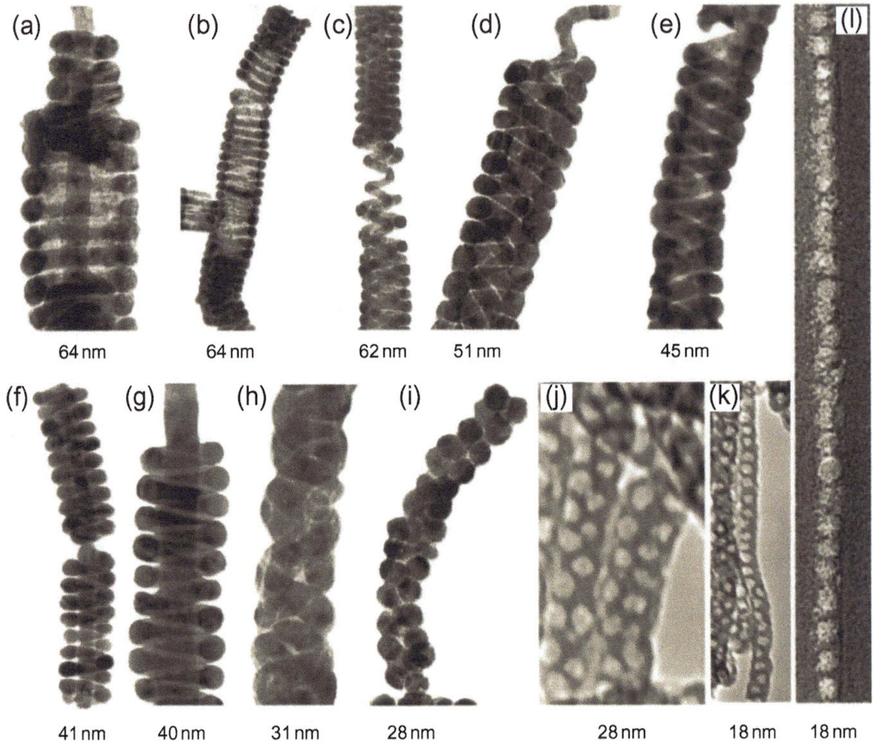

Figure 8.8 TEM images of BCPs confined in cylindrical pores with a variety of diameters Wu et al. (2004).
Reprinted by permission from Macmillan Publishers Ltd: Wu et al. (2004), copyright 2004.

8.3 Aperiodic nanostructures

The templating of complicated patterns with graphoepitaxy differs from the previous reported uniform templating. The above sections discuss ordering of a single BCP morphology, size, and periodicity, in many cases over a large area. They did not discuss complicated patterns and the associated graphoepitaxial parameters. This section delves into research on nonuniform patterns such as placement of particular patterns at specific spots on the substrate, changing the BCP to a geometry or morphology that is different from the natural shape, and the proximal impact of locally varying template patterns. In conclusion, some of the implications and outlooks of complex pattern templating are discussed.

Graphoepitaxy has the ability to locally control the BCP self-assembly in order to direct many aperiodic features such as bends, junctions, and terminations. This work allowed for placement of these features at particular positions on the substrate based on either intuitive or nonintuitive design rules. Both experimental work and theoretical modeling has helped progress this approach from simple trial-and-error to model-based engineering. Experimentally, BCP geometry was matched by EBL posts that were roughly 10 nm diameter and coated with a polymer brush to attract the minority cylindrical block.

Yang et al. showed the first examples of aperiodic structure fabrication using sets of dashed and double templating posts to intuitively direct PDMS cylinders into different complicated patterns. An example SEM of this work, a meandering PDMS cylinder array, is shown in Figure 8.9a, specifically directed by the closely spaced HSQ posts at the bends (Yang et al., 2010). In this approach, PDMS cylinders preferentially aligned parallel to HSQ dashes or double posts, even if this new orientation had a periodicity that was further from the equilibrium spacing, L_0, than other possible orientations. Other complicated patterns included nested *L*s, jogs, and zig-zags. This was an intuitive, solve-forward approach to designing and achieving complicated patterns simply based on the dimensions and spacing of the templating features. To improve upon the above work, self-consistent field theory (SCFT) was carried out by Zhang et al. to explain how elliptical nanoposts, instead of spherical posts, at the ideal height, more effectively templated complicated aperiodic features (Zhang et al., 2014). The elliptical

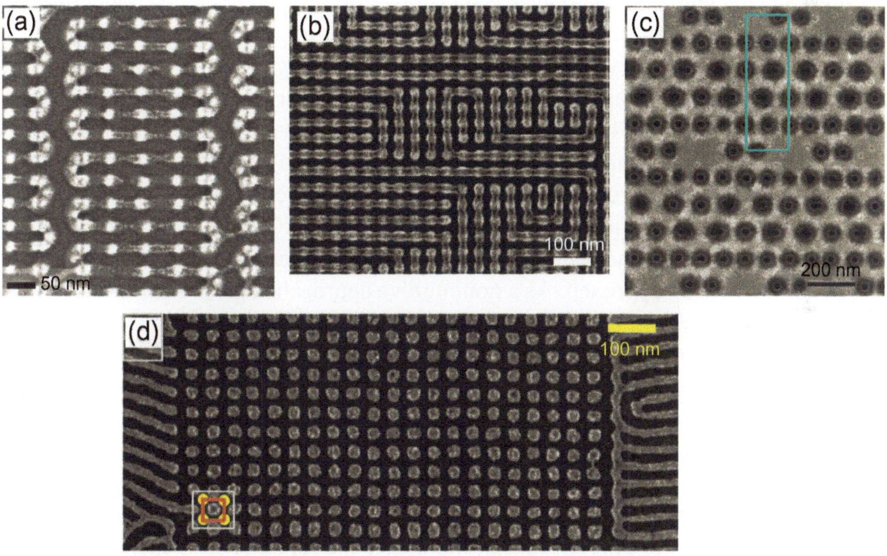

Figure 8.9 SEMs of complicated BCP patterns by graphoepitaxial templating. (a) EBL-fabricated HSQ posts templated meandering PDMS cylinders (Yang et al., 2010). (b) EBL-fabricated HSQ single- and double posts templated the bends and junctions of PS-*b*-PDMS cylinders (Chang et al., 2014). (c) Holes, fabricated with 193i optical lithography, templated the self-assembly of PS-*b*-PMMA cylinders for hole shrinking (Yi et al., 2012). (d) EBL-fabricated PMMA post arrays templated the change in BCP morphology from cylinders (outside the template) to spheres (inside the template), and the posts were subsequently removed (represented by yellow dots connected by red lines) (Tavakkoli et al., 2012b). Reprinted by permission from Macmillan Publishers Ltd: Yang et al. (2010), copyright 2010. Yi et al. (2012), copyright (c) 2012 John Wiley & Sons. Tavakkoli et al. (2014), copyright (c) 2014 Wiley Materials. Reprinted by permission from Macmillan Publishers Ltd: Chang et al. (2014), copyright 2014.

nanoposts have the ability to guide the BCP orientation to achieve the well-aligned nanostructures better than simple posts, and the height of the posts was important for achieving this control. Both the ideal height of the posts and density of elliptical posts showed low defectivity and low line-edge roughness. Furthermore, they determined that posts are not necessary at the apices of bends and T-junctions.

The above approaches focused on creating complicated patterns with intuitive placement of templating posts, but Chang et al. developed a tiling pattern for rule-based templating of pattern bends and junctions (Chang et al., 2014). They used a set of topographic template tiles in a square array at a pitch of L_0. Each 3×3-post tile included 8 single posts and an X- or Y-oriented double-dot post at the center. The alignment of adjacent PDMS cylinders was a function of the orientation of both 2 and 4 adjacent double-dot posts. From this analysis, design rules were determined in order to control the position of bends and terminations without relying on intuition, but instead on statistically derived design rules. An example of well-controlled complicated nanopatterns is shown in Figure 8.9b. While this approach was able to effectively engineer a desired complicated pattern, it was not reduced to the minimum number of features, and the template could have been simpler or more complicated depending on the desired final pattern.

Modeling completed by Hannon et al. has shown progress in this direction with the inverse prediction of the required templating posts for a given desired BCP pattern (Hannon et al., 2013, 2014). A computational algorithm was developed to systematically test different post template arrangements compared to a fidelity function, the difference between target structure density and solution structure density. This allowed for variation of post size and post arrangement, different BCP fractions, minimization of the number of posts, and removal of posts that were spaced closer than the resolution of the fabrication approach. As a nonintuitive approach, a given solution was only one of a set of possible solutions.

Irregularly shaped and spaced templating holes have been filled with BCP to shrink the template hole size for contact hole fabrication (Yi et al., 2012 , 2013a,b; Bao et al., 2011). In all reports, a BCP of poly(styrene-*b*-methyl methacrylate) (PS-*b*-PMMA) was used for shrinking of template holes that were patterned by 193 nm immersion optical lithography. Out-of-plane PMMA cylinders assembled to effectively reduce the diameter of the template hole and, in some cases, multiply the number of contact holes per template hole. This approach is similar to the self-assembly in nanopores, as discussed in the previous section. Figure 8.9c shows a prime example of the achieved hole shrinking, in which the contact holes for a 22 nm SRAM design cell were fabricated (Yi et al., 2012). This sublithographic patterning approach enabled extension of optical lithography beyond its limits and self-assembly of the natural geometry of the BCP but for nonperiodic features. There were many geometric benefits of this approach: one to four nanoholes could be produced per templating hole at a resolution much smaller than the optical lithography, the standard deviation of the nanohole registration was as small as 1 nm, and the size of the contact hole could be precisely controlled (Yi et al., 2012). In the end, these factors were mainly dependent on template hole size, BCP film thickness, and templating hole density. The strong boundary confinement allowed for stabilization of either a single contact hole in a single guiding template, or multiple closely spaced contact holes in larger templates. The latter approach reduced template resolution requirements but also reduced contact hole position control (Yi et al., 2012). SCFT modeling replicated the experimentally achieved results, though suggested that, in order for the contact holes, that is, the PMMA cylinders, to reach the substrate, certain conditions were required. The substrate needed to be preferential to the majority block or chemically neutral, the template wall needed to be preferential to the minority block, and the BCP thickness needed to be thinner than the template hole depth for adequate filling (Yi et al., 2013a). To showcase the above results, contact hole designs for SRAM, half-adder, NAND, and DRAM cells with stringent resolution requirements were fabricated with the BCP hole shrinking method (Yi et al., 2012; Bao et al., 2011). Even higher resolution and smaller holes were finally achieved with EBL-fabricated templates and a lower molecular weight BCP (Yi et al., 2013a).

Spherical diblock copolymers do not typically form square-packed arrays of features, though square arrays of templating posts have experimentally and theoretically stabilized a square packing of BCP features. While square arrays of BCP are not naturally forming, Feng et al. developed an SCFT model in which a square arrangement was stabilized with the appropriate packing and spacing of a post templating array (Feng et al., 2011). The incommensurate energy from the square template array, specifically at a pitch of 1.61 and $1.94 \times L_0$, was important to induce this geometric arrangement. As shown in Figure 8.9d, square packing PS-*b*-PDMS BCP features was achieved with post template arrays (Tavakkoli et al., 2012b, 2014) when the arrays of majority-block-functionalized posts were patterned in a rectangular arrangement with varying pitches, holes and spheres of PDMS were fabricated in square and rectangular arrays. The stabilization of this geometric rearrangement was replicated with SCFT modeling.

While most BCP DSA research only uses a single morphology for a single substrate, it is possible to change the stable self-assembled morphology in different parts of a substrate via physical templating (Park et al., 2012; Tavakkoli et al., 2012b, 2014). Tavakkoli et al. produced many different morphologies from naturally cylinder-forming PS-*b*-PDMS BCP (45.5 kg/mol and natural L_0 of 35 nm) with the use of EBL-fabricated post template arrays (Tavakkoli et al., 2012b). The majority-brush-coated templating posts were patterned in HSQ or negative-tone PMMA, and the PMMA templates were removed with oxygen reactive-ion etching after annealing. While the morphology of BCPs has been controlled with other processing conditions (Son et al., 2011a; Orilall and Wiesner, 2011; Bosworth et al., 2009; Jung and Ross, 2009; Park et al., 2006; Gotrik et al., 2012), strain- and physical-template-induced morphology changes can be specifically registered to designed graphoepitaxial templates. The induced morphologies were directed cylinders, undulating cylinders, spheres, ellipsoids, periodic superstructure, perforated lamellae, and perforated lamellae with generated holes. The control over morphology was because of repulsion of the PDMS minority block from the majority-block-functionalized posts and incommensurate periodicity of the posts. Furthermore, at larger post periods, natural BCP patterns could be controlled with commensuration effects. All of the varying morphologies and templating phenomenon were simulated and supported with SCFT. As another example of graphoepitaxial change of morphology, Park et al. made used of a sinusoidal undulating template to spatially vary the BCP film thickness and therefore the BCP morphology (Park et al., 2012). For thinner films at the peaks of the template, in-plane cylinders and perforated lamellae was formed. For thicker films in the valleys, in-plane and out-of-plane lamellae were formed.

Graphoepitaxial templating has been able to change the arrangement of BCP features in other ways. Jung et al. showed the wrapping of PDMS cylinders into circles inside circular physical templates and the reduction of normalized cylinder spacing when the BCP radius of curvature was below two L_0 (Jung et al., 2008b). Chang et al. reported that the change in templating feature size could controllably change the BCP feature size (Chang et al., 2010), and Doerk et al. showed that large masking features could be used to rectify the periodic BCP pattern, thus customizing the 2D layout (Doerk et al., 2014). Last, Bao et al. reported that strong-confinement physical templating could orient cylinders out-of-plane, as opposed to in-plane, without the typical use of a substrate neutralization layer (Bao et al., 2011). Overall, the ability to control the BCP morphology by physical templating could be useful for applications like integrated circuit and memory fabrication or multifunctional sensor fabrication.

In conclusion, the local shape, size, orientation, and geometry of BCP self-assembly have been influenced by graphoepitaxy. Based on the current trend in research, the control over complex ordering with physical templates will continue to improve with future research. The discussed research has shown that physical templates allow for direct, local control over self-assembly, inducing aperiodic patterns and shapes. Graphoepitaxy may be the route toward understanding local control of proximal microdomains and possibly individual molecules. Furthermore, as suggested by Chang et al. and Yi et al., design rules may be developed to allow for reverse engineering of desired patterns into the necessary templates (Yang et al., 2010; Yi et al., 2012). One significant challenge in the future will be scaling up the typically nonuniform or serial template fabrication methods into manufacturable processes.

8.4 Multilayer structures

Most of the discussion on graphoepitaxial templating in the previous sections was limited to controlling the morphology and pattern of a single layer of BCP. The following section explains the fabrication and control of the BCP microdomains in multilayers (Ross et al., 2014). Multilayer structures are interesting for application in cross-point memory, metamaterials, photovoltaics, and interconnects. The following section highlights the physical templating mechanisms on BCP microdomains in multilayers.

Multilayer structures can be fabricated in several ways. Jeong et al. formed multilayer structures by sequential patterning (Jeong et al., 2012) with a combination of nanotransfer printing and BCP self-assembly. Before transferring, each layer of BCP was self-assembled by solvent annealing and directed by topographic trenches of molded PDMS. They sequentially transferred two printed ordered BCP films onto a silicon substrate for a final nanostructure of a bilayer of cross-linked SiO_x nanowires with different in-plane alignment angles. While this method was able to effectively stack aligned nanowire arrays, alignment between top and bottom layers was difficult.

Tavakkoli et al. showed generation of multilayer structures in a single step rather than sequential patterning by using rectangular arrays of EBL-fabricated posts (Tavakkoli et al., 2012a). A majority-block brush was coated on the topographical posts, rendering them attractive to the majority block and repulsive to the minority block of cylindrical microdomains. They showed by both SCFT simulation and experiment that the simple arrays of posts could control bilayers of BCP cylindrical microdomains. A range of morphologies and patterns were achieved depending on the commensurability of the templates with the BCP in the x-, y-, or diagonal directions. Such patterns include aligned cylinders, perpendicular and angled mesh, cylinders on ellipsoids, cylinders on spheres, cylinders on perforated lamellae, and periodic superstructure. Figure 8.10a shows an example of a bilayer film templated by a rectangular array of posts, resulting in mesh-shaped structures. Also, they showed that by simply changing the post periods or arrangement, each layer of cylindrical patterns could be controlled locally, independently, and simultaneously with bends and junctions. This method has the potential to make multilevel interconnects or intralevel contacts.

In order to better understand the above results, Gotrik et al. analyzed BCP bilayers with TEM tomography (Gotrik et al., 2014). They posited that top-down SEM and TEM could not characterize and visualize the morphology and connectivity of multilayer structures. Three-dimensional TEM tomography could characterize and image the interior of 3D structures without etching, which was otherwise necessary for SEM imaging. This method was straightforward and, unlike grazing-incidence X-ray scattering, did not require conversion of scattering data into the structure profiles. Figure 8.10b shows an example of 3D visualization of the mesh-shaped structure from TEM tomography corresponding to the similar results in Figure 8.10a (Gotrik et al., 2014). TEM tomography showed that there was a connection between top- and bottom-layer cylinders. Also top-layer cylinders, in comparison to bottom-layer

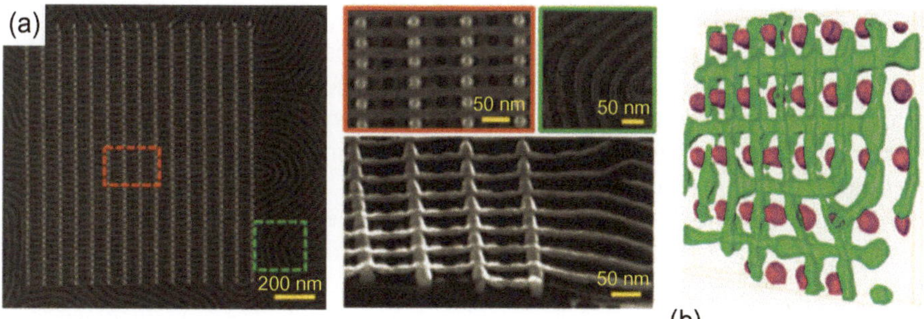

Figure 8.10 (a) SEM images of bilayer mesh-shaped structures from EBL post templating (Tavakkoli et al., 2012a). (b) 3D visualization of the mesh-shaped structure using TEM tomography (Gotrik et al., 2014).
From Tavakkoli et al. (2012a). Reprinted with permission from AAAS. Gotrik et al. (2014), copyright (c) 2014 Wiley Materials.

cylinders, were thinner and their cross-sections were rounder. They showed by SCFT simulation that the top and bottom surfaces, during the annealing process, were neutral to either block.

One of the limitations in the above reports is that the posts were incorporated in to the final BCP nanopatterns, as shown in Figure 8.10a. In order to avoid the incorporation of the post templates into the final multilayer results, Tavakkoli et al. developed sacrificial-post templating method (Tavakkoli et al., 2013, 2014). They used PMMA posts, instead of HSQ posts, which were removed during the postannealing etch of the PS matrix. Sacrificial templating with PMMA relieved the potential pattern transfer problem as a result of dissimilar etch rates between a physical template and the final BCP pattern, as was similarly the case for single-layer templated self-assembly (Tavakkoli et al., 2014).

Chuang et al. studied multilayers of spherical BCP in a V-groove silicon template (Chuang et al., 2006). The poly(styrene-*block*-ferrocenyl dimethylsilane) (PS-*b*-PFS) BCP typically formed a BCC sphere array, but the template forced an FCC structure because of topographical constraint. In other words, they achieved square symmetry on the top surface, parallel to the substrate, unlike the hexagonal symmetry typically observed in a BCP monolayer. The size of the spheres depended on the commensurability between the BCP period and the film thickness in the groove.

There is a need for more complicated templates to control interlayer connections. More research on characterization and metrology of multilayer patterns will overcome the insufficiencies of top-down SEM and TEM.

8.5 Conclusion

In this chapter, we showed that by using topographical templates like lines, dots, and other pattern motifs, BCP self-assembly could be controlled. Fabrication of these physical templates has been accomplished with different lithography methods such as PL, IL, EBL, NIL, and also self-forming methods such as BCP, surface reconstruction, and porous materials. As discussed, commensurability, lateral ordering, and confinement effects play key roles in graphoepitaxy. As a result, the templating method is capable of large-area ordering and alignment of cylindrical, spherical, lamellar, and other morphologies. Furthermore, a wide range of complex patterns, such as bends, junctions, and helices, can be fabricated. In the third dimension, graphoepitaxy is also capable of controlling multilayer self-assembly.

Although there has been significant progress in the area of graphoepitaxial templating during the last decade, current research is still far from achieving the industrial requirements of defect-free mass production. In order to meet this demand, graphoepitaxy will need to employ rapid or complex templating methods such as NIL, advanced lithography, tip-based techniques (Tseng, 2011; Chai et al., 2010; Tseng et al., 2008; Wang and Zou, 2011), and roll-to-roll processes (Ahn and Guo, 2008; Guo and Ahn, 2011; Stuart and Chen, 2009). Furthermore, advancing inverse design methods will help realize better-optimized templates for the fabrication of complicated structures.

Acknowledgments

The authors would like to acknowledge Caroline A. Ross for useful discussion and advising. This material is based upon work supported by the National Science Foundation under Grant Nos. 1234169 and 1246740. Authors thank Taiwan Semiconductor Manufacturing Corporation for financial support.

References

Ahlhorn, T., Pohlmann, H., Kotthaus, J.P., 1989. Holographic generation of gratings with periodicities below 150nm with an excimer laser. In: Excimer Lasers and Applications. Proceedings of SPIE, Vol. 1023.

Ahn, S.H., Guo, L.J., 2008. High-speed roll-to-roll nanoimprint lithography on flexible plastic substrates. Adv. Mater. 20 (11), 2044–2049.

Ahn, M., Heilmann, R.K., Schattenburg, M.L., 2008. Fabrication of 200nm period blazed transmission gratings on silicon-on-insulator wafers. J. Vac. Sci. Technol. B 26 (6), 2179–2182.

Arsenault, A.C., Rider, D.A., Tétreault, N., Chen, J.I.-L., Coombs, N., Ozin, G.A., Manners, I., 2005. Block copolymers under periodic, strong three-dimensional confinement. J. Am. Chem. Soc. 127 (28), 9954–9955.

Bang, J., Jeong, U., Ryu, D.Y., Russell, T.P., Hawker, C.J., 2009. Block copolymer nanolithography: translation of molecular level control to nanoscale patterns. Adv. Mater. 21 (47), 4769–4792.

Bao, X.-Y., Yi, He, Bencher, C., Chang, L.-W., Dai, H., Chen, Y., Chen, P.-T.J., Wong, H.-S.P., 2011. SRAM, NAND, DRAM contact hole patterning using block copolymer directed self-assembly guided by small topographical templates. In: International Electronic Devices Meeting, Vol. 11. pp. 7.7.1–7.7.4.

Bita, I., Yang, J.K.W., Jung, Y.S., Ross, C.A., Thomas, E.L., Berggren, K.K., 2008. Graphoepitaxy of self-assembled block copolymers on Two-dimensional periodic patterned templates. Science 321 (5891), 939–943.

Black, C.T., 2005. Self-aligned self assembly of multi-nanowire silicon field effect transistors. Appl. Phys. Lett. 87 (16), 163116.

Black, C.T., Bezencenet, O., 2004. Nanometer-scale pattern registration and alignment by directed diblock copolymer self-assembly. IEEE Trans. Nanotechnol. 3 (3), 412–415.

Borah, D., Shaw, M.T., Holmes, J.D., Morris, M.A., 2013a. Sub-10nm feature size PS-b-PDMS block copolymer structures fabricated by a microwave-assisted solvothermal process. ACS Appl. Mater. Interfaces 5 (6), 2004–2012.

Borah, D., Simao, C.D., Senthamaraikannan, R., Rasappa, S., Francone, A., Lorret, O., Salaun, M., Kosmala, B., Kehagias, N., Zelsmann, M., Sotomayor-Torres, C.M., Morris, M.A., 2013b. Soft-graphoepitaxy using nanoimprinted polyhedral oligomeric silsesquioxane substrates for the directed self-assembly of PS-b-PDMS. Eur. Polym. J. 49 (11), 3512–3521.

Bosworth, J.K., Black, C.T., Ober, C.K., 2009. Selective area control of self-assembled pattern architecture using a lithographically patternable block copolymer. ACS Nano 3 (7), 1761–1766.

Broers, A.N., Hoole, A.C.F., Ryan, J.M., 1996. Electron beam lithography—resolution limits. Microelectron. Eng. 32 (1–4), 131–142.

Bruccoleri, A., Mukherjee, P., Heilmann, R. K., Yam, J., Schattenburg, M. L., DiPiazza, F., 2012. Fabrication of nanoscale, high throughput, high aspect ratio freestanding gratings. J. Vac. Sci. Technol. B 30 (6), 06FF03-1.

Chai, J., Huo, F., Zheng, Z., Giam, L.R., Shim, W., Mirkin, C.A., 2010. Scanning probe block copolymer lithography. Proc. Natl. Acad. Sci. U. S. A. 107 (47), 20202–20206.

Chang, L.-W., Wong, H.-S.P., 2006. Diblock copolymer directed self-assembly for CMOS device fabrication. In: Design and Process Integration for Microelectronic Manufacturing IV. In: Proceedings of SPIE, Vol. 6156. 615611–615611-6.

Chang, L.-W., Lee, T.L., Wann, C.H., Chang, C.Y., Wong, H.-S.P., 2009. Top-gated FETs/inverters with diblock copolymer self-assembled 20nm contact holes. In: International Electronic Devices Meeting, 2009, pp. 1–4.

Chang, L.-W., Bao, Xinyu, Bencher, C., Philip Wong, H.-S., 2010. Experimental demonstration of aperiodic patterns of directed self-assembly by block copolymer lithography for random logic circuit layout. In: International Electronic Devices Meeting, Vol. 10. pp. 33.2.1–33.2.4.

Chang, J.-B., Son, J.G., Hannon, A.F., Alexander-Katz, A., Ross, C.A., Berggren, K.K., 2012. Aligned Sub-10-nm block copolymer patterns templated by post arrays. ACS Nano 6 (3), 2071–2077.

Chang, J.-B., Choi, H.K., Hannon, A.F., Alexander-Katz, A., Ross, C.A., Berggren, K.K., 2014. Design rules for self-assembled block copolymer patterns using tiled templates. Nat. Commun. 5.

Cheng, J.Y., Ross, C.A., Thomas, E.L., Smith, H.I., Vancso, G.J., 2002. Fabrication of nanostructures with long-range order using block copolymer lithography. Appl. Phys. Lett. 81 (19), 3657.

Cheng, J.Y., Ross, C.A., Thomas, E.L., Smith, H.I., Vancso, G.J., 2003. Templated self-assembly of block copolymers: effect of substrate topography. Adv. Mater. 15 (19), 1599–1602.

Cheng, J.Y., Mayes, A.M., Ross, C.A., 2004a. Nanostructure engineering by templated self-assembly of block copolymers. Nat. Mater. 3 (11), 823–828.

Cheng, J.Y., Jung, W., Ross, C.A., 2004b. Magnetic nanostructures from block copolymer lithography: hysteresis, thermal stability, and magnetoresistance. Phys. Rev. B 70 (6), 064417.

Cheng, J.Y., Zhang, F., Smith, H.I., Vancso, G.J., Ross, C.A., 2006a. Pattern registration between spherical block-copolymer domains and topographical templates. Adv. Mater. 18 (5), 597–601.

Cheng, J.Y., Zhang, F., Chuang, V.P., Mayes, A.M., Ross, C.A., 2006b. Self-assembled one-dimensional nanostructure arrays. Nano Lett. 6 (9), 2099–2103.

Cheng, J.Y., Rettner, C.T., Sanders, D.P., Kim, H.-C., Hinsberg, W.D., 2008. Dense self-assembly on sparse chemical patterns: rectifying and multiplying lithographic patterns using block copolymers. Adv. Mater. 20 (16), 3155–3158.

Cheng, J.Y., Sanders, D.P., Truong, H.D., Harrer, S., Friz, A., Holmes, S., Colburn, M., Hinsberg, W.D., 2010. Simple and versatile methods to integrate directed self-assembly with optical lithography using a polarity-switched photoresist. ACS Nano 4 (8), 4815–4823.

Cheng, H.-H., Keen, I., Yu, A., Chuang, Y.-M., Blakey, I., Jack, K.S., Leeson, M.J., Younkin, T.R., Whittaker, A.K., 2012. EUVL compatible LER solutions using functional block copolymers. In: Alternative Lithographic Technologies IV. Proceedings of SPIE, Vol. 8323, 83231O–83231O-11.

Chou, S.Y., Krauss, P.R., Renstrom, P.J., 1996. Nanoimprint lithography. J. Vac. Sci. Technol. B 14 (6), 4129–4133.

Chu, C.-J., Chung, P.-Y., Chi, M.-H., Kao, Y.-H., Chen, J.-T., 2014. Three-dimensional block copolymer nanostructures by the solvent-annealing-induced wetting in anodic aluminum oxide templates. Macromol. Rapid Commun. 35 (18), 1598–1605.

Chuang, V.P., Cheng, J.Y., Savas, T.A., Ross, C.A., 2006. Three-dimensional self-assembly of spherical block copolymer domains into V-shaped grooves. Nano Lett. 6 (10), 2332–2337.

Chuang, V.P., Gwyther, J., Mickiewicz, R.A., Manners, I., Ross, C.A., 2009. Templated self-assembly of square symmetry arrays from an ABC triblock terpolymer. Nano Lett. 9 (12), 4364–4369.

Do, H.W., Chang, J.-B., Berggren, K.K., 2014. Three-dimensional nanofabrication using hydrogen silsesquioxane/poly(methylmethacrylate) bilayer resists. J. Vac. Sci. Technol. B 32 (6), 06F501.

Dobriyal, P., Xiang, H., Kazuyuki, M., Chen, J.-T., Jinnai, H., Russell, T.P., 2009. Cylindrically confined diblock copolymers. Macromolecules 42 (22), 9082–9088.

Doerk, G.S., Cheng, J.Y., Singh, G., Rettner, C.T., Pitera, J.W., Balakrishnan, S., Arellano, N., Sanders, D.P., 2014. Enabling complex nanoscale pattern customization using directed self-assembly. Nat. Commun. 5, 5805.

Fasolka, M., Harris, D., Mayes, A., Yoon, M., Mochrie, S., 1997. Observed substrate topography-mediated lateral patterning of diblock copolymer films. Phys. Rev. Lett. 79 (16), 3018–3021.

Feng, J., Cavicchi, K.A., Heinz, H., 2011. Control over self-assembly of diblock copolymers on hexagonal and square templates for high area density circuit boards. ACS Nano 5 (12), 9413–9420.

Fucetola, C.P., Korre, H., Berggren, K.K., 2009. Low-cost interference lithography. Journal of Vacuum Science & Technology B: Microelectronics and Nanometer Structures 27 (6), 2958.

Gotrik, K.W., Hannon, A.F., Son, J.G., Keller, B., Alexander-Katz, A., Ross, C.A., 2012. Morphology control in block copolymer films using mixed solvent vapors. ACS Nano 6 (9), 8052–8059.

Gotrik, K.W., Lam, T., Hannon, A.F., Bai, W., Ding, Y., Winterstein, J., Alexander-Katz, A., Liddle, J.A., Ross, C.A., 2014. 3D TEM tomography of templated bilayer films of block copolymers. Adv. Funct. Mater. 24 (48), 7689–7697.

Grigorescu, A.E., Hagen, C.W., 2009. Resists for sub-20-nm electron beam lithography with a focus on HSQ: state of the art. Nanotechnology 20 (29), 292001.

Gu, X., Liu, Z., Gunkel, I., Chourou, S.T., Hong, S.W., Olynick, D.L., Russell, T.P., 2012. High aspect ratio Sub-15 nm silicon trenches from block copolymer templates. Adv. Mater. 24 (42), 5688–5694.

Guo, L.J., 2007. Nanoimprint lithography: methods and material requirements. Adv. Mater. 19 (4), 495–513.

Guo, L.J., Ahn, S.H., 2011. Roll to roll nanoimprint lithography. US8027086 B2, September 27.

Hammond, M.R., Kramer, E.J., 2006. Edge effects on thermal disorder in laterally confined diblock copolymer cylinder monolayers. Macromolecules 39 (4), 1538–1544.

Han, E., Kang, H., Liu, C.-C., Nealey, P.F., Gopalan, P., 2010. Graphoepitaxial assembly of symmetric block copolymers on weakly preferential substrates. Adv. Mater. 22 (38), 4325–4329.

Hannon, A.F., Gotrik, K.W., Ross, C.A., Alexander-Katz, A., 2013. Inverse design of topographical templates for directed self-assembly of block copolymers. ACS Macro Lett. 2 (3), 251–255.

Hannon, A.F., Ding, Y., Bai, W., Ross, C.A., Alexander-Katz, A., 2014. Optimizing topographical templates for directed self-assembly of block copolymers via inverse design simulations. Nano Lett. 14 (1), 318–325.

Henschel, W., Georgiev, Y.M., Kurz, H., 2003. Study of a high contrast process for hydrogen silsesquioxane as a negative tone electron beam resist. J. Vac. Sci. Technol. B 21 (5), 2018–2025.

Hong, S.W., Gu, X., Huh, J., Xiao, S., Russell, T.P., 2011. Circular nanopatterns over large areas from the self-assembly of block copolymers guided by shallow trenches. ACS Nano 5 (4), 2855–2860.

Hong, S.W., Voronov, D.L., Lee, D.H., Hexemer, A., Padmore, H.A., Xu, T., Russell, T.P., 2012a. Controlled orientation of block copolymers on defect-free faceted surfaces. Adv. Mater. 24 (31), 4278–4283.

Hong, S.W., Huh, J., Gu, X., Lee, D.H., Jo, W.H., Park, S., Xu, T., Russell, T.P., 2012b. Unidirectionally aligned line patterns driven by entropic effects on faceted surfaces. Proc. Natl. Acad. Sci. U.S.A. 109 (5), 1402–1406.

Hosaka, S., Akahane, T., Huda, M., Zhang, H., Yin, Y., May 2014. Controlling of 6 nm sized and 10 nm pitched Dot arrays ordered along narrow guide lines using PS–PDMS self-assembly. ACS Appl. Mater. Interfaces 6 (9), 6208–6211.

Ilievski, F., Ross, C.A., 2010. Graphoepitaxy of block copolymers using selectively removable templates. J. Vac. Sci. Technol. B 28 (1), 42.

Jackson, E.A., Hillmyer, M.A., 2010. Nanoporous membranes derived from block copolymers: from drug delivery to water filtration. ACS Nano 4 (7), 3548–3553.

Jeong, S.-J., Kim, J.E., Moon, H.-S., Kim, B.H., Kim, S.M., Kim, J.B., Kim, S.O., 2009. Soft graphoepitaxy of block copolymer assembly with disposable photoresist confinement. Nano Lett. 9 (6), 2300–2305.

Jeong, S.-J., Moon, H.-S., Kim, B.H., Kim, J.Y., Yu, J., Lee, S., Lee, M.G., Choi, H., Kim, S.O., 2010. Ultralarge-area block copolymer lithography enabled by disposable photoresist prepatterning. ACS Nano 4 (9), 5181–5186.

Jeong, J.W., Park, W.I., Do, L.-M., Park, J.-H., Kim, T.-H., Chae, G., Jung, Y.S., 2012. Nanotransfer printing with sub-10 nm resolution realized using directed self-assembly. Adv. Mater. 24 (26), 3526–3531.

Jung, Y.S., Ross, C.A., 2007. Orientation-controlled self-assembled nanolithography using a polystyrene – polydimethylsiloxane block copolymer. Nano Lett. 7 (7), 2046–2050.

Jung, Y.S., Ross, C.A., 2009. Solvent-vapor-induced tunability of self-assembled block copolymer patterns. Adv. Mater. 21 (24), 2540–2545.

Jung, Y.S., Jung, W., Tuller, H.L., Ross, C.A., 2008a. Nanowire conductive polymer Gas sensor patterned using self-assembled block copolymer lithography. Nano Lett. 8 (11), 3776–3780.

Jung, Y.S., Jung, W., Ross, C.A., 2008b. Nanofabricated concentric ring structures by templated self-assembly of a diblock copolymer. Nano Lett. 8 (9), 2975–2981.

Jung, Y.S., Chang, J.B., Verploegen, E., Berggren, K.K., Ross, C.A., 2010. A path to ultranarrow patterns using self-assembled lithography. Nano Lett. 10 (3), 1000–1005.

Kim, H.-C., Rettner, C.T., Sundström, L., 2008a. The fabrication of 20 nm half-pitch gratings by corrugation-directed self-assembly. Nanotechnology 19 (23), 235301.

Kim, S., Lee, J., Jeon, S.-M., Lee, H.H., Char, K., Sohn, B.-H., May 2008. Orientation of lamellar nanostructures in the patterned thin films of a diblock copolymer. Macromolecules 41 (10), 3401–3404.

Kim, S., Shin, D.O., Choi, D.-G., Jeong, J.-R., Mun, J.H., Yang, Y.-B., Kim, J.U., Kim, S.O., Jeong, J.-H., May 2012. Graphoepitaxy of block-copolymer self-assembly integrated with single-step ZnO nanoimprinting. Small 8 (10), 1563–1569.

Kim, B.H., Choi, Y., Kim, J.Y., Shin, H., Kim, S., Son, S.-W., Kim, S.O., Kim, P., 2014. Wrinkle-directed self-assembly of block copolymers for aligning of nanowire arrays. Adv. Mater. 26 (27), 4665–4670.

Li, H.-W., Huck, W.T.S., 2004. Ordered block-copolymer assembly using nanoimprint lithography. Nano Lett. 4 (9), 1633–1636.

Li, L., Yokoyama, H., 2005. Aligning single-layer cylinders of block copolymer nanodomains using soft molding. Adv. Mater. 17 (11), 1432–1436.

Man, X., Andelman, D., Orland, H., Thébault, P., Liu, P.-H., Guenoun, P., Daillant, J., Landis, S., 2011. Organization of block copolymers using nanoimprint lithography: comparison of theory and experiments. Macromolecules 44 (7), 2206–2211.

Milliron, D.J., Raoux, S., Shelby, R.M., Jordan-Sweet, J., May 2007. Solution-phase deposition and nanopatterning of GeSbSe phase-change materials. Nat. Mater. 6 (5), 352–356.

Moon, S.I., McCarthy, T.J., 2003. Template synthesis and self-assembly of nanoscopic polymer 'pencils,'. Macromolecules 36 (12), 4253–4255.

Moon, H.-S., Shin, D.O., Kim, B.H., Jin, H.M., Lee, S., Lee, M.G., Kim, S.O., 2012. Large-area, highly oriented lamellar block copolymer nanopatterning directed by graphoepitaxially assembled cylinder nanopatterns. J. Mater. Chem. 22 (13), 6307.

Mukherjee, P., Bruccoleri, A., Heilmann, R. K., Schattenburg, M. L., Kaplan, A. F., Guo L. J., 2010. Plasma etch fabrication of 60:1 aspect ratio silicon nanogratings with 200 nm pitch. J. Vac. Sci. Technol. B 28 (6) C6P70-75.

Naito, K., Hieda, H., Sakurai, M., Kamata, Y., Asakawa, K., 2002. 2.5-inch disk patterned media prepared by an artificially assisted self-assembling method. IEEE Trans. Magn. 38 (5), 1949–1951.

Onses, M.S., Song, C., Williamson, L., Sutanto, E., Ferreira, P.M., Alleyne, A.G., Nealey, P.F., Ahn, H., Rogers, J.A., 2013. Hierarchical patterns of three-dimensional block-copolymer films formed by electrohydrodynamic jet printing and self-assembly. Nat. Nanotechnol. 8 (9), 667–675.

Orilall, M.C., Wiesner, U., 2011. Block copolymer based composition and morphology control in nanostructured hybrid materials for energy conversion and storage: solar cells, batteries, and fuel cells. Chem. Soc. Rev. 40 (2), 520.

Ouk Kim, S., Solak, H.H., Stoykovich, M.P., Ferrier, N.J., de Pablo, J.J., Nealey, P.F., 2003. Epitaxial self-assembly of block copolymers on lithographically defined nanopatterned substrates. Nature 424 (6947), 411–414.

Park, I., Park, S., Park, H.-W., Chang, T., Yang, H., Ryu, C.Y., 2006. Unexpected hexagonally perforated layer morphology of PS-*b*-PMMA block copolymer in supported thin film. Macromolecules 39 (1), 315–318.

Park, S.-M., Stoykovich, M.P., Ruiz, R., Zhang, Y., Black, C.T., Nealey, P.F., 2007. Directed assembly of lamellae- forming block copolymers by using chemically and topographically patterned substrates. Adv. Mater. 19 (4), 607–611.

Park, S., Lee, D.H., Xu, J., Kim, B., Hong, S.W., Jeong, U., Xu, T., Russell, T.P., 2009. Macroscopic 10-terabit-per-square-inch arrays from block copolymers with lateral order. Science 323 (5917), 1030–1033.

Park, S.-M., Liang, X., Harteneck, B.D., Pick, T.E., Hiroshiba, N., Wu, Y., Helms, B.A., Olynick, D.L., 2011. Sub-10nm nanofabrication *via* nanoimprint directed self-assembly of block copolymers. ACS Nano 5 (11), 8523–8531.

Park, S., Tsarkova, L., Hiltl, S., Roitsch, S., Mayer, J., Böker, A., 2012. Guiding block copolymers into sequenced patterns via inverted terrace formation. Macromolecules 45 (5), 2494–2501.

Phillip, W.A., O'Neill, B., Rodwogin, M., Hillmyer, M.A., Cussler, E.L., 2010. Self-assembled block copolymer thin films as water filtration membranes. ACS Appl. Mater. Interfaces 2 (3), 847–853.

Rockford, L., Liu, Y., Mansky, P., Russell, T., Yoon, M., Mochrie, S., 1999. Polymers on nano-periodic, heterogeneous surfaces. Phys. Rev. Lett. 82 (12), 2602–2605.

Ross, C.A., Jung, Y.S., Chuang, V.P., Ilievski, F., Yang, J.K.W., Bita, I., Thomas, E.L., Smith, H.I., Berggren, K.K., Vancso, G.J., Cheng, J.Y., 2008. Si-containing block copolymers for self-assembled nanolithography. J. Vac. Sci. Technol. B 26 (6), 2489–2494.

Ross, C.A., Berggren, K.K., Cheng, J.Y., Jung, Y.S., Chang, J.-B., 2014. Three-dimensional nanofabrication by block copolymer self-assembly. Adv. Mater. 26 (25), 4386–4396.

Ruiz, R., Kang, H., Detcheverry, F.A., Dobisz, E., Kercher, D.S., Albrecht, T.R., de Pablo, J.J., Nealey, P.F., 2008. Density multiplication and improved lithography by directed block copolymer assembly. Science 321 (5891), 936–939.

Salaün, M., Kehagias, N., Salhi, B., Baron, T., Boussey, J., Sotomayor Torres, C.M., Zelsmann, M., 2011. Direct top-down ordering of diblock copolymers through nanoimprint lithography. J. Vac. Sci. Technol. B 29 (6), 06F208.

Salaün, M., Zelsmann, M., Archambault, S., Borah, D., Kehagias, N., Simao, C., Lorret, O., Shaw, M.T., Sotomayor Torres, C.M., Morris, M.A., 2013. Fabrication of highly ordered sub-20nm silicon nanopillars by block copolymer lithography combined with resist design. J. Mater. Chem. C 1 (22), 3544.

Sanders, D.P., Cheng, J., Rettner, C.T., Hinsberg, W.D., Kim, H.-C., Trung, H., Fritz, A., Harrer, S., Holmes, S., Colburn, M., 2010. Integration of directed self-assembly with 193nm lithography. J. Photopolym. Sci. Technol. 23 (1), 11–18.

Savas, T.A., 1996. Large-area achromatic interferometric lithography for 100nm period gratings and grids. Journal of Vacuum Science & Technology B: Microelectronics and Nanometer Structures 14 (6), 4167.

Schift, H., 2008. Nanoimprint lithography: an old story in modern times? A review. J. Vac. Sci. Technol. B 26 (2), 458–480.

Segalman, R.A., Yokoyama, H., Kramer, E.J., 2001. Graphoepitaxy of spherical domain block copolymer films. Adv. Mater. 13 (15), 1152–1155.

Segalman, R.A., Schaefer, K.E., Fredrickson, G.H., Kramer, E.J., Magonov, S., 2003. Topographic templating of islands and holes in highly asymmetric block copolymer films. Macromolecules 36 (12), 4498–4506.

Shi, A.-C., Li, B., 2013. Self-assembly of diblock copolymers under confinement. Soft Matter 9 (5), 1398–1413.

Shin, K., 2004. Curving and frustrating flatland. Science 306 (5693), 76.

Simão, C., Khunsin, W., Kehagias, N., Francone, A., Zelsmann, M., Morris, M.A., Sotomayor Torres, C.M., 2014. Nanoimprint-assisted directed self-assembly of low-molecular weight block copolymers: a route for 3D and multilevel nanostructures. In: Micro- and Nanotechnology Sensors, Systems, and Applications VI. In: Proceedings of SPIE, Vol. 9083. p. 90832S.

Son, J.G., Chang, J.-B., Berggren, K.K., Ross, C.A., 2011a. Assembly of sub-10-nm block copolymer patterns with mixed morphology and period using electron irradiation and solvent annealing. Nano Lett. 11 (11), 5079–5084.

Son, J.G., Hannon, A.F., Gotrik, K.W., Alexander-Katz, A., Ross, C.A., 2011b. Hierarchical nanostructures by sequential self-assembly of styrene-dimethylsiloxane block copolymers of different periods. Adv. Mater. 23 (5), 634–639.

Stewart-Sloan, C.R., Thomas, E.L., 2011. Interplay of symmetries of block polymers and confining geometries. Eur. Polym. J. 47 (4), 630–646.

Stoykovich, M.P., Müller, M., Kim, S.O., Solak, H.H., Edwards, E.W., de Pablo, J.J., Nealey, P.F., 2005. Directed assembly of block copolymer blends into nonregular device-oriented structures. Science 308 (5727), 1442–1446.

Stoykovich, M.P., Kang, H., Daoulas, K.C., Liu, G., Liu, C.-C., de Pablo, J.J., Müller, M., Nealey, P.F., 2007. Directed self-assembly of block copolymers for nanolithography: fabrication of isolated features and essential integrated circuit geometries. ACS Nano 1 (3), 168–175.

Stuart, C., Chen, Y., 2009. Roll in and roll out: a path to high-throughput nanoimprint lithography. ACS Nano 3 (8), 2062–2064.

Sundrani, D., Darling, S.B., Sibener, S.J., 2004. Guiding polymers to perfection: macroscopic alignment of nanoscale domains. Nano Lett. 4 (2), 273–276.

Switkes, M., Bloomstein, T.M., Rothschild, M., 2000. Patterning of sub-50 nm dense features with space-invariant 157 nm interference lithography. Appl. Phys. Lett. 77 (20), 3149.

Tang, Q., Ma, Y., 2010. High density multiplication of graphoepitaxy directed block copolymer assembly on two-dimensional lattice template. Soft Matter 6 (18), 4460–4465.

Tavakkoli, K.G., A. Piramanayagam, S.N., Ranjbar, M., Sbiaa, R., Chong, T.C., 2011. Path to achieve sub-10-nm half-pitch using electron beam lithography. J. Vac. Sci. Technol. B 29 (1), 011035.

Tavakkoli, K.G., A. Gotrik, K.W., Hannon, A.F., Alexander-Katz, A., Ross, C.A., Berggren, K.K., 2012a. Templating three-dimensional self-assembled structures in bilayer block copolymer films. Science 336 (6086), 1294–1298.

Tavakkoli, K.G., A. Hannon, A.F., Gotrik, K.W., Alexander-Katz, A., Ross, C.A., Berggren, K.K., 2012b. Rectangular symmetry morphologies in a topographically templated block copolymer. Adv. Mater. 24 (31), 4249–4254.

Tavakkoli, K.G., A. Ranjbar, M., Piramanayagam, S.N., Wong, S.K., Poh, W.C., Sbiaa, R., Chong, T.C., 2012c. Reverse nanoimprint lithography for fabrication of nanostructures. Nanosci. Nanotechnol. Lett. 4 (8), 835–838.

Tavakkoli, K.G., A. Nicaise, S.M., Berggren, K.K, Gotrik, K.W., Ross, C.A., 2013. Removable templates for directed self assembly. US20130244439 A1, September 19.

Tavakkoli, K.G., A. Nicaise, S.M., Hannon, A.F., Gotrik, K.W., Alexander-Katz, A., Ross, C.A., Berggren, K.K., 2014. Sacrificial-post templating method for block copolymer self-assembly. Small 10 (3), 493–499.

Thébault, P., Niedermayer, S., Landis, S., Chaix, N., Guenoun, P., Daillant, J., Man, X., Andelman, D., Orland, H., 2012. Tailoring nanostructures using copolymer nanoimprint lithography. Adv. Mater. 24 (15), 1952–1955.

Thurn-Albrecht, T., Schotter, J., Kästle, G.A., Emley, N., Shibauchi, T., Krusin-Elbaum, L., Guarini, K., Black, C.T., Tuominen, M.T., Russell, T.P., 2000. Ultrahigh-density nanowire arrays grown in self-assembled diblock copolymer templates. Science 290 (5499), 2126–2129.

Tseng, A.A. (Ed.), 2011. Tip-based nanofabrication: fundamentals and applications. Springer, New York.

Tseng, A.A., Jou, S., Notargiacomo, A., Chen, T.P., May 2008. Recent developments in Tip-based nanofabrication and its roadmap. J. Nanosci. Nanotechnol. 8 (5), 2167–2186.

Vieu, C., Carcenac, F., Pépin, A., Chen, Y., Mejias, M., Lebib, A., Manin-Ferlazzo, L., Couraud, L., Launois, H., 2000. Electron beam lithography: resolution limits and applications. Appl. Surf. Sci. 164 (1–4), 111–117.

Voet, V.S.D., Pick, T.E., Park, S.-M., Moritz, M., Hammack, A.T., Urban, J.J., Ogletree, D.F., Olynick, D.L., Helms, B.A., 2011. Interface segregating fluoralkyl-modified polymers for high-fidelity block copolymer nanoimprint lithography. J. Am. Chem. Soc. 133 (9), 2812–2815.

Wagner, D.J., Jayatissa, A.H., 2005. Nanoimprint lithography: review of aspects and applications. In: Nanofabrication: Technologies, Devices, and Applications II. In: Proceedings of SPIE, Vol. 6002, 60020R-1–60020R-9.

Wang, H., Zou, Q., 2011. Development of high-throughput control techniques for tip-based nanofabrication. In: Tseng, A.A. (Ed.), Tip-Based Nanofabrication. Springer, New York, pp. 325–355.

Wu, Y., Cheng, G., Katsov, K., Sides, S.W., Wang, J., Tang, J., Fredrickson, G.H., Moskovits, M., Stucky, G.D., 2004. Composite mesostructures by nano-confinement. Nat. Mater. 3 (11), 816–822.

Xiang, H., Shin, K., Kim, T., Moon, S.I., McCarthy, T.J., Russell, T.P., 2004. Block copolymers under cylindrical confinement. Macromolecules 37 (15), 5660–5664.

Xiang, H., Shin, K., Kim, T., Moon, S., Mccarthy, T.J., Russell, T.P., 2005. The influence of confinement and curvature on the morphology of block copolymers. J. Polym. Sci. B 43 (23), 3377–3383.

Xiao, S., Yang, X., Edwards, E.W., La, Y.-H., Nealey, P.F., 2005. Graphoepitaxy of cylinder-forming block copolymers for use as templates to pattern magnetic metal dot arrays. Nanotechnology 16 (7), S324–S329.

Yan, N., Sheng, Y., Liu, H., Zhu, Y., Jiang, W., 2015. Templated self-assembly of block copolymers and morphology transformation driven by the rayleigh instability. Langmuir 31 (5), 1660–1669.

Yang, J.K.W., Berggren, K.K., 2007. Using high-contrast salty development of hydrogen silsesquioxane for sub-10-nm half-pitch lithography. J. Vac. Sci. Technol. B 25 (6), 2025.

Yang, J.K.W., Cord, B., Duan, H., Berggren, K.K., Klingfus, J., Nam, S.-W., Kim, K.-B., Rooks, M.J., 2009. Understanding of hydrogen silsesquioxane electron resist for sub-5-nm-half-pitch lithography. J. Vac. Sci. Technol. B 27 (6), 2622–2627.

Yang, J.K.W., Jung, Y.S., Chang, J.-B., Mickiewicz, R.A., Alexander-Katz, A., Ross, C.A., Berggren, K.K., 2010. Complex self-assembled patterns using sparse commensurate templates with locally varying motifs. Nat. Nanotechnol. 5 (4), 256–260.

Yang, X., Xiao, S., Hu, W., Hwu, J., van de Veerdonk, R., Wago, K., Lee, K., Kuo, D., 2014. Integration of nanoimprint lithography with block copolymer directed self-assembly for fabrication of a sub-20 nm template for bit-patterned media. Nanotechnology 25 (39), 395301.

Yi, H., Bao, X.-Y., Zhang, J., Bencher, C., Chang, L.-W., Chen, X., Tiberio, R., Conway, J., Dai, H., Chen, Y., Mitra, S., Wong, H.-S.P., 2012. Flexible control of block copolymer directed self-assembly using small, topographical templates: potential lithography solution for integrated circuit contact hole patterning. Adv. Mater. 24 (23), 3107–3114.

Yi, H., Latypov, A., Wong, H.-S.P., 2013a. Computational simulation of block copolymer directed self-assembly in small topographical guiding templates. In: Alternative Lithographic Technologies V. Proceedings of SPIE, Vol. 8680, 86801 L.

Yi, H., Bao, X.-Y., Tiberio, R., Wong, H.-S.P., 2013b. Design strategy of small topographical guiding templates for sub-15 nm integrated circuits contact hole patterns using block copolymer directed self assembly. In: Alternative Lithographic Technologies V. In: Proceedings of SPIE, Vol. 8680, p. 868010.

Yu, B., Sun, P., Chen, T., Jin, Q., Ding, D., Li, B., Shi, A.-C., 2006. Confinement-induced novel morphologies of block copolymers. Phys. Rev. Lett. 96 (13), 138306.

Yu, J., Geng, C., Zeng, Y., Yan, Q., Wang, X., Shen, D., 2011. Confined self-assembly of asymmetric diblock copolymers within silica nanobowl arrays. ACS Macro Lett. 1 (1), 62–66.

Zailer, I., Frost, J.E.F., Chabasseur-Molyneux, V., Ford, C.J.B., Pepper, M., 1996. Crosslinked PMMA as a high-resolution negative resist for electron beam lithography and applications for physics of low-dimensional structures. Semicond. Sci. Technol. 11 (8), 1235.

Zankovych, S., Hoffmann, T., Seekamp, J., Bruch, J.-U., Torres, C.M.S., 2001. Nanoimprint lithography: challenges and prospects. Nanotechnology 12 (2), 91.

Zhang, L., Wang, L., Lin, J., 2014. Harnessing anisotropic nanoposts to enhance long-range orientation order of directed self-assembly nanostructures via large cell simulations. ACS Macro Lett. 3 (8), 712–716.

Part Three

Application of directed self-assembly in nanomanufacturing

The inverse directed self-assembly problem

9

A. Latypov, T.H. Coskun
GLOBALFOUNDRIES Inc., Santa Clara, CA, USA

9.1 Introduction

Directed self-assembly (DSA) utilizing melts or solutions of block copolymers (BCP) is presently drawing attention as another option for next-generation lithography (Herr, 2011). Unlike other nanofabrication methods where the size of the smallest manufacturable feature is determined, for example, by the capability of an external field to generate a spatial variation of a characteristic dimension of this smallest feature, DSA utilizes the interactions between BCP macromolecules resulting in their self-assembly into the features of characteristic sizes of these macromolecules ranging from tens to a few nanometers.

The particular family of self-assembling patterns that can be naturally formed in a freely annealing BCP melt or solution is dictated by the composition of the BCP molecule. Lamella structures and hexagonally arranged cylinders formed by diblock BCP molecules are usually considered to be suitable candidates to form, respectively, lines and spaces and vias needed for typical integrated circuits (ICs). Chemoepitaxy prepatterns or graphoepitaxy confinement wells patterned at a larger scale using optical/EUV or e-beam lithography-based processes provide guidance to DSA that can be used to shape the DSA morphologies into patterns usable in a fabrication of ICs. For example, DSA lamellae can be bent, isolated, or terminated by properly selected graphoepitaxy prepatterns to form lines and spaces of ICs, and DSA cylinders can be properly positioned and grouped to form arrays of contact holes and vias.

Predicting DSA patterns resulting from given chemoepitaxy or graphoepitaxy directing pattern is a well-known *direct* problem of computational DSA simulations (Fredrickson, 2006). In this chapter, we concentrate on an *inverse problem of DSA*—finding directing graphoepitaxy or chemoepitaxy patterns, resulting in given desired DSA patterns on a wafer. We also refer to this problem and the algorithms for its solution as *DSA Process/Proximity Correction (DSA PC) problem or algorithms.*

This chapter discusses several approaches to the formulation and computational solution of the inverse DSA problem, particularly, approaches based on a linearization of the DSA model and minimizing the objective function ensuring the formation of the desired DSA patterns. We illustrate these approaches by presenting the results of their application to a practically useful problem (Yi et al., 2012) of forming arrays of DSA contact holes using graphoepitaxy guiding templates.

Directed Self-assembly of Block Copolymers for Nano-manufacturing. http://dx.doi.org/10.1016/B978-0-08-100250-6.00009-2

In this chapter, without losing the generality, we demonstrated an approach to the solution of the inverse DSA problem using the self-consistent field theory (SCFT) DSA model (Fredrickson, 2006; Takahashi et al., 2012; Matsen, 1997; Ceniceros and Fredrickson, 2004), however, this approach can be applied to other models of both graphoepitaxy and chemoepitaxy DSA processes.

9.2 DSA model and inverse DSA problem

9.2.1 SCFT DSA model

The DSA model used in this work is based on the following SCFT Hamiltonian for a compressible diblock copolymer melt (Fredrickson, 2006; Takahashi et al., 2012):

$$H\left[w_A,w_B\right]=\rho_0\int \mathbf{dx}\left[\frac{1}{\chi_{AB}}\left(\frac{w_B-w_A}{2}\right)^2+\left(\frac{2\bar{\chi}_{\mathrm{w}}}{\chi_{AB}+2\kappa}\frac{\rho_{\mathrm{w}}}{\rho_0}-\frac{2\kappa}{\chi_{AB}+2\kappa}\left(1-\frac{\rho_{\mathrm{w}}}{\rho_0}\right)\right)\frac{w_A+w_B}{2}\right.$$
$$\left.-\frac{1}{\left(\chi_{AB}+2\kappa\right)}\left(\frac{w_A+w_B}{2}\right)^2-2\frac{\chi_{\mathrm{w}}}{\chi_{AB}}\frac{\rho_{\mathrm{w}}}{\rho_0}\frac{w_B-w_A}{2}\right]-n\ln Q\left[w_A,w_B\right], \tag{9.1}$$

where $w_A=w_A(\mathbf{x})$ and $w_B=w_B(\mathbf{x})$ are the self-consistent fields (potentials) acting on the phases A and B of the diblock copolymer, respectively, and n is a total number of the BCP molecules in the melt. Finite compressibility is modeled by a quadratic penalty term $0.5\kappa\left(\rho_A+\rho_B+\rho_{\mathrm{w}}-\rho_0\right)^2$, where ρ_0 is the constant total density parameter and κ is a compressibility parameter. $\rho_{\mathrm{w}}=\rho_{\mathrm{w}}(\mathbf{x},\mathbf{d})$ is the confining wall material density, given as a function of the spatial coordinates $\mathbf{x}$ and the wall shape design variables $\mathbf{d}$. $\rho_K=\rho_K\left[w_A,w_B\right]$, where K is either A or B, are the spatially varying densities of the BCP phases A and B, which are known nonlocal functions of the self-consistent fields. The parameter χ_{AB} is the Flory parameter characterizing the interaction of the BCP phases with each other, $\chi_{\mathrm{w}}=\left(\chi_{\mathrm{w}B}-\chi_{\mathrm{w}A}\right)/2$ and $\bar{\chi}_{\mathrm{w}}=\left(\chi_{\mathrm{w}A}+\chi_{\mathrm{w}B}\right)/2$ are the Flory-like parameters, where $\chi_{\mathrm{w}K}$ ($K=A$ or B) characterize the interaction of the BCP phases with the wall material. $Q[w_A, w_B]$ is a single polymer chain partition function, which is a known functional of the self-consistent fields.

Using the mean-field approximation, we are looking for the self-consistent fields providing a saddle point of the SCFT Hamiltonian (9.1) (Fredrickson, 2006). The nonlocal dependence of the BCP phase densities on the fields is found by solving Fokker–Planck reaction-diffusion equations for propagators of diblock BCP in given potential fields (Fredrickson, 2006).

Equating the variations of the SCFT Hamiltonian (9.1) with respect to the SCFT potential fields w_A and w_B to zero and combining the resulting two stationary (saddle point) conditions, results in the following nonlinear and nonlocal SCFT equations:

$$\rho_A\left[w_A,w_B\right]+\rho_B\left[w_A,w_B\right]-\frac{\rho_0}{\left(\chi_{AB}+2\kappa\right)}\left(w_A+w_B\right)$$
$$=\frac{2\kappa}{\chi_{AB}+2\kappa}\rho_0-\frac{2\overline{\chi}_{\mathrm{w}}+2\kappa}{\chi_{AB}+2\kappa}\rho_{\mathrm{w}}\left(\mathbf{x},\mathbf{d}\right),$$
$$\rho_B\left[w_A,w_B\right]-\rho_A\left[w_A,w_B\right]+\frac{\rho_0}{\chi_{AB}}\left(w_B-w_A\right) \tag{9.2}$$
$$=\frac{2\chi_{\mathrm{w}}}{\chi_{AB}}\rho_{\mathrm{w}}\left(\mathbf{x},\mathbf{d}\right).$$

The SCFT model (9.2) based on the Hamiltonian (9.1) can be used to simulate the graphoepitaxy DSA for a compressible or incompressible BCP melt in a confinement well (Takahashi et al., 2012; Matsen, 1997). Alternatively, treating $\rho_{\mathrm{w}}(\mathbf{x},\mathbf{d})$ as a fixed density approximation of the polymer brush or the cross-linked polymer mat material allows us to use Equation (9.2) as a model for chemical epitaxy DSA.

In many cases, one can neglect the compressibility of the BCP melt, which corresponds to setting $\kappa\to+\infty$ in Equation (9.2).

Given the values of the design variables **d** and the corresponding density of confinement walls, chemoepitaxy prepattern brushes, or cross-linked polymer mats, $\rho_{\mathrm{w}}(\mathbf{x},\mathbf{d})$, Equation (9.2) are solved iteratively with respect to the unknown functions $w_A=w_A\left(\mathbf{x}\right)$ and $w_B=w_B\left(\mathbf{x}\right)$ (Fredrickson, 2006; Matsen, 1997; Ceniceros and Fredrickson, 2004). After the solution is found, the equilibrium distribution of the BCP densities, $\rho_K=\rho_K\left[w_A,w_B\right]$, where K is either A or B, can be found from the solution of the Fokker–Planck partial differential equations for propagators of diblock BCP in given SCFT potential fields (Fredrickson, 2006; Matsen, 1997; Ceniceros and Fredrickson, 2004).

9.2.2 Inverse DSA problem, DSA PC

In inverse DSA problems, the desired target positions and sizes of DSA features are given, and we aim to find the placement and shape of manufacturable directing patterns (chemoepitaxy or graphoepitaxy) resulting in the formation of DSA features as close as possible to the given targets. We also refer to this family of problems as DSA PC problems and the algorithm for its solution as DSA PC algorithm.

Many concepts and approaches previously used in the solution of the optical proximity correction (OPC) problems in computational lithography (Ma and Arce, 2010) can be applied to DSA PC problems.

In general, one can separate the inverse DSA problem from the inverse litho problem and perform each problem solutions individually. In this two-step approach, the inverse DSA step is performed first and followed by the inverse litho step. The goal of the inverse DSA step is to find a litho target, and the goal of the inverse litho step is to find a mask shape to generate this litho target. Another, single-step, approach is to combine DSA and litho forward simulations to find a mask shape that will ultimately result in DSA targets. Each of these methods has some accuracy limitations and implementation complexities. For example, the two-step method may be viewed as relatively simple, compared to the single-step approach, due to decomposition of co-optimization of DSA and litho into two consecutive steps. However, such decomposition limits the solution domain and involves two-stage error accumulations. The single-step approach, on the other hand, combines the effects of two differently behaving processes and sensitivities into one cost function that might be unstable in an optimization algorithm.

Another method proposed (Wang et al., 2014) involves a two-step approach, which combines the simplicity of the two-step approach with the power of extending the solution domain. In this method, a rule-based educated initial confinement well guess is followed by a hybrid co-optimization.

In the following sections, we concentrate on the inverse DSA part of the problem, since the inverse litho part has been well discussed in the literature for general litho problems. However, one can take the advantage of DSA process in litho correction by making the litho correction DSA-aware. One example of this is shown in Section 9.2.7 where the assist features (AF) are placed and optimized by considering the following DSA process. A similar approach can also be applied to illumination source optimization where the source can evolve to a better solution due to larger solution domain, when DSA helps healing the AF and side-lobe prints.

A full chip correction flow needs to consider both litho and DSA corrections. In this flow, one needs to consider that the interaction range of the litho and DSA process steps are different. For example, in the graphoepitaxy DSA, the DSA proximity effects strongly affect the shape and placement of DSA features adjacent to each other inside the same confinement well. In addition to this, if the confinement wells are overfilled with the BCP material, they become "connected" and the outcomes of the DSA processes in the adjacent confinement wells may be influenced by DSA proximity effects between these adjacent wells. The presence of the confinement well overfill and the strength of the DSA proximity effects between the adjacent confinement wells is expected to depend on the parameters of the DSA process characterizing the deposition the BCP material onto the wafer (e.g., the parameters of the spin-coating process), as well as the local density of the confinement wells and the spatial proximity of the adjacent confinement wells. In the examples shown in Sections 9.2.6 and 9.2.7, we assume that the graphoepitaxy confinement wells are not overfilled, and the DSA processes in each confinement well are independent.

The corrections for the proximity effects in optical lithography step should consider multiple confinement well targets where the interaction between them cannot be ignored. This domain difference needs to be carefully considered especially in co-optimization approaches where the litho and DSA forward simulations are effectively combined together. To solve this complication, one can apply a hybrid approach where each correction is performed individually in an iterative manner until a converging solution is obtained. This approach involves a group-based correction by ignoring the interaction between groups in intermediate iterations. It has been shown that one can achieve reasonable accuracy by gradually decreasing the max fragment movement as the number of iterations increases (Wang et al., 2014).

As in the OPC problems, another aspect of the inverse DSA problem is the initial guess of the confinement well. This initial guess also depends on how the DSA targets are grouped together to form a confinement well. Even though the grouping strategy

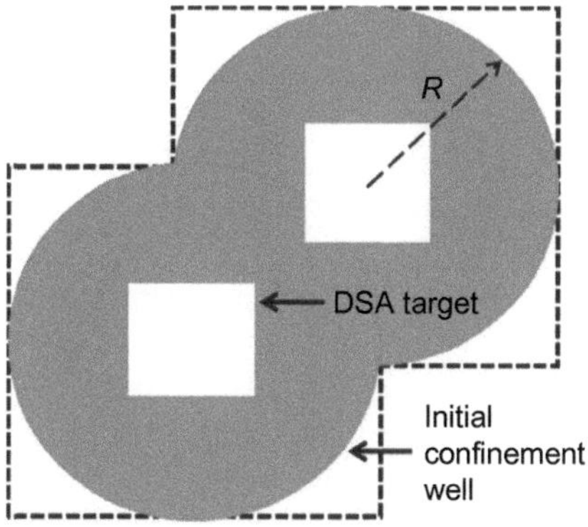

Figure 9.1 Determination of initial confinement well (litho target).

might be part of design problem, we found that a simple approach is quite efficient to generate confinement well groups (Wang et al., 2014). In this approach, a circle around the center of each DSA target is placed, as illustrated in Figure 9.1. The radius of the circle is optimized on some test patterns by using a simple searching algorithm. The generated circles are merged together to form group boundaries. These group boundaries are also used as an educated guess to the initial confinement well shape and litho target to be optimized further. One can also use the manhattanized version of these initial litho targets to generate initial mask and mask fragmentations to be used in the litho correction. This is illustrated by dashed lines in Figure 9.1.

9.2.3 Inverse DSA problem as a constrained optimization problem

The problem to be solved by the DSA PC algorithm can be formulated as the following constrained optimization problem:

$$\begin{gathered}\text{Find } \mathbf{d} \text{ such that}\\ C(\rho[\mathbf{w}]) \text{ is minimal}\\ \text{subject to } \mathbf{S}[\mathbf{w}] = \mathbf{f}(\mathbf{d}) \text{ and } \mathbf{M}(\mathbf{d}) \leq \mathbf{0}.\end{gathered} \tag{9.3}$$

Here, in addition to the notations introduced in Section 9.2.1, $C(\boldsymbol{\rho}[\mathbf{w}])$ is a cost function quantifying the proximity of the DSA result to the given targets; $\rho = (\rho_A, \rho_B)^{\mathrm{T}}$ are the average densities of the BCP components; $\mathbf{w} = (w_A, w_B)^{\mathrm{T}}$ are the SCFT fields; and $\mathbf{S}[\mathbf{w}] = \mathbf{f}(\mathbf{d})$ are the SCFT equations (9.2). The *manufacturability constraints* $\mathbf{M}(\mathbf{d}) \leq \mathbf{0}$ are dictated by the restrictions imposed to account for the limitations of the process used to fabricate the directing patterns. For example, if an optical lithography process is used to fabricate the directing patterns, such constraints may include the minimal possible distance between the directing pattern features, minimal size of the feature of the directing pattern, and so forth. In these notations, for a given set of the design variables **d**, Equation (9.2) can be solved to find the SCFT fields, **w**, and to calculate the corresponding BCP phase average densities, $\boldsymbol{\rho}$. The value of the cost function can then be calculated from these phase densities. Having this procedure in mind, we can also write the cost function as $C = C(\mathbf{d})$.

Etch transfer of the DSA patterns from the BCP film into the substrate is usually done using an etching process, or processes that first predominantly remove one of the BCP phases (similarly to the resist development process in optical or EUV lithography), and then etch the substrate through the deprotected areas in the remaining BCP phase. Based on these processes, several options for the cost function C can be used in the DSA PC problem formulation.

For instance, a possible cost function can be based on the edge placement error (EPE) (Wang et al., 2014). In this approach, target edge positions are specified, and then a DSA etching process model is applied to the phase density distribution resulting from the solution of the SCFT equations (9.2) in order to calculate the current edge positions, as shown in Figure 9.2. These current positions of the edges are compared to the target positions, and the differences between these positions at certain points (EPE check points) are calculated to find the current EPE. Finally, some norm of EPE is calculated and used as a cost function. One simple approximate way to determine the edge positions is to apply a constant threshold to the density of one of the BCP components. As an example, minority phase can be thresholded to generate the edge positions in Figure 9.2.

Another way to formulate a cost function of the DSA PC problem involves specifying a target density distribution for one of the BCP components and then comparing the current density distribution of the same component to this target density distribution. Some suitable norm of the difference of these density distributions can then be used as a cost function.

Similar to OPC problems (Ma and Arce, 2010), the DSA PC problem (9.3) is a constrained nonlinear optimization problem. In the same manner as in many approaches to the solution of OPC problems, inequality manufacturability constraints can be incorporated into the cost function using a penalty method, that is, by using a modified cost function $C(\mathbf{d}) + \mathbf{s}^{\mathrm{T}}\mathbf{G}(\mathbf{d})$ with constraint violation penalties, where $\mathbf{s}$ is the given positive penalty coefficients and $\mathbf{G}(\mathbf{d}) = \max(0, \mathbf{M}(\mathbf{d}))^2$ is one example

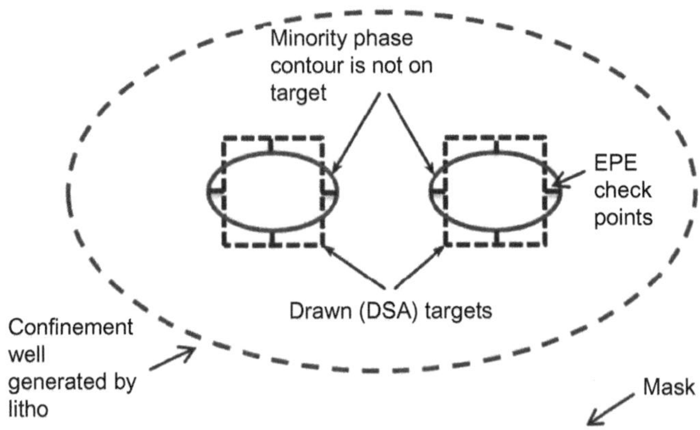

Figure 9.2 EPE-based cost function construction.

of the penalty functions. In the following, we will assume that the constraints are incorporated into the cost function and the optimization problem is unconstrained, with each evaluation of the cost function $C = C(\mathbf{d})$ requiring a solution of the direct DSA problem (9.2).

9.2.4 Iterative methods to solve inverse DSA problem

As in the OPC problem solutions (Ma and Arce, 2010), several families of iterative algorithms can be used to solve the DSA PC problem (9.3). In the *steepest descent method* (Ma and Arce, 2010), the optimization algorithm updates the estimate to the solution iteratively by advancing in the direction opposite to the gradient of the cost function:

$$\mathbf{d}^{(k+1)} = \mathbf{d}^{(k)} - t_k C_{\mathrm{d}}^{(1)}\left(\mathbf{d}^{(k)}\right), \quad k = 0,1,2,\ldots$$

where $t_k > 0$ is the iterative step parameter and $C_{\mathrm{d}}^{(1)}\left(\mathbf{d}^{(k)}\right)$ denotes the vector of the first derivatives of the cost function by the design variables $\mathbf{d}$, evaluated at $\mathbf{d}^{(k)}$. Another important family of methods is the *Newton method* (Ma and Arce, 2010):

$$\mathbf{d}^{(k+1)} = \mathbf{d}^{(k)} - t_k \left(C_{\mathrm{d}}^{(2)}\left(\mathbf{d}^{(k)}\right)\right)^{-1} C_{\mathrm{d}}^{(1)}\left(\mathbf{d}^{(k)}\right), \quad k = 0,1,2,\ldots$$

where $C_{\mathrm{d}}^{(2)}\left(\mathbf{d}^{(k)}\right)$ denotes a matrix of the second-order partial derivatives of the cost function with respect to the design variables, evaluated at $\mathbf{d}^{(k)}$. Other important families and variations of iterative optimization algorithms include quasi-Newton methods, the conjugate gradient method, and a trust region family of methods (Ma and Arce, 2010).

All of these smooth optimization methods require evaluation of the partial derivatives of the cost function. In the absence of analytical expressions for these partial derivatives, their evaluations in the course of iterative optimization algorithm necessitates solving the direct DSA problem (9.2) multiple times for each evaluation of partial derivatives. For instance, in case of N_{d} design variables, if the first-order finite-difference formulae are used to evaluate the gradient of the cost function, $C_{\mathrm{d}}^{(1)}\left(\mathbf{d}^{(k)}\right)$, the cost function needs to be evaluated $N_{\mathrm{d}} + 1$ times, for different values of the design variable vector $\mathbf{d}$, with each evaluation of the cost function requiring a solution of the direct DSA problem (9.2). Because of a relatively high computational cost of the solution of the latter problem, alternative computationally efficient ways of evaluating partial derivatives of the cost function are needed.

9.2.5 Linearization of the SCFT equations

One approach to a computationally efficient way of evaluating the partial derivatives of the DSA PC cost function, with respect to the design variables, is based on a linearized form of the SCFT equations. The complexity of the SCFT equations (9.2) and a high computational cost of their solution are both due the nonlinear and nonlocal dependence of the BCP phase densities on the SCFT fields, $\rho_K = \rho_K\left[w_A, w_B\right]$, where $K = A$ or B. One approach to linearization of this dependence is to use a weak inhomogeneity expansion (Fredrickson, 2006; Ceniceros and Fredrickson, 2004). In this expansion, the fields variations are viewed as a weak perturbation of a constant field values: $w_K(\mathbf{x}) = w_K^{(0)} + w_K^{(1)}(\mathbf{x})$, where $K = A$ or B and $w_K^{(1)}(\mathbf{x}) \ll w_K^{(0)}$. In these assumptions, the BCP densities can be linearized as follows (Ceniceros and Fredrickson, 2004):

$$\rho_A[w_A, w_B] \approx \rho_0\left(f + Ng_{AA} \otimes w_A^{(1)} + Ng_{AB} \otimes w_B^{(1)}\right),$$

$$\rho_B[w_A, w_B] \approx \rho_0\left((1-f) + Ng_{AB} \otimes w_A^{(1)} + Ng_{BB} \otimes w_B^{(1)}\right),$$

where f is the average volume fraction of the phase A, N is a degree of polymerization, g_{AA}, g_{AB}, g_{BB} are certain isotropic linearization kernels, with a known closed-form analytical expressions for their Fourier transformations in terms of modified Debye functions (Ceniceros and Fredrickson, 2004).

Although the validity of weak inhomogeneity expansion may be questioned for the SCFT field variations occurring in the DSA problems of practical significance, such linearization has been used with good results in approximating the terms in Equation (9.2) for relatively small changes of the SCFT fields occurring during the iterative solution of the SCFT equations. Based on a partial linearization, semi-implicit terms have been introduced into these iterations in the semi-implicit Seidel (SIS) method (Fredrickson, 2006; Ceniceros and Fredrickson, 2004), leading to a substantial improvement of a convergence speed.

Acting in a similar way, one may linearize the SCFT equations, $\mathbf{S}[\mathbf{w}] = \mathbf{f}(\mathbf{d})$, and use the resulting linearized equations to evaluate the derivatives of the cost function with respect to the design variables. For instance, linearization of the left-hand side of the SCFT equations and BCP densities with respect to a small change in the SCFT fields, $\Delta\mathbf{w}$, caused by a small change in the design variables, $\Delta\mathbf{d}$, yields

$$\mathbf{S}[\mathbf{w} + \Delta\mathbf{w}] = \mathbf{S}[\mathbf{w}] + \mathbf{L}_1\Delta\mathbf{w} + O\left(|\Delta\mathbf{w}|^2\right),$$

$$\rho[\mathbf{w} + \Delta\mathbf{w}] = \rho[\mathbf{w}] + \mathbf{L}_2\Delta\mathbf{w} + O\left(|\Delta\mathbf{w}|^2\right),$$

where $\mathbf{L}_1$ and $\mathbf{L}_2$ are the linear operators that can be evaluated and inverted relatively computationally inexpensively, compared to the computational cost of the iterative solution of the SCFT equations (9.2). Using such approximating linearizations, the change in value of the cost function caused by a small change in the design variables, $\Delta\mathbf{d}$, can be evaluated by applying the linear operations $(\mathbf{L}_1)^{-1}$ and $\mathbf{L}_2$, precalculated for the current solution. Using this result, the gradient of the DSA PC cost function, $C_{\mathbf{d}}^{(1)}$, can be calculated much faster, compared to the straightforward approach utilizing the finite-difference formulae and the iterative solution of the SCFT equations (9.2).

Although the above presentation applies a complete linearization in both SCFT equations and the density dependence on the SCFT fields, many variations of the linearization approach are possible. For instance, because of the relatively low computational cost of the single evaluation of BCP densities corresponding to the changed SCFT fields, compared to the iterative solution of the SCFT equations, densities may be re-evaluated without linearization, leading to a more accurate evaluation of derivatives without a substantial relative increase in the number of floating point operations. Also, similarly to a partial linearization of the SCFT equations in the SIS iterations (Fredrickson, 2006; Ceniceros and Fredrickson, 2004), only some terms in the SCFT equations (9.2) may be linearized, with other terms treated nonlinearly, with these partially linearized SCFT equations still solved noniteratively (Fredrickson, 2006; Ceniceros and Fredrickson, 2004).

Along with allowing speeding up the calculation of the cost function derivatives, the above linearization provides some insight into the formal mechanism of frequency multiplication in the DSA process. To see that, consider the incompressible melt ($\kappa \to +\infty$) of the diblock BCP with an equal average volume fraction of both BCP phases ($f = 0.5$). We will also use a simplified "directing field" chemoepitaxy direction model, in which the chemoepitaxy brushes or cross-linked mats are first modeled as a prescribed density distribution $\rho_w(\mathbf{x})$ in (9.2), with an additional simplification, neglecting the volumetric displacement due to these brushes or mats, but taking into account their interaction with the BCP through the "directing fields." In such simplification, the SCFT equations reduce to the following equations:

$$\rho_A\left[w_A, w_B\right] + \rho_B\left[w_A, w_B\right] = \rho_0,$$
$$\rho_B\left[w_A, w_B\right] - \rho_A\left[w_A, w_B\right] + \frac{\rho_0}{\chi_{AB}}\left(w_B - w_A\right) = \frac{\rho_0}{\chi_{AB}} w^{(d)},$$

where $w^{(d)} = w^{(d)}(\mathbf{x})$ is an externally imposed directing field. Using the weak inhomogeneity expansion in these equations leads to a system of linear equations with respect to the SCFT fields.

These linear equations can be explicitly solved in the Fourier space, leading to the following expressions for the Fourier transformations of the BCP densities:

$$\tilde{\rho}_A = \frac{1}{2}\rho_0\left(\delta(\mathbf{k}) - N\tilde{G}(\mathbf{k})\tilde{w}^{(d)}(\mathbf{k})\right),$$
$$\tilde{\rho}_B = \frac{1}{2}\rho_0\left(\delta(\mathbf{k}) + N\tilde{G}(\mathbf{k})\tilde{w}^{(d)}(\mathbf{k})\right),$$

where tilde denotes a Fourier transformation, and a Fourier transformation of the solution kernel is

$$\tilde{G}(\mathbf{k}) = \frac{1}{\chi_{AB}N}\frac{\tilde{g}_{AA}(\mathbf{k}) - \tilde{g}_{AB}(\mathbf{k})}{\tilde{g}_{AA}(\mathbf{k}) - \tilde{g}_{AB}(\mathbf{k}) + \dfrac{1}{\chi_{AB}N}}.$$

Because the functions $\tilde{g}_{AA}, \tilde{g}_{AB}, \tilde{g}_{BB}$ depend only (Ceniceros and Fredrickson, 2004) on $|\mathbf{k}|$, the Fourier transformation of the solution kernel is also a function of $|\mathbf{k}|$, as plotted on Figure 9.3 (a blue line). $\tilde{G}(\mathbf{k})$ has a distinct maximum at the spatial frequency, corresponding to a spatial wavelength of approximately 3.22 R_g, where R_g is the unperturbed radius of gyration of the BCP. The effect of application of such kernel to the externally imposed directing field is amplification of the

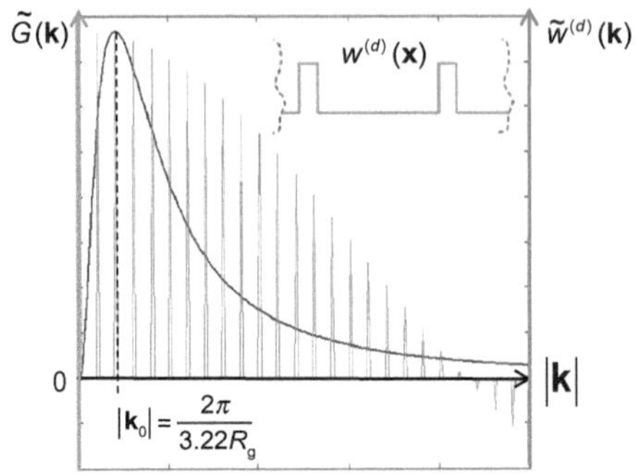

Figure 9.3 The solution kernel (blue) of the linearized SCFT equations, in Fourier space, and a Fourier transformation (green) of the directing field, imposed as a periodized rectangular pulse.

spatial frequencies close to the frequency corresponding to the natural period of this lamella-forming BCP. For example, the green plot on Figure 9.3 shows a Fourier transformation of an external field applied as a periodized rectangular pulse with a spatial period, approximately corresponding to twice the natural period of the given BCP. The effect of the solution kernel of the linearized model in this case is amplification of the external field harmonics corresponding to the natural pitch of the BCP, eventually leading to the frequency doubling in the solution of the full system of the SCFT equations (9.2).

It should be noted, that the above straightforward linearization does not lead to an accurate SCFT model, because it fails to capture many important qualitative properties of the DSA process, which are due to its nonlinearity. Contrary to what is observed in simulations and in reality, the above linearized model predicts a linear dependence of the stationary BCP density variations on the externally imposed directing field. In particular, in the absence of the directing field, the linearized model yields uniform distribution of BCP densities, failing to predict the "fingerprint" lamellae patterns occurring in a bulk BCP melt. The above result from the linearized model can still be viewed (Fredrickson, 2006) as a result pertaining to the linearized stability analysis of the homogeneous BCP density distributions, demonstrating a peak instability at the spatial wavelengths close to the natural pitch of the BCP, which should eventually lead to a periodic lamellae solution.

9.2.6 Examples

Using the DSA PC problem formulation (9.3), several examples of the DSA PC algorithm application have been produced, and they are documented below. These examples are formulated for an important problem of placing arrays of vias or contact holes, fabricated using a graphoepitaxy confinement well DSA process. A direct problem was treated as a 2D problem of DSA in a thin film of a BCP melt. We used a BCP density-based cost function, using target BCP cylinders density profiles from a SCFT solution for hexagonally arranged cylinders.

To solve the optimization problem, we used two approaches: (1) the MATLAB Optimization Toolbox function based on a finite-difference evaluation of the partial derivatives and an active-set algorithm to account for constraints and (2) our unconstrained steepest descent algorithm utilizing linearized SCFT equations to evaluate a gradient of the cost function.

The SCFT parameters used in our examples are approximating those for a PS-b-PMMA diblock copolymer with a molecular weight of 46.1–21 kg/mol. The volume fraction of the PS phase was 0.7, and the unperturbed radius of gyration R_g was 10 nm. The Flory–Huggins parameter $\chi_{AB}N$ was 20.5, and compressibility parameter κN was set to 50. The wall affinity parameter $\chi_w N$ was set to (−17). The negative value of this parameter corresponds to the walls of the confinement well having higher affinity to the minority phase, resulting in a formation of the minority phase "wetting layer" near the wall, along with the cylinders in the interior of the confinement well. The step size of the spatial grid used in the SCFT solver was 0.25 R_g.

Figure 9.4 introduces the design variable and parameterization conventions used in this DSA PC algorithm. The actual shape of the confinement well boundary is given by a smooth curve, which is parameterized by a Manhattan polygon. The initial shape of this parameterizing polygon (shown in blue) is specified by the user, who also specifies the end points of the straight line segments forming the edges of this initial polygon. The design variables **d** are defined as the displacements applied to each segment of the initial polygon, shifting these segments, while leaving them parallel to their

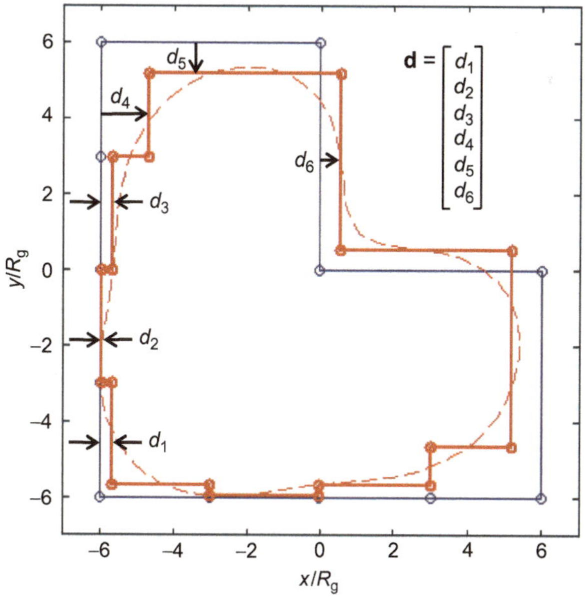

Figure 9.4 Design variables and parameterization conventions. Blue: initial, reference parameterizing polygon; solid red: parameterizing polygon, corresponding to the segment displacements, given by the design variables **d**; dashed red: the actual shape of the confinement well boundary, corresponding to the parameterizing polygon.

initial orientation. Positive components of **d** correspond to the shifts of these segments into the interior of the polygon. One example of a parameterizing polygon is shown on Figure 9.4 in solid red. In this example, mirror symmetry with respect to the bisector of the first main quadrant in the *XY* plane is enforced, resulting in the displacements of the 12 segments specified by 6 design variables. The smooth curve, corresponding to the actual shape of the confinement well (shown in dashed red), can be calculated using one of many possible algorithms, such as a spline interpolation.

Example 1 Two-parameter confinement well for fabrication of two vias

In this example, the confinement well is parameterized by a rectangle, which is defined by two parameters—height and width. The purpose of the DSA PC algorithm is to ensure a symmetric placement of two vias at a distance of 3.3 R_g between their centers. Figure 9.5 (left, top row) shows the evolution of the majority phase density field through the course of the DSA PC iterations. The plots at the bottom row show the initial polygon (solid blue rectangle), the current parameterizing rectangle (solid red), the current shape of the confinement well boundary (dotted red), the minority phase density contour, corresponding to the normalized density value of 0.4 (solid blue smooth contour), and the target density contour (solid black circle), corresponding to the same normalized density value of 0.4. The latter contour is plotted only for the left cylinder. These plots show that the BCP density converges to the target density with satisfactory accuracy within the first 9–15 iterations of the DSA PC algorithm.

Figure 9.5 (right) shows the comparison of the cost function value versus elapsed time for two versions of the DSA PC algorithm. The red line corresponds to the MATLAB Optimization Toolbox based optimization, utilizing finite-difference evaluation of the derivatives of the cost function. The blue line corresponds to our steepest descent optimization algorithm utilizing a linearized form of the SCFT equations to evaluate the gradient of the cost function, as outlined in Section 9.2.5. Both algorithms

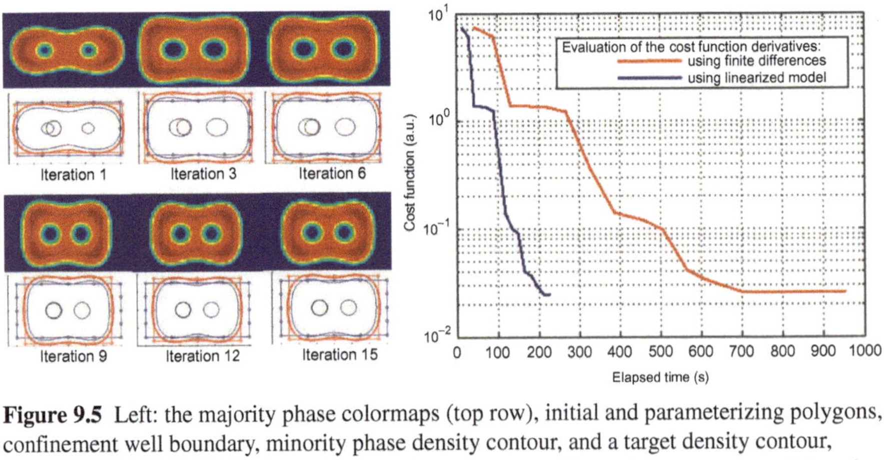

Figure 9.5 Left: the majority phase colormaps (top row), initial and parameterizing polygons, confinement well boundary, minority phase density contour, and a target density contour, though iterations of the DSA PC algorithm. Right: cost function value versus elapsed time for two versions of the DSA PC algorithm.

demonstrate a similar convergence history in terms of the cost function versus the number of iterations, but our steepest descent algorithm demonstrates about four times improvement in convergence time, due to faster evaluation of the partial derivatives of the cost function.

Example 2 Three-parameter confinement well for fabrication of two vias, varying the distance between via centers

In this example, we used our DSA PC algorithm to calculate the shapes of the confinement wells needed to place the minority phase cylinders at distances between their centers ranging from 2.7 to 4.5 R_g. The confinement well parameterizing polygon has three parameters in this example—the width and height of the rectangle and the displacement of the central segment of the longer side of this rectangle. The initial rectangle dimensions are 6 $R_g \times 12$ R_g. The length of the central segment along the longer side is always 6 R_g, and this segment can shift inward and outward, according to the sign and value of the third design variable.

Figure 9.6 shows the results obtained from our DSA PC algorithm, for each prescribed value of the center-to-center distance. The top-row plots show the colormaps of the majority phase density. The bottom-row plots show the initial polygon (a solid blue rectangle), the current parameterizing polygon (in solid red), the current shape of the confinement well boundary (dotted red), the minority phase density contour corresponding to the normalized density value of 0.4 (solid blue smooth contours), and the target density contour (a solid black circle) corresponding to the same normalized density value of 0.4. The latter contour is plotted only for the left cylinder.

The solutions obtained from the DSA PC algorithm show a qualitative trend; to bring the cylinders closer together, compared to their natural distance of 3.3 R_g, the algorithm needs to employ an outward notch and push the left and right walls closer together. To move the minority phase cylinders further apart, the opposite trend is used; the left and right walls are moved further apart and an inward notch is employed. The algorithm fails to achieve a sufficiently small value of a cost function for the minimal center-to-center distance between the minority phase cylinders of 2.7 R_g, compared to the final values of the cost function achieved for other tested values of center-to-center distances. This is evident from the noticeable discrepancy between the actual minority phase 0.4 value contour and the target density contour, for the case $\Delta R = 2.7$ R_g. One contributing factor to this degradation in convergence is that our target density profile

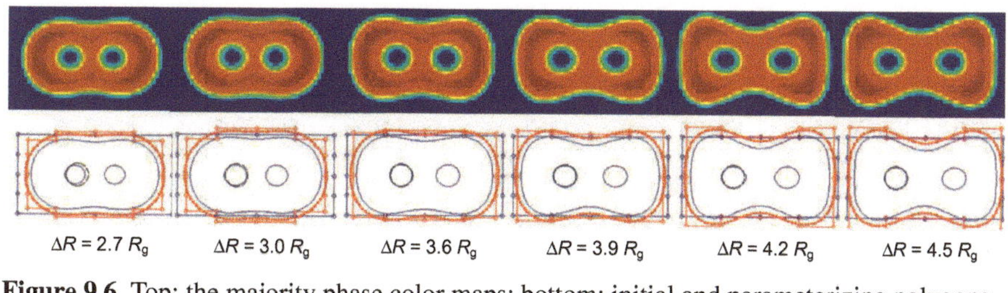

Figure 9.6 Top: the majority phase color maps; bottom: initial and parameterizing polygons, confinement well boundary, minority phase density contour, and a target density contour.

fails to account for the proximity effects between two minority phase cylinders pushed close together in the case $\Delta R = 2.7\ R_g$.

Example 3 Six-parameter confinement well for fabrication of three vias

In this example, we used our DSA PC algorithm to place the centers of the minority phase cylinders in the vertices of the right-angle triangle with the lengths of its legs (catheti) equal to 3.3 R_g. The parameterizing polygon has 12 moving segments, with the enforcement of mirror symmetry with respect to the bisector of the first main quadrant in the XY plane leading to 6 design variables. These parameters and other conventions for this example are shown in Figure 9.4.

The result of this DSA PC run is shown in Figure 9.7.

The plot on the left of Figure 9.7 shows the initial polygon (solid blue), the current parameterizing polygon (solid red), the current shape of the confinement well boundary (dotted red), the minority phase density contour, corresponding to the normalized density value of 0.4 (solid blue smooth contours), and the target density contours (solid black circles), corresponding to the same normalized density value of 0.4. The latter contours are plotted only for two of three minority phase cylinders. The plot on the right of the same figure shows the majority phase color map.

In this example, although the placement of the centers of the minority phase cylinders is achieved with an accuracy of a fraction of a nanometer, the discrepancy between the contours corresponding to the normalized density value of 0.4 (actual versus target) reaches 2–3 nm. We attribute this residual discrepancy to the coarse fragmentation of our parameterizing polygon. A finer parameterizing of the shape of the confinement well may be attempted to improve the final result, however, one should keep in mind the specific capabilities of the traditional lithography method (DUV / EUV / e-beam) used to pattern the confinement well.

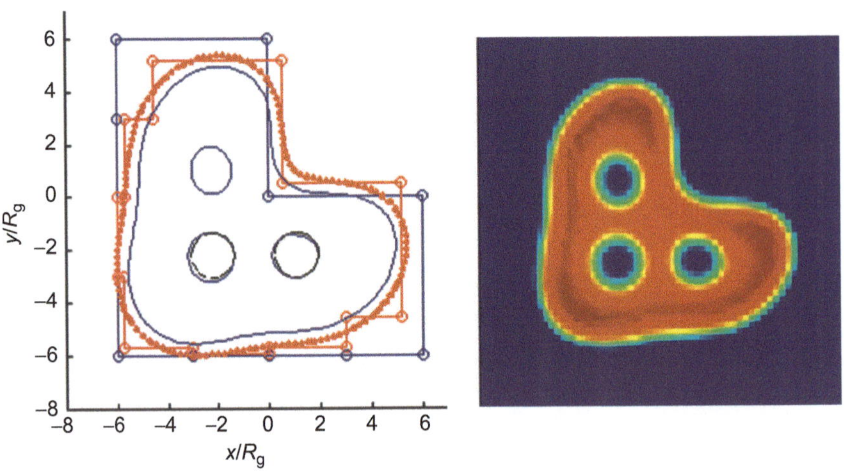

Figure 9.7 DSA PC result for Example 3. Left: initial and parameterizing polygons, confinement well boundary, minority 0.4 phase density contour, and a target 0.4 density contour. Right: majority phase density color map.

9.2.7 DSA-aware assist features: Using DSA to enhance the advantages of the optical lithography assist features

The RET/OPC steps of the optical lithography mask data preparation for DSA confinement wells may use traditional AF, improving both the optical lithography process windows and the PV bands of these DSA confinement wells. Such traditional AF need to be properly positioned and sized, using either rule-based or model-based approaches to ensure the desired improvements and to ensure nonprintability of these AF on the wafer. The fact that the optical lithography step is followed by the DSA process step provides an option to enhance the advantages of traditional AF.

An important property of a confinement well graphoepitaxy DSA process in a confinement well is that, depending on the size and the shape of the confinement well and other parameters of the DSA process, the final DSA result may be lacking etch-transferrable cylinders. In such cases, BCP is "sealing" the confinement well and preventing it from being etch transferred into the substrate during the etching step. This property can be illustrated by the results of one of our earlier parametric studies described in more detail in Latypov et al. (2013) and summarized in Figure 9.8, showing the majority phase density colormaps in the plane normal to the substrate and passing through the axis of symmetry of the confinement wells, for confinement wells of dimensions ranging from 55 to 90 nm.

These results show that the DSA processes in single cylinder confinement wells with sizes ranging between 60 and 80 nm, resulted in a minority phase cylinder at the center of the well. Such minority phase cylinder will be selectively removed during the development step of the DSA process and the resulting opening will be etch transferred into the substrate during the etching step. For confinement wells of sizes less than 60 nm, for example, for a 55-nm confinement well, such minority phase cylinder does not form. For a range of confinement wells dimensions, including 85 and 90 nm confinement wells, either an incomplete cylinder or a disk-shaped cavity form. The latter two DSA results will not be etch transferable into the substrate.

The availability of ranges of confinement well dimensions and shapes, leading to DSA results that are not etch transferable into a substrate, makes it possible to enhance the capabilities of traditional AF. We use this property to introduce the *DSA-aware* AF, as illustrated on a schematic diagram in Figure 9.9, where the presence of an etch-transferable DSA cylinder is indicated by blue circles. Like traditional AF,

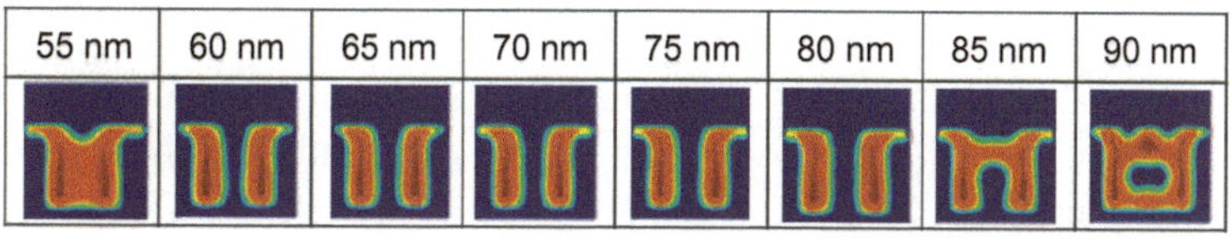

Figure 9.8 The majority phase density colormap in the plane normal to the substrate and passing through the axis of symmetry of the confinement wells, for confinement wells of dimensions ranging from 55 to 90 nm. The DSA process details and parameters can be found in Latypov et al. (2013).

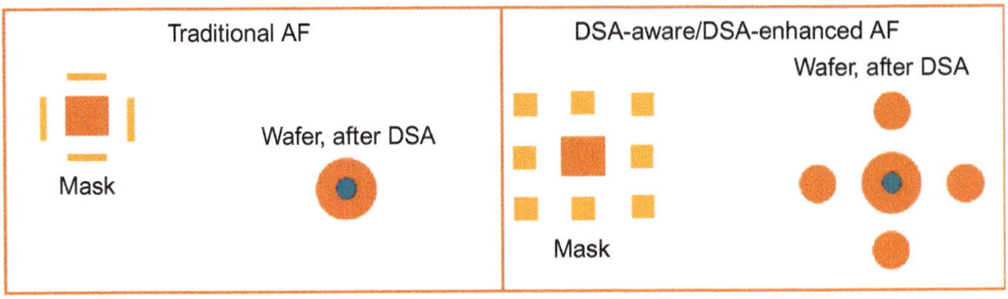

Figure 9.9 The concepts of traditional assist features (AF) and DSA-aware assist features.

DSA-aware AF, shown in Figure 9.9 (right), serve to improve the optical lithography process window and the PV band of the confinement well. Unlike the traditional AF, the DSA-aware AF can print on the wafer; as long as they do not result in etch-transferable DSA cylinders in them. Allowing DSA-aware AF to print on the wafer, significantly relaxes the constraints imposed on traditional AF, thus helping them to further optimize the process window and the PV band. Another advantage of DSA-aware AFs might be a mask making cost reduction and an improvement in the accuracy of the mask making process, both due to the fact that DSA-aware AFs can be made larger on the mask as compared to the traditional AFs.

Figure 9.10 shows one example of process window improvement due to DSA-aware AF in case of a DSA contact hole shrinkage process. Note that the application of DSA-aware AF is not limited to the DSA contact hole shrinkage cases. The contact hole shrinkage example is chosen here because of its simplicity to demonstrate the basic concept of DSA-aware AF. A similar technique can be applied to confinement wells with multiple DSA cylinders and to simultaneous source and mask optimization.

In this example, an immersion tool with 1.35 NA and annular illumination is used. First, we simulated the confinement well generation procedure using the traditional AF. We optimized the AF locations to have the maximum process window while constraining their sizes so as they don't print through the process variations. This case is labeled as "Traditional AF" in Figure 9.10. Following that, we enabled the DSA-aware AF and optimized their locations to get the maximum process window, but also let them print through the process window. This case is labeled as "DSA-aware AF" in Figure 9.10. In both cases, the isolated contact size is optimized to print at 50 nm at nominal condition. Figure 9.10c and d shows the litho contours at nominal and defocus conditions. As shown in Figure 9.10c, AF print at nominal as well as defocus conditions when DSA-aware AF are enabled. Figure 9.10e shows the results of DSA simulations demonstrating that AF of this size and smaller are not etch transferable in the DSA process step. In this DSA graphoepitaxy example, the minority phase etches faster than the majority phase creating a hole where minority phase is localized. Traditional AF insertion resulted in 73 nm DOF at 5% EL, whereas we were able to get 89 nm DOF at 5% EL with DSA-aware AF as shown in Figure 9.10a. Figure 9.10b shows the cross-sections of confinement wells and minority phase density distributions of the main feature with DSA-aware AF and with traditional AF at different optical lithography focus conditions. These simulation results show etch-transferable

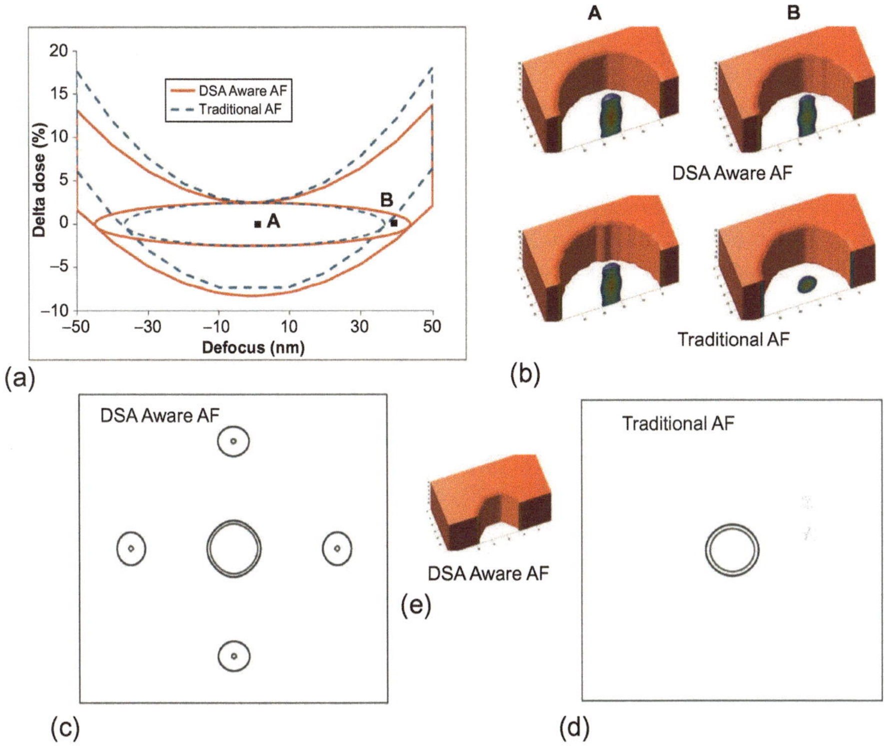

Figure 9.10 Process window enhancement by using DSA-aware assist features. (a) Optical lithography process window with DSA-aware assist features and with traditional AF. (b) Cross-sections of confinement wells and minority phase density distributions with DSA-aware AF and with traditional AF at different focus conditions (A and B). (c) and (d) Litho contours at nominal (A) and defocus (B) conditions with DSA-aware AF and with traditional AF. (e) The cross-section of a confinement well corresponding to the DSA-aware AF printed at nominal conditions. The absence of the minority phase feature at the center of this confinement well indicates that the DSA result in it is not etch transferrable into the substrate.

DSA pattern survived at the edge of the process window (marked as B), when DSA-aware AF were used, whereas it became not etch transferable when traditional non-printing AF were used. In this example, we were able to increase DOF by 16 nm by using DSA-aware AF.

9.3 Conclusions and future directions

We present an approach to the solution of the inverse DSA problem, based on the SCFT DSA solver. The inverse DSA problem is formulated as a smooth constrained nonlinear optimization problem, where evaluation of the cost function involves a solution of the SCFT equations at each iteration. To speed up the evaluation of the partial derivatives of the cost function with respect to the design variables, we propose to use a linearized version of the SCFT equations, obtained, for example, by using the weak inhomogeneity expansion.

We present several proof-of-the-concept examples, demonstrating applications of our implementations of the proposed DSA PC algorithms to the DSA PC problems of designing the shapes of confinement wells used in vias/contact holes fabrication process based on graphoepitaxy DSA.

A number of directions can be identified as future work.

First, a faster DSA model is desirable to speed up the calculations. Different levels of accuracies are needed from the DSA model based on its application. The most applicable model to inverse DSA problem might be a compact model similar to compact resist models used in OPC tools (see Chapter 11). In recent years, several research groups presented their work on the models of the DSA process, which claimed to be sufficiently fast to be used for the inverse DSA problem solution on the scale of the whole wafer or a single IC chip, within reasonable time. In Liu et al. (2012) and Lai et al. (2013, 2014), along with the rigorous DSA models, the results of applications of several faster simplified DSA models are presented, without presenting the detailed background of these simplifications, other than saying that these simplifications are based on a 2D phenomenological phase field model, requiring a solution of the Cahn–Hilliard type of equation. A similar approach to a simplified DSA model leading to the Cahn–Hilliard equation has been used in Qin et al. (2013), where such model has been employed in an evolutionary algorithm used to solve a family of inverse DSA problems, determining the optimal placements of circular dots guiding a lamellae-forming BCP to the desired target patterns. Another similar approach to a faster and simplified DSA model, the Ohta–Kawasaki model (Ohta and Kawasaki, 1986) has been used for direct DSA simulations in Yoshimoto and Taniguchi (2013) and Yoshimoto et al. (2014). In Torres et al. (2014), along with the discussion of the IC design aspects stemming from a possibility to use DSA to fabricate certain IC layers, both rigorous and compact DSA model development is mentioned, and the compact model accuracy requirements are discussed. In Latypov (2014), the approach to fast DSA model, based on a linearization of the dependence of BCP densities on the SCFT fields (see Section 9.2.5) has been presented and tested.

The fast DSA model should be accurate enough to predict the center location of the DSA-formed features and their shapes. They should also allow calibration against experimental observations.

It should be kept in mind that these models do not necessarily need to predict the 3D variations and dependency on the confinement side wall angle. However, if the fast DSA model lacks such 3D prediction capability, the process and solution domain need to be optimized by using more accurate rigorous simulations or experiments. For example, Figure 9.11 demonstrates one example of the DSA process optimization for the two-cylinder confinement well designed using inverse DSA solver, where the thickness of the BCP film is optimized to avoid the 3D variations in the DSA solution, which may negatively affect the etch transfer of the DSA result from the BCP film into the substrate.

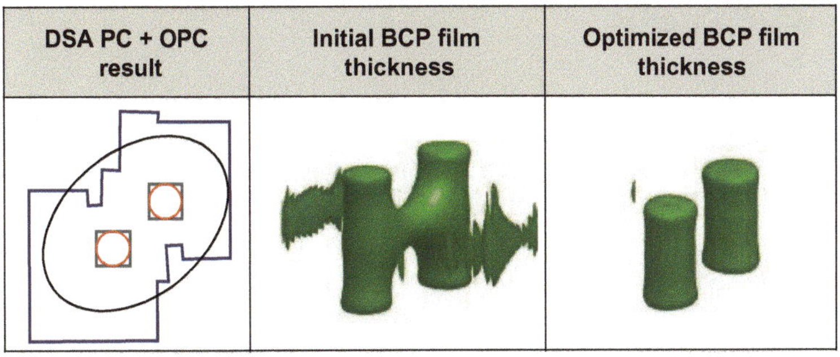

Figure 9.11 Effect of BCP film thickness optimization to derive a more stable solution.

For such verification and optimization, the variation of the BCP densities in the direction normal to the substrate needs to be accounted for in a 3D DSA model to ensure that the final result has the BCP density variations usable in the etch transfer into the substrate. Another form of verification model could be a compact model, which can sample the variations in the direction normal to the substrate. This type of models can be generated by calibration against either rigorous simulations or experimental data. A further simplification to these models can be done to provide an estimate of "good" or "bad" patterns in terms of whether it can be etched through or not.

Yet another potential future challenge for the practical inverse DSA problem solution comes from the effects occurring due to an overfill of the graphoepitaxy confinement wells and the well-to-well DSA proximity effects, discussed in Section 9.2.2. These effects need to be explored and treated in a computationally efficient way.

Although the inverse DSA problem solution examples presented in this chapter deal with graphoepitaxy DSA processes in confinement wells, the inverse problem formulation and solution approaches presented in Sections 9.2.3–9.2.5 are applicable to other topographical direction approaches and chemoepitaxy DSA as well. For instance, the similarly formulated inverse DSA problems for lamellae-forming BCPs directed by topographical poles or circular dots have been treated in Qin et al. (2013) and Hannon et al. (2014). One aspect of chemoepitaxy DSA relevant to inverse DSA problems is that chemoepitaxy patterns can be designed to change the local orientation of the BCP microdomains in the thin film allowing the formation of isolated features. For instance, a formation of an isolated line in a thin lamella-forming BCP film can be achieved by blocking a formation of some of the lamellae perpendicular to the substrate using an extended chemoepitaxy patch, resulting in some lamellae running parallel to the substrate of the BCP film. Such chemoepitaxy DSA has been demonstrated experimentally (Stoykovich et al., 2007; Xu et al., 2011) and through simulations (Latypov et al., 2013), for isolated lamellae arrays and cylinders. A design of the sizes, shapes, and placements of graphoepitaxy patches, controlling the orientation of the BCP microdomains in the thin BCP film and allowing to pattern a broader variety of patterns in the substrate (e.g., isolated groups of controllably bent and terminated lines), is yet another important future application for inverse DSA problems.

Acknowledgments

The authors would like to thank Roel Gronheid of IMEC and also our colleagues at GLOBALFOUNDRIES—Luigi Capodieci, Deniz Civay, Vito Dai, Nicholas Eib, Richard Farrell, Rani Ghaida, Sarah McGowan, Moshe Preil, Gerard Schmid, Christopher Spence, Keith Standiford, Edward Teoh, Wei-Long Wang, Ji Xu, He Yi, Yang Yue, and Yi Zou—for reviewing the results presented in this chapter and for many useful suggestions.

This chapter is based, in part, on the recently published SPIE papers (Latypov, 2013; Latypov et al., 2013, 2014) coauthored by authors. The permission from SPIE to reproduce the material (text and figures) from these SPIE papers is hereby gratefully acknowledged.

References

Ceniceros, H.D., Fredrickson, G.H., 2004. Numerical solution of polymer self-consistent field theory. SIAM, Multiscale Model. Simul. 2, 452–474.

Fredrickson, G.H., 2006. The Equilibrium Theory of Inhomogeneous Polymers. Oxford University Press, Oxford.

Hannon, A.F., Ding, Y., Bai, W., Ross, C.A., Alexander-Katz, A., 2014. Optimizing topographical templates for directed self-assembly of block copolymers via inverse design simulations. Nano Lett. 14 (1), 318–325.

Herr, D., 2011. Directed block copolymer self-assembly for nanoelectronics fabrication. J. Mater. Res. 26 (02), 122–139.

Lai, K., Liu, C.-c., Pitera, J., Dechene, D.J., Schepis, A., Abdallah, J., Tsai, H., Guillorn, M., Cheng, J., Doerk, G., Tjio, M., Rettner, C., Odesanya, O., Ozlem, M., Lafferty, N., 2013. Computational aspects of optical lithography extension by directed self-assembly. In: Optical Microlithography XXVI, 868304, San Jose, California, 2013. Proceedings of SPIE, vol. 8683.

Lai, K., Ozlem, M., Pitera, J.W., Liu, C.-c., Schepis, A., Dechene, D., Krasnoperova, A., Brue, D., Abdallah, J., Tsai, H., Guillorn, M., Cheng, J., Doerk, G., Tjio, M., Topalogu, R., Fahkry, M., Lafferty, N., 2014. Computational lithography platform for 193i-guided directed self-assembly. In: Optical Microlithography XXVII, 90521A, San Jose, California. Proceedings of SPIE, vol. 9052.

Latypov, A., 2013. Computational solution of inverse directed self-assembly problem. In: Alternative Lithographic Technologies V, 86800Z, San Jose, California, 2013. Proceedings of the SPIE, vol. 8680.

Latypov, A., 2014. Towards the fast DSA: improvement of computational performance of the DSA model. In: Alternative Lithographic Technologies VI, 904923, San Jose, California, 2014. Proceedings of SPIE, vol. 9049.

Latypov, A., Preil, M., Schmid, G.M., Xu, J., Yi, H., Yoshimoto, K., Zou, Y., 2013. Exploration of the directed self-assembly based nano-fabrication design space using computational simulations. In: Alternative Lithographic Technologies V, 868013, San Jose, 2013. Proceedings of SPIE, vol. 8680.

Latypov, A., Coskun, T.H., Garner, G., Preil, M., Schmid, G., Xu, J., Zou, Y., 2014. Simulations of spatial DSA morphology, DSA-aware assist features and block copolymer-homopolymer blends. In: Alternative Lithographic Technologies VI, 904908, San Jose, California, 2014. Proceedings of SPIE, vol. 9049.

Liu, C.-C., Pitera, J., Lafferty, N., Lai, K., Rettner, C., Tjio, M., Arellano, N., Cheng, J., 2012. Progress towards the integration of optical proximity correction and directed self-assembly of block copolymers with graphoepitaxy. In: Alternative Lithographic Technologies IV, 83230X, San Jose, California, 2012. Proceedings of SPIE, vol. 8323.

Ma, X., Arce, G.R., 2010. Computational Lithography. John Wiley & Sons, New York.

Matsen, M.W., 1997. Thin films of block copolymer. J. Chem. Phys. 106, 7781–7791.

Ohta, T., Kawasaki, K., 1986. Equilibrium morphology of block copolymer melts. Macromolecules 19 (10), 2621–2632.

Qin, J., Khaira, G.S., Su, Y., Garner, G.P., Miskin, M., Jaeger, H.M., de Pablo, J.J., 2013. Evolutionary pattern design for copolymer directed self-assembly. Soft Matter 9, 11467.

Stoykovich, M.P., Kang, H., Daoulas, K.C., Liu, G., Liu, C.-C., de Pablo, J.J., Muller, M., Nealey, P.P., 2007. Directed self-assembly of block copolymers for nanolithography: fabrication of isolated features and essential integrated circuit geometries. ACS Nano 1 (3), 168–175.

Takahashi, H., Laachi, N., Delaney, K.T., Hur, S.-M., Weinheimer, C.J., Shykind, D., Fredrickson, G.H., 2012. Defectivity in laterally confined lamella-forming diblock copolymers: thermodynamic and kinetic aspects. Macromolecules 45 (15), 6253–6265.

Torres, J.A., Sakajiri, K., Fryer, D., Granik, Y., Ma, Y., Krasnova, P., Fenger, G., Nagahara, S., Kawakami, S., Rathsack, B., Khaira, G., de Pablo, J., Ryckaert, J., 2014. Physical verification and manufacturing of contact/via layers using grapho-epitaxy DSA processes. In: Design-Process-Technology Co-optimization for Manufacturability VIII, 90530R, San Jose, California, 2014. Proceedings of SPIE, vol. 9053.

Wang, W.-L., Latypov, A., Zou, Y., Coskun, T., 2014. A full-chip DSA correction framework. In: Alternative Lithographic Technologies VI, 90491 J, San Jose, 2014. In: Proceedings of SPIE, vol. 9049.

Xu, J., Russell, T.P., Ocko, B., Checco, A., 2011. Confined block copolymer assembly in chemical patterned squares. Soft Matter. 7, 3915–3919.

Yi, H., Bao, X., Zhang, J., Tiberio, R., Conway, J., Chang, L.-W., Mitra, S., Wong, H.-S.P., 2012. Contact hole patterning for random logic circuits using block copolymer directed self-assembly. In: SPIE 2012, Conferences on Alternative Lithographic Technologies IV, 8323-31, San Jose.

Yoshimoto, K., Taniguchi, T., 2013. Large-scale dynamics of directed self-assembly defects on chemically pre-patterned surface. In: Alternative Lithographic Technologies V, 86801I, San Jose, California, 2013. In: Proceedings of SPIE, vol. 8680.

Yoshimoto, K., Fukawatase, K., Ohshima, M., Naka, Y., Maeda, S., Tanaka, S., Morita, S., Aoyama, H., Mimotogi, S., 2014. Optimization of directed self-assembly hole shrink process with simplified model. J. Micro/Nanolith. MEMS MOEMS 13 (3), 031305.

Directed self-assembly guiding template design for contact hole patterning

10

H. Yi, H.-S.P. Wong
Stanford University, Stanford, CA, USA

10.1 Flexible control of directed self-assembly using physical guiding templates

Patterning contact holes for very large-scale integrated (VLSI) circuits is one of the most difficult challenges for lithography. It requires very precise control of both the size and the position of cylindrical structures. The current state-of-the-art optical lithography can reach the 3-sigma values for critical dimension (CD) control and overlay (registration of one layer to another layer of lithography) accuracy at 2.0 and 4.0 nm, respectively (International Technology Roadmap for Semiconductors, 2013). However, for the 10 nm technology node and beyond, the required lithography 3-sigma values for CD control and overlay accuracy are below 1.0 and 3.2 nm. Among all the emerging patterning approaches, block copolymer directed self-assembly (DSA) is one of the most promising candidates for patterning contact holes due to its low cost, high throughput, and well-controlled sub-20 nm features; it has demonstrated potential to scale well below 14 nm (Cheng et al., 2008; Black et al., 2007; Bita et al., 2008; Ruiz et al., 2008).

Apart from the requirement of overlay accuracy and contact sizes, an additional challenge for contact hole patterning is that most contacts in integrated circuits (ICs), especially those in standard cells of logic gates, are distributed in an aperiodic order. To generate aperiodic block copolymer self-assembly patterns with precise size and position control, researchers have explored the use of DSA using lithographically defined guiding templates (Bita et al., 2008; Segalman et al., 2001). The chemical guiding template approaches use chemical patterns of preferential affinity on the substrate surface to form aperiodic patterns (Detcheverry et al., 2010; Stoykovich et al., 2005). To date, most studies on chemical guiding templates have been focused on generating lamella patterns for lines and spaces (Ruiz et al., 2008; Tang et al., 2008). In order to generate aperiodic cylindrical patterns for holes, physical guiding templates approaches have proven promising. Due to the strong confinement from physical guiding templates, the block copolymer's natural symmetry is altered and aperiodic cylindrical patterns, therefore, readily form inside the templates. Because the DSA pattern position is determined by position of the guiding templates, DSA patterns can be almost arbitrarily placed as required (Figure 10.1) (Bita et al., 2008; Bao et al., 2011).

Directed Self-assembly of Block Copolymers for Nano-manufacturing. http://dx.doi.org/10.1016/B978-0-08-100250-6.00010-9

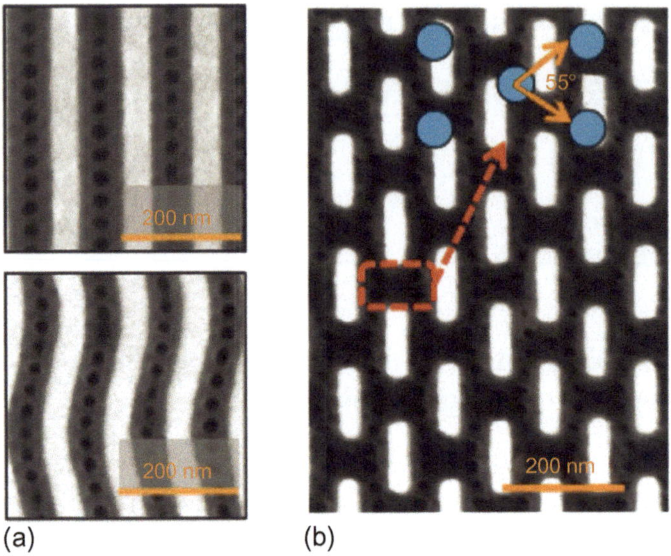

Figure 10.1 (a) DSA patterns inside the trench. Physical confinement forces self-assembly to rearrange according to prepattern placement. (b) DSA patterns guided by the separated islands. The inherently hexagonal closed pack structure of the block copolymer is not followed due to the physical confinement by the guiding template.

By controlling the template shape and block copolymer film thickness, various types of cylindrical structures are studied and placed as required in ICs exactly, using a simple 70:30 PS-*b*-PMMA diblock copolymer system as a model material system. This type of PS-*b*-PMMA block copolymer is known to naturally generate hexagonally ordered cylindrical microdomains with diameter of ~20 nm and pitch of ~40 nm (Bao et al., 2011). A series of small square templates were fabricated using conventional optical lithography and etching process on 300 mm Si wafer with sizes ranging from 75 to 140 nm and depth around 50 nm. After spin coating and annealing at 180 °C, the wafers were exposed to deep UV light before acetic acid development to remove PMMA from the DSA holes. The DSA patterns were characterized by high-resolution scanning electron microscopy (SEM). This DSA process inside the small guiding templates does not require the homopolymer or the random copolymer coating that was used to neutralize the substrate surface. We demonstrated that the self-assembled features can almost be arbitrarily placed as required by circuit fabrication and not limited to regular patterns, by combining templates of different types on one wafer (Yi et al., 2012a).

Figure 10.2 summarizes some of the major IC-layout relevant DSA patterns obtained in small topographical templates. Single-hole patterns (Figure 10.2a and b) can be formed in square templates from 60 to 100 nm. The 4-hole patterns with square symmetry can be formed in the relatively large template size window between 110 and 130 nm (Figure 10.2c). In rectangular templates (width from 60 to 80 nm, length from 100 to 150 nm), 2-hole and 3-hole patterns can be formed. Besides hole size distribution, we also study the overlay accuracy of DSA patterns. We calculate the

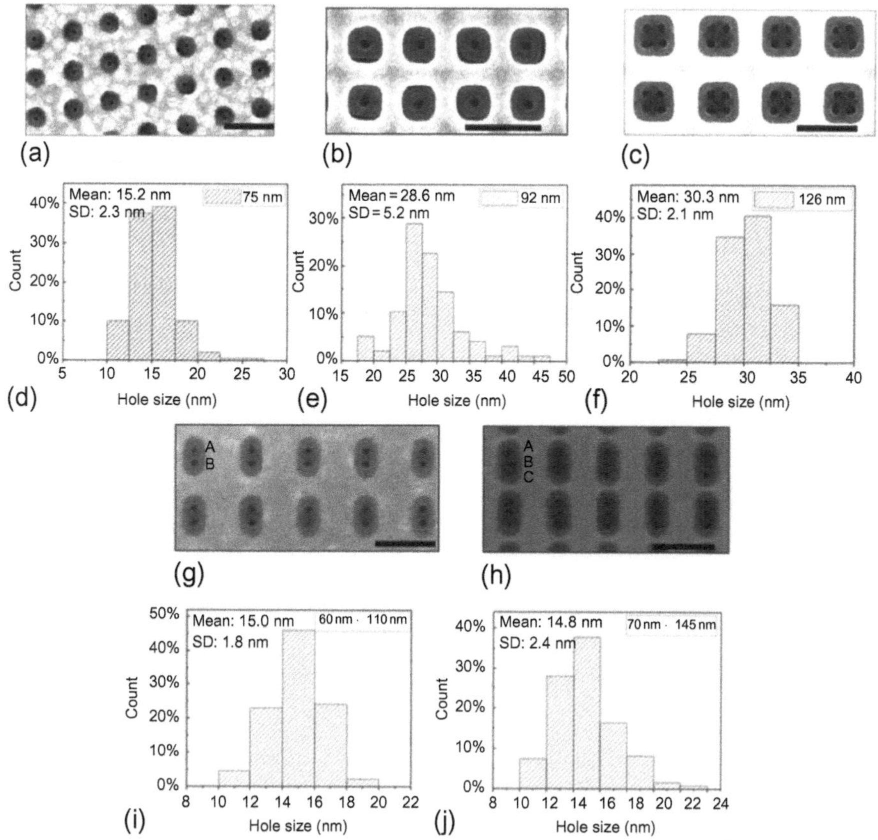

Figure 10.2 SEM images of DSA patterns confined by templates with different sizes and the corresponding hole size distribution. (a) Single hole in 75 nm squares. (b) Single hole in 92 nm squares. (c) 4-Hole patterns in 126 nm square templates. (d) Hole size distribution of (a). (e) Hole size distribution of (b). (f) Hole size distribution of (c). (g) 2-Hole patterns in 60 nm × 110 nm rectangle templates. (h) 3-Hole pattern in 70 nm × 145 nm rectangle templates. (i) Hole size distribution of (g). (j) Hole size distribution of (h). Scale bars: 200 nm. Reprinted (adapted) with permission from Yi et al. (2012a). Copyright 2015 John Wiley and Sons.

average position of DSA patterns in guiding templates of the same size, and obtain the overlay accuracy of the self-assembled features in terms of average positional deviation of self-assembled holes from the average position. It is shown in Figure 10.3 that the average positional deviations for both 2-hole and 3-hole patterns are less than 2 nm, indicating a high repeatability and accurate placement. For template size larger than 140 nm, the pattern formation becomes complicated and uncontrollable due to the reduced confinement effect from the guiding templates and the internal force of the block copolymers becoming dominating. As shown in Figure 10.3, the average DSA pattern size deviated from the naturally self-assembled PS-*b*-PMMA (~20 nm). The

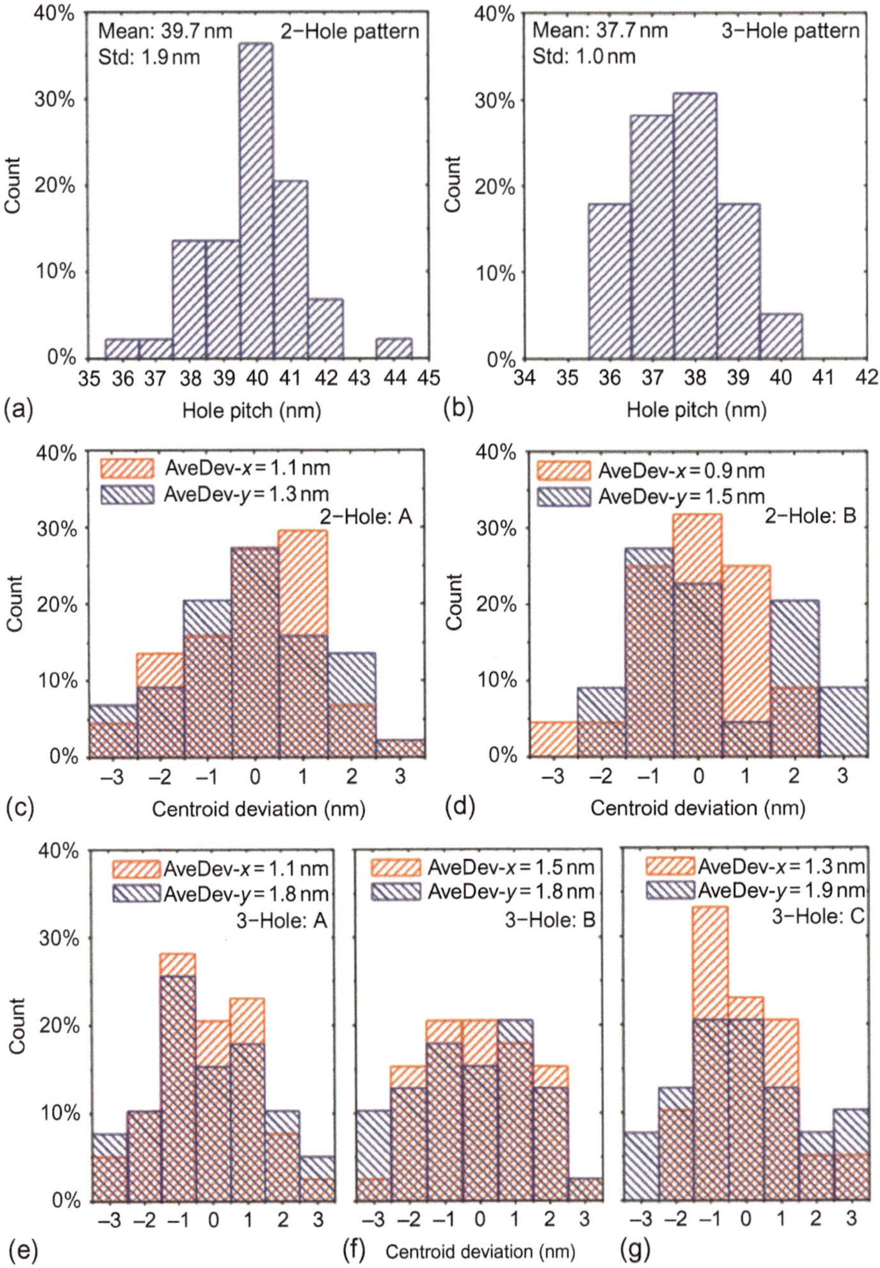

Figure 10.3 The pitch and overlay accuracy analysis for 2-hole and 3-hole patterns. (a) Hole pitch distribution for the 2-hole pattern. (b) Hole pitch distribution for the 3-hole pattern. (c) Overlay accuracy analysis for upper hole A in the 2-hole pattern. (d) Overlay accuracy analysis for lower hole B in the 2-hole pattern. (e) Overlay accuracy analysis for upper hole A in the 3-hole pattern. (f) Overlay accuracy analysis for middle hole B in the 3-hole pattern. (g) Overlay accuracy analysis for lower hole C in the 3-hole pattern.
Reprinted (adapted) with permission from Yi et al. (2012a). Copyright 2015 John Wiley and Sons.

size deviation and symmetry change suggested that both the PS and PMMA polymer chains were stretched or compressed during the self-assembly with the strong lateral confinement.

Our experiment reveals that DSA pattern formation can be controlled by varying the block copolymer film thickness in the template (Yi et al., 2012a). It is important to note that block copolymer film thickness inside the guiding templates is not only controlled by the spin speed in DSA process, but also by the size and density of guiding templates. The size and density of templates can cause variation of the final polymer thickness in the template well after reflowing during annealing at elevated temperatures. For the case in which one template has multiple DSA holes, the hole pitch could depend on the size of the template as well. The relation between the DSA hole pitch and the size and shape of the templates can be considered as constituting a design space for the block copolymer self-assembly into cylindrical patterns. The DSA design space defines the pattern geometries as a function of spin speed, template size, and density, with which a set of conditions can be determined to generate a specific pattern. It is important to note that the chemical affinity of the substrate and the depth of the template also affect the DSA pattern. However, the templates in one contact layer must share the same depth and substrate material, so for simplicity, these two parameters are not listed in the design space. All of the templates used in these experiments were fabricated on Si and etched to a depth of 50 nm.

The design space of DSA 2-hole and 3-hole pattern is shown in Figure 10.4. We measure the change in pitch for the 2- and 3-hole DSA patterns resulting from increasing the length of the elliptical templates while keeping the width unchanged (Figure 10.4a). There is clear transition from 2-hole to 3-hole pattern as the template length increases. Figure 10.4b shows the required spin speed and correspondent template size and spacing to generate 4-hole pattern, indicating a broad process window for DSA. Proper template designs can avoid conditions at the boundaries of the design space to ensure a wide process window. The design space can be codified as "design rules" for IC layouts and incorporated in electronic design automation (EDA) tools for use in manufacturing (Du et al., 2013; Lai et al., 2013).

10.2 Contact/via patterning using block copolymer DSA—Practical examples

10.2.1 SRAM contact hole patterning for 22 nm node using block copolymer DSA

The flexible control of DSA pattern afforded by small physical guiding templates enables patterning contact holes of practical devices for memory and logic circuitries. In this section, we will illustrate DSA patterning of contacts holes for nanoscale devices. As a first example, we chose 22 nm technology node SRAM as a demonstration vehicle to testify our patterning strategy for contact holes using block copolymer DSA since SRAM is an industry-standard ramp-up vehicle for logic process technology development (Bao et al., 2011). Figure 10.5a shows a device layout directly derived

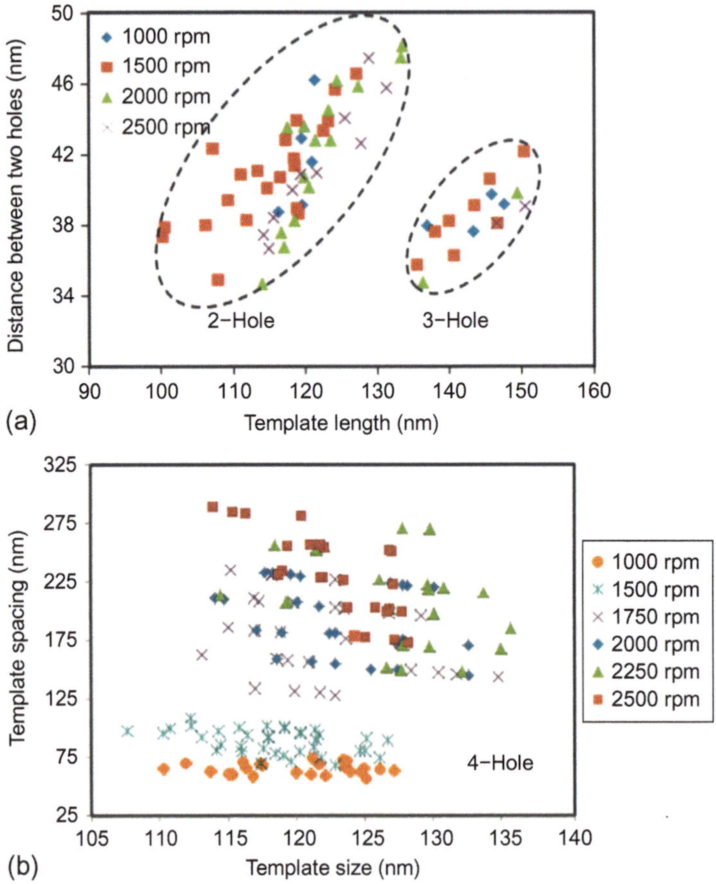

Figure 10.4 Design space for the 2-hole, 3-hole, and 4-hole pattern. Each data point represents a template that successfully generates a desired DSA pattern. (a) The transition between the 2-hole pattern and the 3-hole pattern under different block copolymer coating speed. As template length increases, the distance between holes keep increasing until the 2-hole pattern turns into a 3-hole pattern. (b) The required spin coating speed, template size, and spacing to generate a 4-hole pattern in a square template.
Reprinted (adapted) with permission from Yi et al. (2012a). Copyright 2015 John Wiley and Sons.

from the 22 nm 6T-SRAM cell demonstrated by IBM (Haran et al., 2008). In the layout, the contact holes were designed with multiple pitches and orders, and elongated holes were used to connect the polysilicon gate and the active source/drain diffusion region (Figure 10.5a, rectangular connections). To improve the design regularity, the elongated connections can be replaced by two identical square contact holes with smaller pitch (~45 nm) (Figure 10.5a, indicated with blue circles), which can hardly be patterned by the current optical lithography. The DSA process needs to generate hole patterns with pitches and orders for this SRAM design in a single process step.

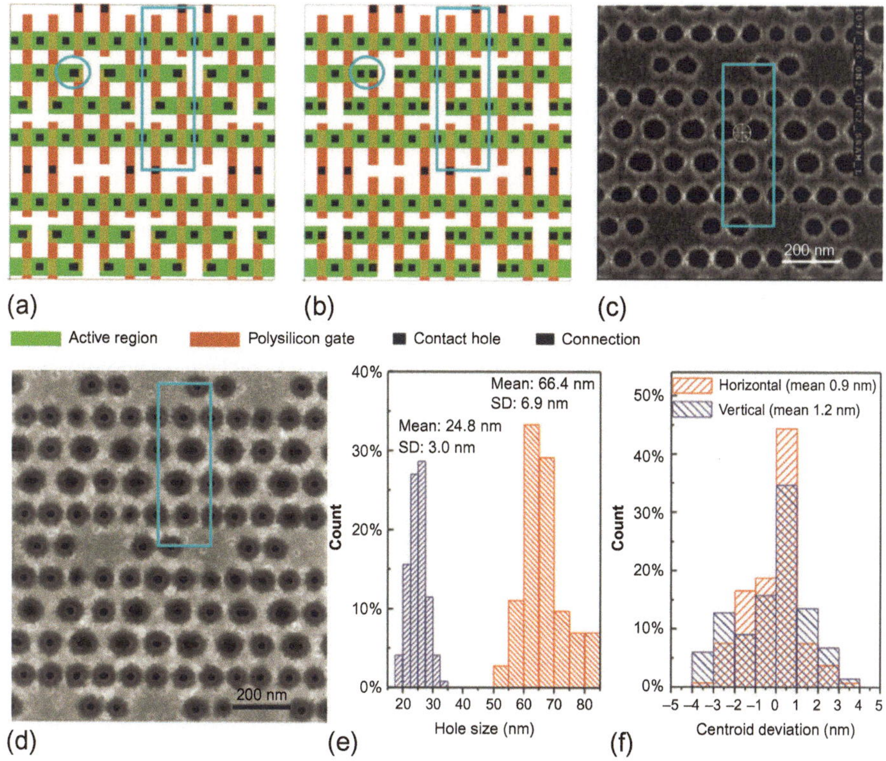

Figure 10.5 Contact hole patterning for 22 nm SRAM cell. (a) Contact hole layout derived directly from the IBM 22 nm 6T-SRAM circuit. (b) Modified layout by replacing the rectangular connections with square holes, and one of the modification is highlighted by blue circles. (c) Guiding templates fabricated with 193 nm immersion optical lithography. (d) PS-*b*-PMMA DSA in the guiding templates. (e) Statistical results for the patterned contact hole size distribution (blue) and the fabricated guiding template size distribution (red). (f) The centroid deviation of the patterned features from the guiding templates.
Reprinted (adapted) with permission from Yi et al. (2012a). Copyright 2015 John Wiley and Sons.

Based on the multiple types of patterns achieved using small topographical templates shown above, we fabricated small guiding templates (shown in Figure 10.5c) according to the DSA-friendly SRAM layout in Figure 10.5a. The templates were patterned on 300 mm Si wafers by 193 nm immersion lithography with single exposure and etched with conventional reactive ion etching process. The fabricated templates have an average size of about 66 nm (analyzed in Figure 10.5e and f). Larger size elliptical templates were fabricated to generate contact hole pairs for the connection between polysilicon gate and active source/drain region. Contact holes with average size of ~25 nm with 3 nm standard deviation were successfully generated after performing the PS-*b*-PMMA DSA process (shown in Figure 10.5d). The DSA generated holes were located at the centroids of the guiding templates with overlay accuracy about 1 nm (Figure 10.5f). The contact hole pairs were generated in the larger elliptical templates with an average pitch of ~30 nm, which are suitable for making the connections for contacting the polysilicon gate and active source/drain region. Thus, contact holes for the 22-nm SRAM devices were successfully patterned using simple PS-*b*-PMMA DSA process with small templates of 66 nm patterned by the current 193 nm immersion optical lithography. The CD limit of the 193 nm immersion optical lithography has been shrunk down by a factor of two to three. This is the first time the contact hole pattern for an industry-standard circuit is demonstrated using DSA combined with 193 nm immersion lithography.

10.2.2 DSA contact patterning for 14 nm random logic circuits

As the second example, we report contact hole patterning for 22 nm technology node "random logic" (Geiger et al., 1990) circuits in an optimized DSA-aware layout design using the same PS-*b*-PMMA DSA process. Patterning for "random logic" circuits is more challenging because contacts are placed more randomly than memory circuits. As shown above, although the small guiding templates are shown to guide the self-assembly off the natural geometry by strong boundary confinement, it is insufficient to simply surround contact holes with guiding templates without optimizing the placement and geometry of the guiding templates. This is because the guiding templates are larger than the desired contact holes, and it may not be possible to pattern the guiding template (with larger feature sizes) using conventional optical lithography due to pitch requirements.

As a starting point, we develop unidirectional DSA-aware designs for a standard cell library. During the course of device scaling in the past decade, industry has realized that simply shrinking the layout geometries from one technology node to the next is no longer viable because of the proximity effects in optical lithography. Instead of having bidirectional structures, such as L-shaped or T-shaped metal lines, IC layouts have now evolved toward gridded unidirectional layouts (Belledent et al., 2012; Smayling et al., 2008; Vaidyanathan et al., 2012, 2013). Unidirectional IC layouts represent an important IC development trend due to their larger process window and higher yield, features for which they have been widely adopted by the semiconductor industry since the 45 nm technology node. Such layouts generally involve an IC layout architecture wherein each device layer consists only of unidirectional lines and contact holes positioned at predetermined grid points (Vaidyanathan et al., 2012, 2013). The bends and junctions of the previous bidirectional layouts are implemented by orthogonal lines on separate layers connected by a contact. The irregularity of the contact distribution is greatly reduced by aligning the contacts to regular grid points. While unidirectional IC design is not a must for DSA, the use of unidirectional layout design could greatly improve the throughput and yield for DSA contact patterning (Yi et al., 2012b). In the scenario that multiple self-assembled holes are generated inside one template, this yield improvement can be maximized when the contact pitch/size matches the natural geometry of the self-assembly. This is because the ideally ordered (natural pitch/size) polymers have the lowest free energy and the DSA process will be most energetically favored, gaining the lowest defectivity. Therefore, DSA-aware unidirectional cell design provides an excellent starting point for DSA patterning optimization by minimizing layout modification; simplifying guiding template design and improving DSA yield.

Figure 10.6a shows a half-adder layout from a 45 nm open cell library.[1] During layout optimization process, the transistor sizes and connections (pin locations) of the logic cells were kept unchanged to minimize any impact on the VLSI (physical) design flow at the full-chip level. We shrank the cell layouts down to 14 nm node with poly gate pitch of 82 nm and metal-1 pitch of 64 nm (shown in Figure 10.6b) (Belledent et al., 2012). In addition, the size and pitch of guiding templates must be consistent with the capabilities of conventional optical lithography that has larger feature sizes. For example, Figure 10.6c shows an improper contact layout design, where the corresponding guiding templates are too closely packed to be printed. To enable patterning of the guiding templates, the positions of two metal lines are switched while

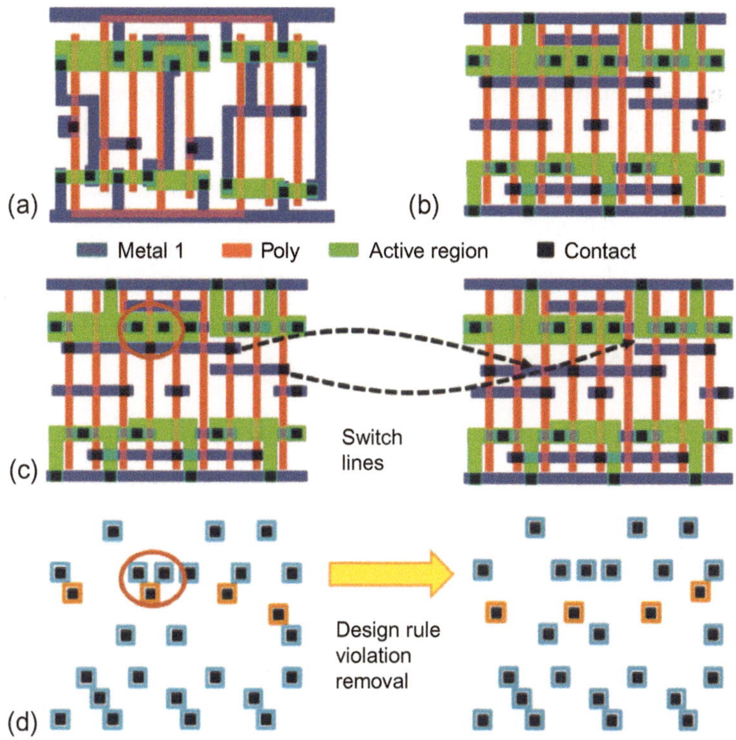

Figure 10.6 DSA-aware design methodology. (a) Conventional half-adder HA-X1 layout from Nangate 45 nm open cell library. (b) Redesigned DSA-aware half-adder HA-X1 layout. (c) The two metal lines switch their positions to remove design rule violation. (d) The contacts (orange boxes) are moved to new locations so they can be patterned using lithography. Reprinted (adapted) with permission from Yi et al. (2012a). Copyright 2015 John Wiley and Sons.

[1] http://www.nangate.com/?page_id=2325.

maintaining the same circuit function and the same cell area as the original layout. Thus, layout optimization must ensure that contact holes can be patterned by DSA and guiding templates are printable by conventional optical lithography (shown in Figure 10.6d). Overall, we redesigned a total of 85 logic cells layouts (including inverters, buffers, NAND, NOR, AOI, OAI, adders, and flip-flops) from the above cell library into DSA-aware layouts. No area penalty is observed for any of these cells, and the full-chip VLSI (physical) design flow is largely unchanged, because all necessary changes occur at the library cell level.

The topographical guiding templates for contact holes were patterned on silicon using e-beam lithography and etched to a depth of ~50 nm (Figure 10.7c). Figure 10.7 shows the contact layout, guiding template, DSA patterns, and statistical analysis for 22-nm half adder. In half-adder HA-X1, contact holes of 14.3 nm size were generated in guiding templates of 51.3 nm size (Figure 10.7d). The overlay accuracy of the patterned features in horizontal and vertical direction is 1.0 nm (Figure 10.7g and f). As shown in Figure 10.7d, DSA also healed the defects: the DSA pattern size variation (1.7 nm) is smaller than that of the template (2.6 nm). It is also important to note that while some small templates merge together during fabrication process (Figure 10.7c), the separate self-assembled holes can still maintain reasonable size and keep overlay accuracy, showing high tolerance for template defects. DSA holes overlay accuracy and size variation are well controlled and not affected by connected templates. This ensures a reasonable process window for DSA.

10.3 Alphabet approach: A general design strategy for DSA contact/via patterning

10.3.1 Introduction to the alphabet approach

This section introduces the alphabet approach that uses a minimal set of small physical templates to pattern all contacts configurations on ICs. A general scalable guiding template design strategy can be developed to link the DSA material properties to the technology node requirements. To demonstrate this strategy, we pattern the same logic cell (one-bit half adder) for three technology nodes: 14, 11, and 7 nm.[2]

The size of the cylindrical DSA pattern and its natural periodicity depends on the chemical properties of the block copolymer. However, the detailed characteristics of the pattern, such as the number of cylinders inside a physical guiding template and the distance between different cylinders, can be controlled by the shape and size of the physical guiding templates. The previous section established that using small physical guiding templates, one can flexibly control the shape and position of the DSA pattern;

[2] The term "technology node" provides a single, simple indicator of overall industry progress in integrated circuit (IC) feature scaling. It was specifically defined as the smallest half-pitch of contacted metal lines on any product. Source: http://www.itrs.net/Links/2007ITRS/IRCPosition.html. Discussions on the use of the term "technology node" can be found at http://www.itrs.net/Links/2009ITRS/2009Chapters_2009Tables/2009_ExecSum.pdf and http://www.itrs.net/Links/2013ITRS/2013Chapters/2013ExecutiveSummary.pdf.

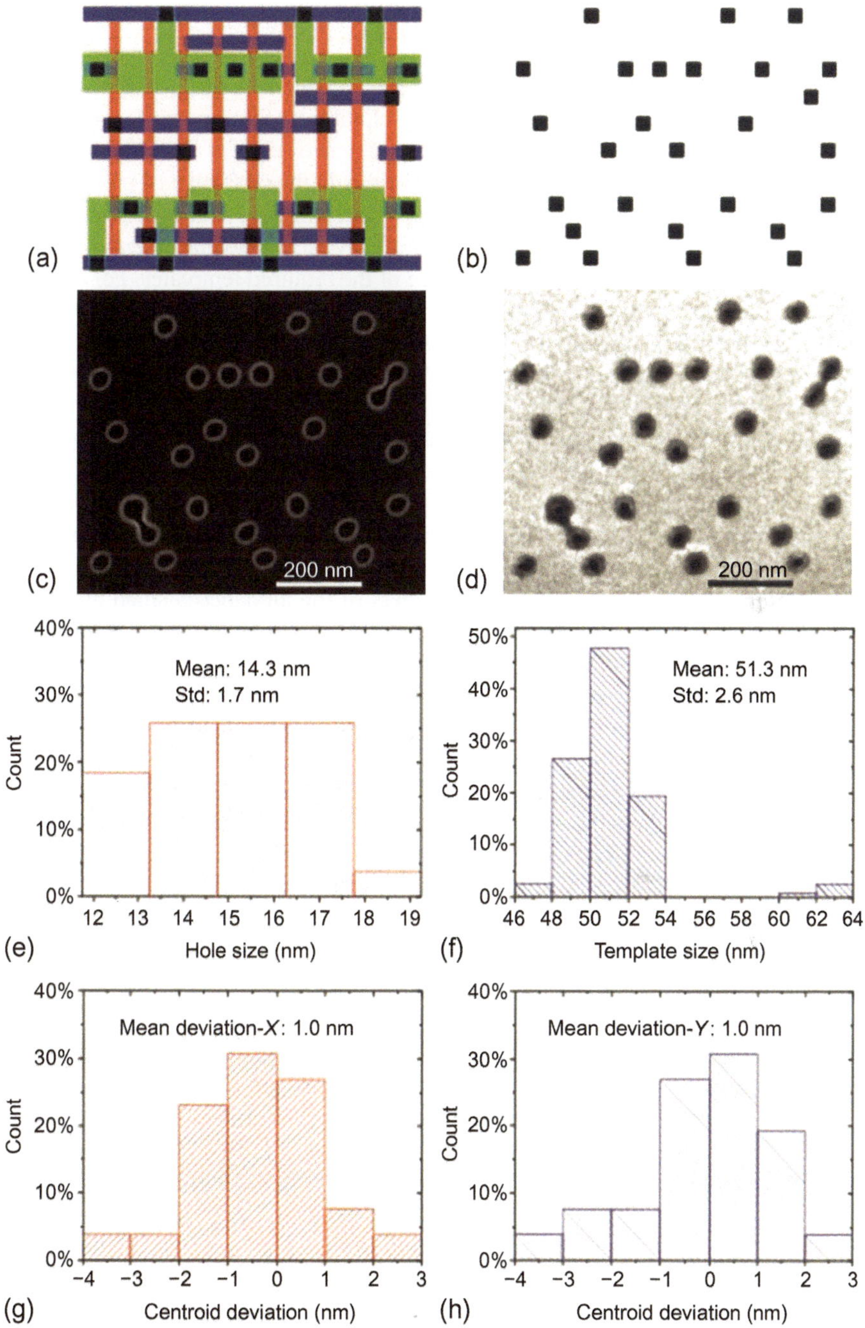

Figure 10.7 Contact hole patterning for 22 nm HA-X1 (one-bit half adder) cell. (a) DSA-aware HA-X1 (half adder) layout. (b) Contact layout. (c) SEM image of half-adder guiding template after etching. Some templates are connected due to proximity effects of e-beam exposure. (d) DSA patterns in the guiding templates. (e) Patterned contact hole size distribution. Scale bar: 200 nm. (f) Fabricated guiding template size distribution. (g) Centroid deviation of the patterned features from the guiding templates in the horizontal direction. (h) Centroid deviation of the patterned features from the guiding templates in the vertical direction.
Reprinted (adapted) with permission from Yi et al. (2012a). Copyright 2015 John Wiley and Sons.

this makes it possible to use 1-hole DSA patterns to pattern contact holes for random logic circuits (Yi et al., 2012b). However, as technology scales down, the contact/via density scales up, which simultaneously opens the possibility of using multiple-hole DSA patterns for contact hole patterning and brings in the challenge of printing guiding templates at a small pitch. Using DSA for patterning IC contacts requires further knowledge of the placement of contacts in an IC layout, as the placement of contacts in the IC layout determines the shape and size of the required templates. For example, an isolated contact could be patterned using a small circular template generating one single DSA hole, and two adjacent contacts could be patterned using an elliptical template generating a pair of DSA holes. While these two examples seem simple and straightforward, the use of these simple templates alone is not sufficient in a practical situation. More types of contact combinations and template shapes are needed to decompose an integrated full-chip contact layer with billions of contacts.

We hypothesize that there exists a set of guiding templates analogous to the letters of an alphabet that can cover all the possible contact hole patterns of a full-chip contact layer (Figure 10.8). This alphabet approach would significantly simplify DSA contact hole patterning when the total number of letters of the alphabet is small and would allow us to focus on fully characterizing the design spaces only for the letters of the alphabet. By positioning these letters in various locations, we would be able to pattern the full-chip contact layer in the same way that the 26 letters of the English alphabet is sufficient to compose an English newspaper. Some of the most basic letters, such as circular templates for 1-hole DSA patterns and elliptical templates for 2- and 3-hole DSA patterns, have been studied extensively (Yi et al., 2012a). To establish a complete alphabet, though, requires the examination of the entire standard cell library as well as the optimization of the layout to further reduce the number of letters in the alphabet as discussed in Du et al. (2013).

As the first step toward implementing the alphabet approach, a systematic design strategy is necessary to translate the placement of contacts into template designs. This

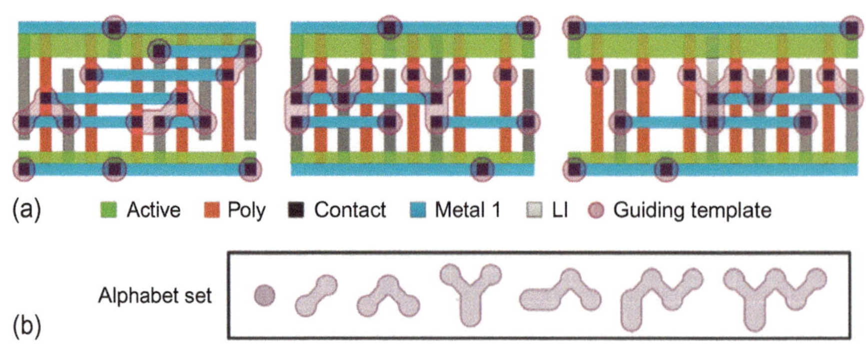

Figure 10.8 (a) Examples of three standard cells (Du et al., 2013). (b) An example of the alphabet set based on the cells in (a). With the templates in this alphabet set, one can pattern all the cells in (a) using DSA.
Reprinted (adapted) with permission from Yi et al. (2015). Copyright 2015 American Chemical Society.

design strategy not only must be applicable to one technology generation, but also must be generally applicable to future technology generations. To develop such a design strategy, we need to consider the material properties of the block copolymer as well as the size, pitch, and overlay accuracy requirements for the target technology nodes.

As mentioned in Section 10.2.2, we choose the unidirectional design as the starting point for exploring general design strategies of guiding templates. This is because the adoption of a unidirectional IC design would fix the pitch between adjacent contacts to a set value determined only by the poly-gate pitch and the metal-line pitch (Figure 10.10). The three smallest pitches between contacts for a range of technology nodes are listed in Table 10.1 (International Technology Roadmap for Semiconductors, 2013; Belledent et al., 2012; Vaidyanathan et al., 2012). As the patterning of the closely but irregularly positioned contact groups is one of the most challenging problems for contact/via printing, it is preferable to limit the number of possible distances between contacts to as few as possible. In this way, the guiding template design only needs to target a few fixed contact pitches and combinations, thus reducing the types of templates required to cover all the contact configurations in a full-chip layer. In other words, one desires to limit the size of the alphabet. A smaller alphabet could lead to simplified template design, improved template printing quality, and higher yield for DSA itself (Du et al., 2013; Xiao et al., 2013, 2014) (Figure 10.8).

Once the distance between closely positioned contacts in certain technology nodes is determined, it is crucial to know whether these distances can be achieved using block copolymer DSA. If the contact pitch is somewhere close to the natural pitch of the block copolymer DSA pattern, then multiple-hole templates corresponding to various letters in the alphabet may be adopted to pattern a group of closely positioned contacts. This set of information could be found in the DSA design space (see Section 10.2.1). Another example of the 2- and 3-hole design space is illustrated in Figure 10.10. These two DSA patterns are generated inside elliptical guiding templates, which have been identified as among the most common letters in the alphabet from examination of a representative set of standard cells in a standard cell library. We measure the change

Table 10.1 Pitch length between adjacent contacts versus technology nodes

Node (nm)	20	14	11	10	7	5
X-pitch (P_X)	90	82	64	58	46	40
Y-pitch (P_Y)	70	64	50	44	36	32
$\sqrt{(P_X/2)^2 + P_Y^2}$	83	76	59	53	43	38
$\sqrt{P_X^2 + P_Y^2}$	114	104	81	73	58	51
CD control (3σ)	2.0	1.4	1.1	1.0	0.7	0.5
Overlay (3σ)	8.0	5.1	3.6	3.2	2.3	1.4

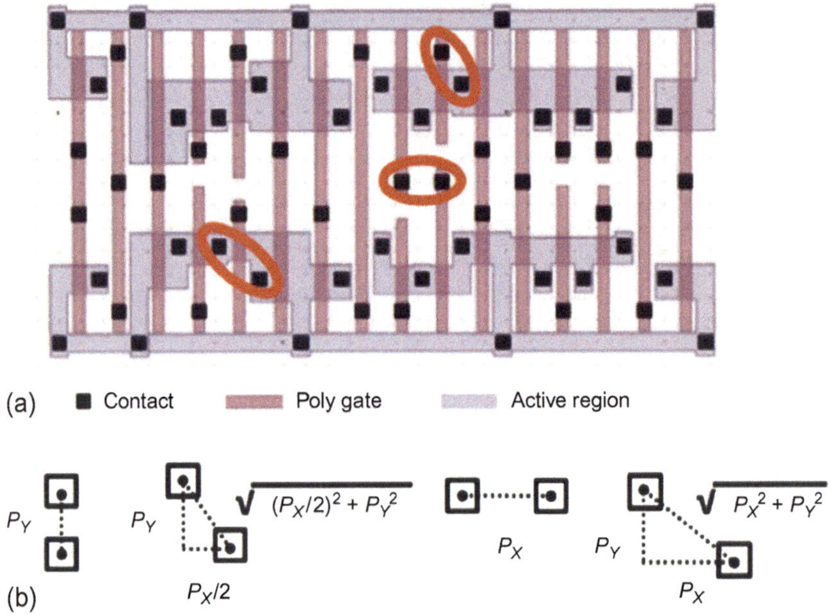

Figure 10.9 (a) An example of cell layout designed in gridded design rules (Smayling et al., 2010). All lines are designed with the same width and pitch, and the contact holes are positioned only at predetermined grid points. The lines of poly gate (purple) and metal-1 (green) layers are perpendicular. Some closely positioned contacts are marked with red circles. (b) Four smallest contact pitches in (a), which correspond to the contact pairs in red circles except for the first pair.
Reprinted (adapted) with permission from Yi et al. (2015). Copyright 2015 American Chemical Society.

in pitch for the 2- and 3-hole DSA patterns resulting from increasing the length of the elliptical templates while keeping the width unchanged (see Figure 10.10). From Figure 10.10, we observe that the pitch of a 2-hole DSA pattern must be between 36 and 50 nm, while the pitch of a 3-hole DSA pattern cannot exceed 48 nm. The importance of developing the DSA design space is that it can directly relate the DSA hole pitch to the polymer film thickness and template size and shape, without requiring knowledge of the chemical properties or process. This allows the design space, once fully characterized, to provide reliable guidance for further template design regardless of the type of block copolymer or DSA process used. For a full-chip design, analogous design spaces for the other letters in the alphabet must be established as well.

10.3.2 A general design strategy for DSA contact hole patterning

The template design strategies described above are built on knowledge of the contact pitch and DSA design space. To determine which strategy to use, we must first

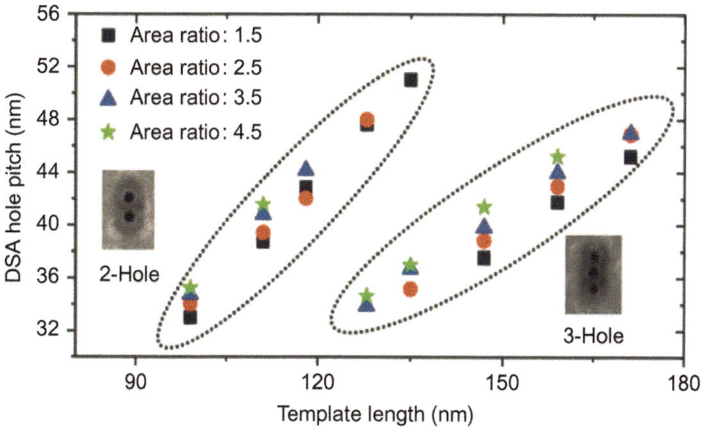

Figure 10.10 2-Hole and 3-hole design space for the same BCP film thickness. Each data point represents one template generating a 2-hole or 3-hole pattern with different template length and area ratios. Area ratio is the top surface area over the template area; a larger area ratio means lower template density. As the template length increases, the distance between holes also increases until the 2-hole patterns turn into 3-hole patterns. It is shown that if the area ratio is larger, either the hole pitch will be larger for the same template length or a 3-hole pattern will replace the 2-hole pattern.
Reprinted (adapted) with permission from Yi et al. (2015). Copyright 2015 American Chemical Society.

compare the contact pitch in a given technology node with the adjustable DSA hole pitch. This comparison, regardless of the technology node and the polymer type, can lead to three different results (Figure 10.11): (1) the minimum contact pitch is larger than the resolution of conventional optical lithography, (2) the minimum contact pitch is larger than the maximum DSA hole pitch but smaller than the resolution of conventional optical lithography, or (3) the minimum contact pitch falls within the range of the DSA hole pitch (Yi et al., 2013). Each result requires the use of a different template design strategy with its own pros and cons. There are also various quality metrics to take into account when evaluating a template design, such as the size distribution and overlay accuracy of the DSA holes as well as the design's tolerance to variations in the lithography process. It is important to note that the overlay accuracy, namely, the registration of the DSA patterns with features already patterned on the wafers, is one of the factors most strongly impacting the DSA pattern quality. As the dimensions of IC circuits are scaled aggressively, the budget for overlay is shrinking quickly as well. The overlay requirement for 14 and 11 nm logic circuits are 5.1 and 3.6 nm, respectively (International Technology Roadmap for Semiconductors, 2013). Therefore, accommodating the overlay accuracy of DSA patterns within the overall overlay budget is of paramount importance for implementing DSA for next generation lithography.

The overlay accuracy of DSA consists of two parts: the intrinsic position error of the DSA pattern and the pattern fidelity of the guiding template. The intrinsic position error of DSA refers to the variation of DSA patterns inside an ideal guiding template and

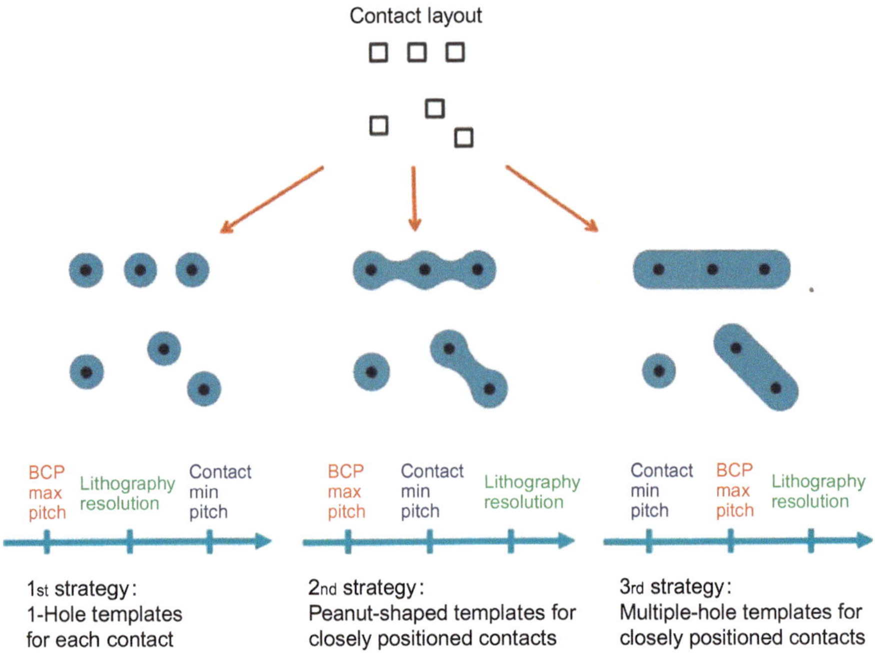

Figure 10.11 A demonstration of three strategies. From left to right: (1) The minimum contact pitch is larger than the resolution of conventional optical lithography. (2) The minimum contact pitch is larger than the maximum DSA hole pitch but smaller than the resolution of conventional optical lithography. (3) The minimum contact pitch falls within the range of DSA hole pitch.

Reprinted (adapted) with permission from Yi et al. (2015). Copyright 2015 American Chemical Society.

depends chiefly on the chemical properties of the block copolymer. The pattern fidelity of the guiding template refers to the position error of the DSA pattern brought about by a nonideal template, which may shift the position of the DSA pattern with respect to the designed location. For instance, a size offset between nonideal templates and ideal targets can be caused by the process variations of conventional lithography, such as focus and dose variations, mask alignment errors, and various lens aberrations (Wuister et al., 2014). Since the DSA pattern is sensitive to the shape and size of the guiding template, such lithography process variations lead to size variations in the guiding templates and ultimately to nontrivial position error in DSA patterns. For the discussion in this chapter, we only focus on the pattern fidelity of guiding templates. The intrinsic DSA position error will be investigated in future works.

When the contact pitch is larger than the resolution of conventional lithography, 1-hole templates for each individual contact should be used. Since the minimum contact pitch in this case is larger than the maximum DSA hole pitch, an attempt to use larger templates to generate multiple holes will either lead to extra DSA holes or DSA holes with the wrong pitch (Figure 10.12). The sole use of 1-hole templates

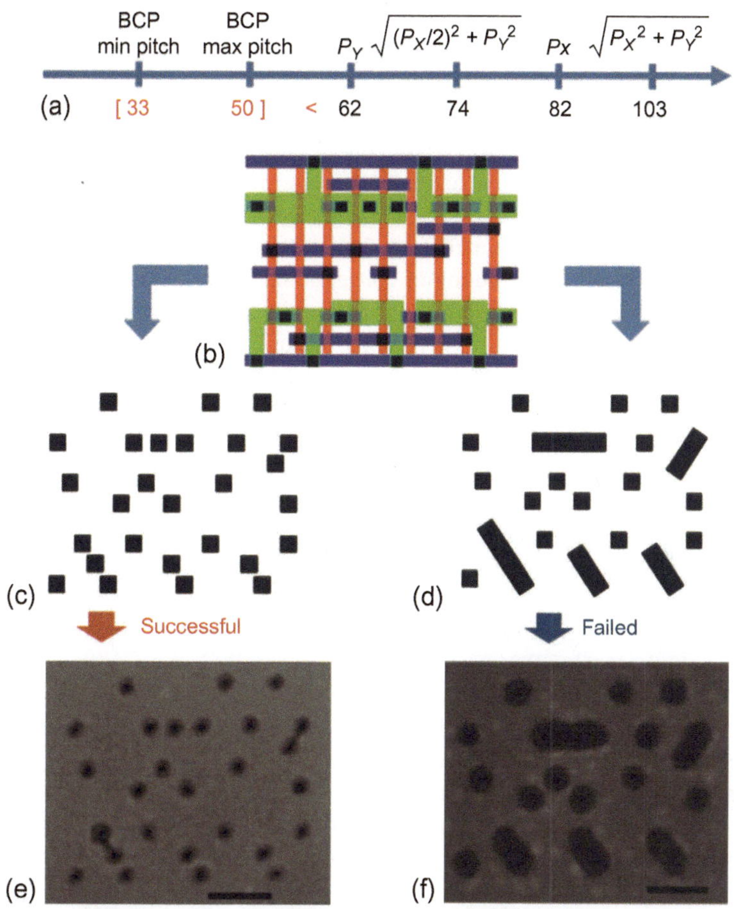

Figure 10.12 (a) An axis showing that the maximum DSA hole pitch is smaller than the smallest pitch length for the 14 nm node. The maximum and minimum pitch values for the BCP are obtained from the design space in Figure 10.10. (b) Circuit layout of one-bit half adder (Yi et al., 2012a). The green blocks represent active regions, blue lines represent the metal-1 layer, red lines represent the poly gate layer and the black squares represent the contact layer. (c) Template design with only 1-hole templates. (d) Template design with larger templates for 2 or 3 holes. (e) Successful DSA patterning using design in (c). (f) Failed DSA patterning using design in (d), as extra holes appear in oversized templates. Scale bar: 200 nm. Reprinted (adapted) with permission from Yi et al. (2015). Copyright 2015 American Chemical Society.

benefits from simplicity and results in high tolerance to process variation. As stated in Yi and Wong (2012), 1-hole DSA patterns can be generated inside circular templates ranging from 55 to 75 nm in diameter, where the DSA hole diameter remains approximately constant at 20 nm and the overlay accuracy at 1 nm. This means that even if the template size varies due to process variations, such as focus and dose variations, a 1-hole DSA pattern can still be generated successfully. On the other hand, this strategy requires that every template be separated, thereby necessitating a design rule check to ensure templates do not merge due to lithography resolution limitations and cause defects. In our example, we modified the layout of a 14 nm one-bit half adder so that the templates of the three adjacent contacts would not connect during the fabrication process due to the proximity effect. Based on the new modified layout, the contacts of the 14 nm one-bit half adder are successfully patterned with DSA using strategy (1) described above (Yi et al., 2013).

Alternatively, if the minimum contact pitch is larger than the maximum DSA hole pitch but smaller than the resolution of conventional optical lithography, one faces a dilemma between using 1-hole templates and multiple-hole templates, such as 2-hole or 3-hole elliptical templates. A very simple example involves patterning a pair of contacts with the pitch of 64 nm, as required in the 11 nm node (Figure 10.13). If a 2-hole elliptical template is adopted, the required contact pitch will be larger than the maximum DSA pitch, thus resulting in an extra hole in the middle. On the other hand, if two 1-hole templates are adopted with each template around 60 nm in diameter, then conventional optical lithography will be unable to resolve the 4 nm gap between these two templates. One possible solution to this conundrum is to instead use a special kind of guiding templates that is peanut-shaped (Figure 10.13) (Yi and Wong, 2013). The peanut shapes are specifically important, as they represent the case where two circular templates designed to be separate merge due to the limited pitch resolution of conventional lithography. Because a 2-hole pair can still form inside the merged template as a useful pattern, the peanut-shaped templates effectively extend the DSA hole pitch. Knowing that the DSA hole pitch inside the peanut-shaped templates is strongly influenced by the "neck" of the peanut, one can use three parameters to characterize the peanut shape: the template length, width, and the width of the connection part. Among these three parameters, the connection width has the most influence on the

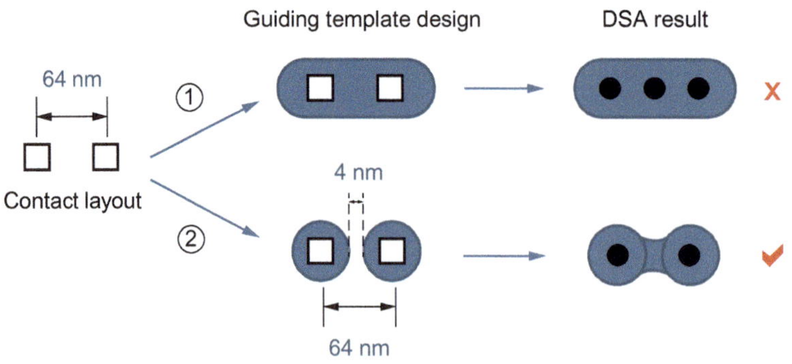

Figure 10.13 Different choices of the guiding template design for the contact pair shown on the left. The first one uses an elongated elliptical template but generates more DSA holes than desired. The second one uses two circular templates. However, due to lithography resolution, the small gap in between could not be resolved. Therefore, the second strategy actually results in a peanut shape.

Reprinted (adapted) with permission from Yi et al. (2015). Copyright 2015 American Chemical Society.

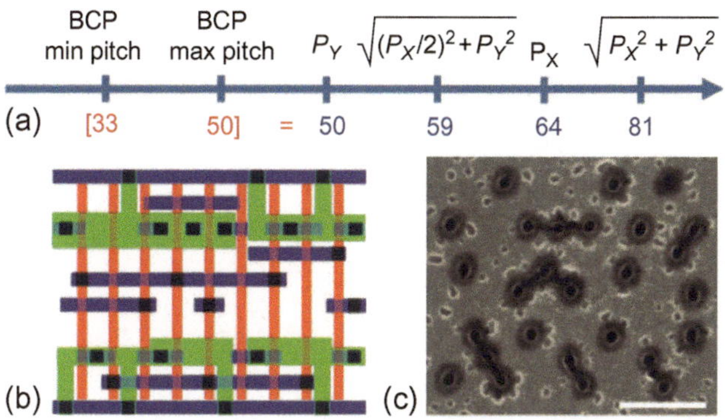

Figure 10.14 (a) An axis showing that the maximum BCP pitch is close to the minimum contact pitch for the 11 nm node but smaller than the lithography resolution (~60 nm). Peanut-shaped templates can be designed for the closely positioned contact pairs. (b) Circuit layout of one-bit half adder given in Figure 10.5(b). (c) 11 nm HA-X1 DSA contact hole patterns with layout given in (b). Scale bar: 200 nm. The DSA process used here is the same as described in Section 10.2.1.

distance variation between the two holes. In a previous study, the DSA design space of peanut-shaped templates was explored (Yi and Wong, 2013). Based on this design space, the templates for DSA contact patterning for the 11 nm one-bit half adder are redesigned, as shown in Figure 10.14.

For the third type of result—namely, the minimum contact pitch falls within the range of DSA hole pitch—one can use multiple-hole templates for generating multiple holes. In this scenario, the pitch of the DSA holes must match the contact pitch in the layout to avoid positional errors. To produce the targeted hole pitch, the geometry of the templates are determined from the corresponding DSA design space. An example of using this strategy to pattern a 7 nm node one-bit half adder is shown in Figure 10.15 (Yi et al., 2013). First, the contact pitch is obtained from the targeted technology node. The appropriate template size and shape can then be found in the DSA design space (Figure 10.10) using the targeted contact pitch. The advantage of this strategy is that the resolution requirement for lithography is relaxed, because the templates can merge as long as their dimensions and shapes can be controlled. Indeed, the stringent resolution requirement for the first design strategy necessitates that we switch the position of two metal lines in the original half-adder layout to prevent the templates from merging due to the lithography resolution limit. When the third design strategy is adopted for 7 nm node, the resolution requirement for the template is relaxed and both the original and modified half-adder layouts can be patterned (Figure 10.16). Though this strategy can flexibly adapt to different layout designs, it is important to note that this strategy also needs to satisfy higher overlay accuracy requirement, as it is often targeted toward smaller technology nodes such as 7 nm node (Table 10.1). Therefore, the pattern fidelity of the guiding templates is very important and must be examined carefully.

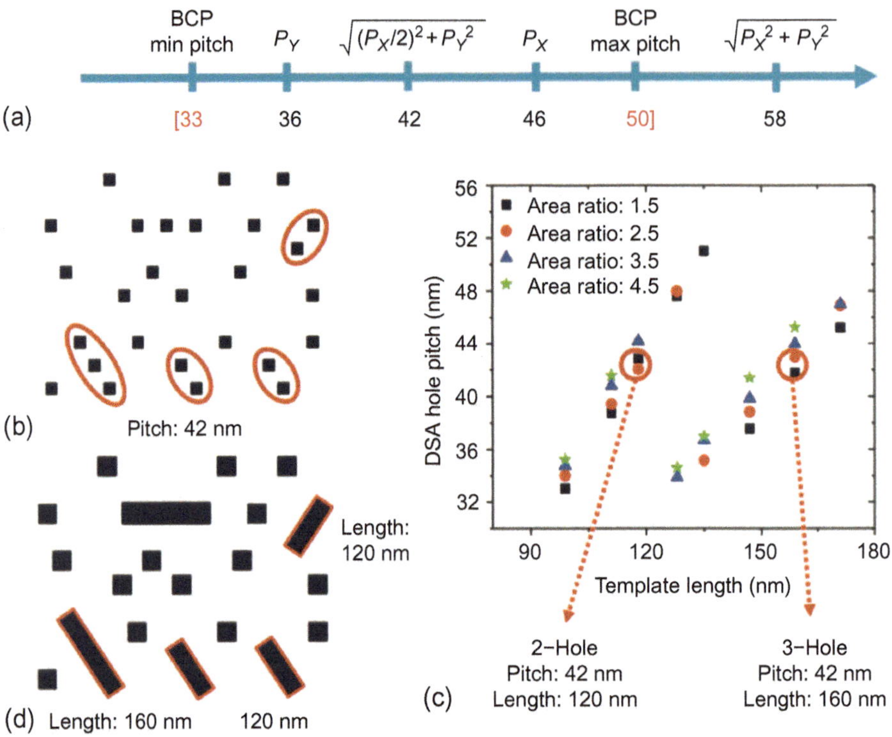

Figure 10.15 (a) An axis showing that the BCP pitch range matches with the smallest set of contact pitches for the 7 nm node. Multiple-hole templates can be designed for all the contact pairs with pitches within the achievable BCP pitch range. (b) Contact layout. All the contact pairs with 42 nm pitch are labeled with red circles. (c) 2-Hole and 3-hole design space from Figure 10.10 From this design space, we can find the template length necessary to achieve 2-hole and 3-hole patterns with 42 nm pitch are 120 and 160 nm, respectively. (d) Template design based on this strategy. Templates that generate holes with 42 nm pitch are outlined red. Reprinted (adapted) with permission from Yi et al. (2015). Copyright 2015 American Chemical Society.

10.3.3 Future directions

In this section, we illustrate a guiding template design strategy to apply DSA to contact hole patterning in IC manufacturing. The strategy couples the requirements for the targeted technology node to the design space of the block copolymer. Three different design solutions are presented and experimentally demonstrated for using PS-*b*-PMMA in DSA contact patterning for 14, 11, and 7 nm node one-bit half adders. This strategy is general and can be applied to technology nodes beyond the experimentally demonstrated 7 nm node, provided the block copolymer used meets the necessary size and pitch requirements. In the face of rapid device scaling, these three design solutions provide a viable path for low-cost DSA technology to extend optical lithography beyond its current limits.

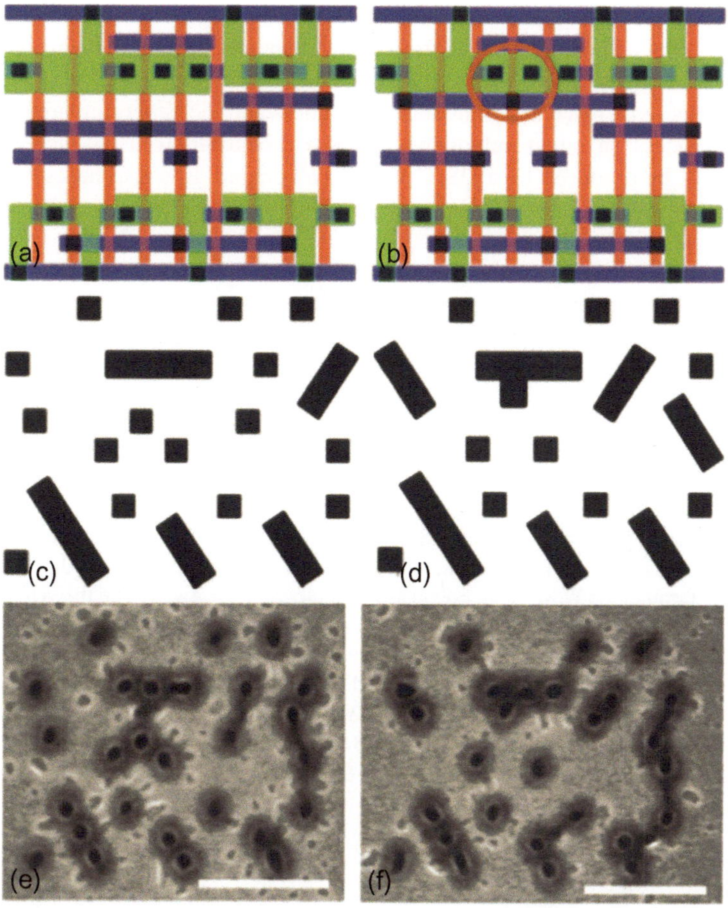

Figure 10.16 (a) DSA-aware HA-X1 (half adder) layout. (b) A similar DSA-aware HA-X1 layout that cannot be implemented by the first strategy due to the difficulty of printing the three closely packed contacts (in the red circle) separately. With the third strategy, the resolution requirement is relaxed and this layout can be patterned. (c) Contact layout design based on the layout in (a). (d) Contact layout design based on the layout in (b). (e) 7 nm HA-X1 DSA contact hole patterns with the layout in (a). (f) 7 nm HA-X1 DSA contact hole patterns with the layout in (b). Scale bar: 200 nm. The DSA process used here is the same as described in Section 10.2.1.

Reprinted (adapted) with permission from Yi et al. (2014). Copyright 2015 American Chemical Society.

This strategy also points to some important research directions: IC layouts can be further optimized to improve the yield and accuracy of DSA process. To implement DSA for full-chip manufacturing, computer aided design (CAD) issues related to DSA must be fully explored, because CAD is the key enabler for manufacturing and is a missing part that has not been studied so far. Research topics, including the minimization and the optimization of the alphabet set (Du et al., 2013), DSA hotspot detection

(Xiao et al., 2013, 2014), and the co-optimization of lithography simulation and template design are receiving attention from researchers and waiting to be explored thoroughly. Finding answers to these open questions would accelerate the development of EDA tools (e.g., place and route tools) to understand the patterning preferences of the DSA process, such that forbidden patterns will be disallowed and more preferred patterns will be produced by a DSA-aware EDA engine. It is also necessary to systematically characterize the design space for DSA that includes interactions between the materials and the ability to print the design template faithfully. Challenges such as optimizing and tuning the template design based on overlay (Wuister et al., 2014), defectivity (Bencher et al., 2012), and the metrology (Sarma et al., 2014) required for 3D DSA structures inspection will need to be further investigated in advance of practical implementation in industry.

Acknowledgments

This work was supported in part by US NSF (CMMI, No. 1029452), the Global Research Collaboration (GRC) of the Semiconductor Research Corporation (SRC), the W.M. Keck Foundation through the W.M. Keck Foundation Faculty Scholar Award, and the Willard R. and Inez Kerr Bell Professorship endowment. Discussions with Y. Du, Z. Xiao, and Prof. Martin Wong (University of Illinois, Urbana-Champaign), Bart Laenens (ASML), James Conway, and Richard Tiberio (Stanford) are much appreciated. We gratefully acknowledge the contributions of L.-W. Chang (Stanford, now with Xilinx), X.-Y. Bao (Stanford, now with Applied Materials), and our collaborators at Stanford, UIUC, UCSB, ASML, Globalfoundries, Applied Materials, IBM, TSMC, IMEC, and Mentor Graphics over the years.

References

Bao, X.-Y., Yi, H., Bencher, C., Chang, L.-W., Dai, H., Chen, Y., Chen, P.-T., Wong, H.-S., 2011. SRAM, NAND, DRAM contact hole patterning using block copolymer directed self-assembly guided by small topographical templates. In: Electron Devices Meeting (IEDM), 2011. IEEE International, pp. 7.7.1–7.7.4.

Belledent, J., Smayling, M., Pradelles, J., Pimenta-Barros, P., Barnola, S., Mage, L., Icard, B., Lapeyre, C., Soulan, S., Pain, L., 2012. Sub-20nm hybrid lithography using optical, pitch-division, and e-beam. In: SPIE Advanced Lithography. International Society for Optics and Photonics, pp. 83230F–83230F-10.

Bencher, C., Yi, H., Zhou, J., Cai, M., Smith, J., Miao, L., Montal, O., Blitshtein, S., Lavi, A., Dotan, K., 2012. Directed self-assembly defectivity assessment (part II). In: SPIE Advanced Lithography. International Society for Optics and Photonics, pp. 83230N–83230N-10.

Bita, I., Yang, J.K., Jung, Y.S., Ross, C.A., Thomas, E.L., Berggren, K.K., 2008. Graphoepitaxy of self-assembled block copolymers on two-dimensional periodic patterned templates. Science 321 (5891), 939–943.

Black, C.T., Ruiz, R., Breyta, G., Cheng, J.Y., Colburn, M.E., Guarini, K.W., Kim, H.-C., Zhang, Y., 2007. Polymer self assembly in semiconductor microelectronics. IBM J. Res. Dev. 51 (5), 605–633.

Cheng, J.Y., Rettner, C.T., Sanders, D.P., Kim, H.C., Hinsberg, W.D., 2008. Dense self-assembly on sparse chemical patterns: rectifying and multiplying lithographic patterns using block copolymers. Adv. Mater. 20 (16), 3155–3158.

Detcheverry, F.A., Nealey, P.F., de Pablo, J.J., 2010. Directed assembly of a cylinder-forming diblock copolymer: topographic and chemical patterns. Macromolecules 43 (15), 6495–6504.

Du, Y., Guo, D., Wong, M.D., Yi, H., Wong, H.-S.P., Zhang, H., Ma, Q., 2013. Block copolymer directed self-assembly (DSA) aware contact layer optimization for 10 nm 1D standard cell library. In: 2013 IEEE/ACM International Conference on Computer-Aided Design (ICCAD). IEEE, pp. 186–193.

Geiger, R.L., Allen, P.E., Strader, N.R., 1990. VLSI Design Techniques for Analog and Digital Circuits. vol. 90. McGraw-Hill, New York.

Haran, B.S., Kumar, A., Adam, L., Chang, J., Basker, V., Kanakasabapathy, S., Horak, D., Fan, S., Chen, J., Faltermeier, J., Seo, S., Burkhardt, M., Burns, S., Halle, S., Holmes, S., Johnson, R., McLellan, E., Levin, T.M., Zhu, Y., Kuss, J., Ebert, A., Cummings, J., Canaperi, D., Paparao, S., Arnold, J., Sparks, T., Koay, C.S., Kanarsky, T., Schmitz, S., Petrillo, K., Kim, R.H., Demarest, J., Edge, L.F., Jagannathan, H., Smalley, M., Berliner, N., Cheng, K., LaTulipe, D., Koburger, C., Mehta, S., Raymond, M., Colburn, M., Spooner, T., Paruchuri, V., Haensch, W., McHerron, D., Doris, B., 2008. 22 nm technology compatible fully functional 0.1 μm 6T-SRAM cell. In: Electron Devices Meeting (IEDM), December 15–17, 2008. IEEE International, pp. 1–4.

International Technology Roadmap for Semiconductors, 2013. http://public.itrs.net (accessed September 9).

Lai, K., Liu, C.-C., Pitera, J., Dechene, D.J., Schepis, A., Abdallah, J., Tsai, H., Guillorn, M., Cheng, J., Doerk, G., 2013. Computational aspects of optical lithography extension by directed self-assembly. In: Proc. SPIE, 2013. pp. 868304.

Ruiz, R., Kang, H., Detcheverry, F.A., Dobisz, E., Kercher, D.S., Albrecht, T.R., de Pablo, J.J., Nealey, P.F., 2008. Density multiplication and improved lithography by directed block copolymer assembly. Science 321 (5891), 936–939.

Sarma, C., Bunday, B., Cepler, A., Dziura, T., Kim, J., Lin, G., Yin, J., 2014. Novel metrology methods for fast 3D characterization of directed self-assembly (DSA) patterns for high volume manufacturing. In: SPIE Advanced Lithography. International Society for Optics and Photonics, pp. 90500O–90500O-10.

Segalman, R.A., Yokoyama, H., Kramer, E.J., 2001. Graphoepitaxy of spherical domain block copolymer films. Adv. Mater. 13 (15), 1152–1155.

Smayling, M.C., Bencher, C., Chen, H.D., Dai, H., Duane, M.P., 2008. APF pitch-halving for 22 nm logic cells using gridded design rules. In: SPIE Advanced Lithography. International Society for Optics and Photonics, pp. 69251E–69251E-8.

Smayling, M.C., Socha, R.J., Dusa, M.V., 2010. 22 nm logic lithography in the presence of local interconnect. In: SPIE Advanced Lithography. International Society for Optics and Photonics, pp. 764019–764019-8.

Stoykovich, M.P., Müller, M., Kim, S.O., Solak, H.H., Edwards, E.W., De Pablo, J.J., Nealey, P.F., 2005. Directed assembly of block copolymer blends into nonregular device-oriented structures. Science 308 (5727), 1442–1446.

Tang, C., Lennon, E.M., Fredrickson, G.H., Kramer, E.J., Hawker, C.J., 2008. Evolution of block copolymer lithography to highly ordered square arrays. Science 322 (5900), 429–432.

Vaidyanathan, K., Ng, S.H., Morris, D., Lafferty, N., Liebmann, L., Bender, M., Huang, W., Lai, K., Pileggi, L., Strojwas, A., 2012. Design and manufacturability tradeoffs in unidirectional and bidirectional standard cell layouts in 14 nm node. In: SPIE Advanced Lithography. International Society for Optics and Photonics, pp. 83270K–83270K-12.

Vaidyanathan, K., Liu, R., Liebmann, L., Lai, K., Strojwas, A., Pileggi, L., 2013. Rethinking ASIC design with next generation lithography and process integration. In: SPIE Advanced Lithography. International Society for Optics and Photonics, pp. 86840C–86840C-15.

Wuister, S., Druzhinina, T., Ambesi, D., Laenens, B., Yi, L.H., Finders, J., 2014. Influence of litho patterning on DSA placement errors. In: SPIE Advanced Lithography. International Society for Optics and Photonics, pp. 90491O–90491O-13.

Xiao, Z., Du, Y., Wong, M.D.F., Zhang, H., 2013. DSA template mask determination and cut redistribution for advanced 1D gridded design. In: SPIE Advanced Lithography. International Society for Optics and Photonics, pp. 888017–888017-8.

Xiao, Z., Du, Y., Tian, H., Wong, M.D.F., Yi, H., Wong, H.-S.P., Zhang, H., 2014. Directed self-assembly (DSA) template pattern verification. In: Proceedings of the 51st Annual Design Automation Conference on Design Automation Conference. ACM, San Francisco, CA, USA, pp. 1–6.

Yi, H., Wong, H.-S.P., 2013. Block copolymer directed self-assembly two-hole pattern inside peanut-shaped templates. In: 57th International Conference on Electron, Ion, and Photon Beam Technology and Nanofabrication (EIPBN).

Yi, H., Bao, X.Y., Zhang, J., Bencher, C., Chang, L.W., Chen, X., Tiberio, R., Conway, J., Dai, H., Chen, Y., 2012a. Flexible control of block copolymer directed self-assembly using small, topographical templates: potential lithography solution for integrated circuit contact hole patterning. Adv. Mater. 24 (23), 3107–3114.

Yi, H., Bao, X.-Y., Zhang, J., Tiberio, R., Conway, J., Chang, L.-W., Mitra, S., Wong, H.-S.P., 2012b. Contact-hole patterning for random logic circuits using block copolymer directed self-assembly. In: SPIE Advanced Lithography. International Society for Optics and Photonics, pp. 83230W–83230W-6.

Yi, H., Bao, X.-Y., Tiberio, R., Wong, H.-S.P., 2013. Design strategy of small topographical guiding templates for sub-15 nm integrated circuits contact hole patterns using block copolymer directed self assembly. In: SPIE Advanced Lithography. International Society for Optics and Photonics, pp. 868010–868010-9.

Yi, H., Bao, X.-Y., Tiberio, R., Wong, H.S.P., 2015. A general design strategy for block copolymer directed self-assembly patterning of integrated circuits contact holes using an alphabet approach. Nano Lett, 15 (2), 805–812.

Yi, H., Wong, H.-S.P., 2012. Design space for one-hole pattern using block copolymer directed self-assembly. In: 56th International Conference on Electron, Ion, and Photon Beam Technology and Nanofabrication (EIPBN).

Modelling and analysis of large-scale, template self-assembly manufacturing techniques

11

J. Andres Torres
Mentor Graphics Corporation, Wilsonville, OR, USA

Moving a technology from the research lab to full volume manufacturing is not an easy task. Directed self-assembly (DSA) has been a research topic since the early turn of the twenty-first century. It all started with the characterisation of the self-assembly process of diblock-copolymers, but what really sparked the interest of the technology in industrial applications was the ability to guide the assembly process with low spatial resolution systems to form useful layout configurations in the design and manufacture of electronic devices.

The work by Kim et al. (2003) and In et al. (2006) demonstrated the possibility of using traditional lithographic processes to generate the guiding patterns required to self-assemble cylindrical configurations at pre-determined positions. At that point, the main challenge was to synthesize enough block copolymer (BCP) material with well-defined characteristics, especially in terms of polydispersity, as it has been demonstrated that polydispersity and chi-N are directly responsible for edge roughness (Peters et al., 2013) and density of dislocation defects.

Novel polymerization techniques (Jarnagin, 2013) have allowed the volume production of high-quality low-polydispersity BCPs, which in turn have permitted large-scale characterisation of processes based on DSA (Gronheid et al., 2014).

At this point, most of the basic components required to support volume manufacturing are available.

- A method to create relevant design structures at desired locations (grapho- or chemo-epitaxy).
- Availability of large quantities of high-quality BCP materials with low polydispersity.
- Full-chip capable models for physical verification and mask synthesis.

It is the latter of these components that is the main focus of this chapter.

11.1 Large-scale modelling requirements

There has been ample work in the area of DSA modelling. Molecular dynamics and Monte Carlo simulators (Wang et al., 2000, 2003) are first-principle models that follow a coarse-grained approach representing polymer segments as individual beads with pre-defined potentials. Because of the computational complexity of these models, others have been proposed, including self-consistent field theory (SCFT) methods

Directed Self-assembly of Block Copolymers for Nano-manufacturing. http://dx.doi.org/10.1016/B978-0-08-100250-6.00011-0

(Reilly et al., 2013). Most of these first-principle models allow probing the assembly process across the thickness of the resist, and evaluating with a high degree of confidence different BCP compositions and guiding pattern styles. Unfortunately, the fastest implementation of these methods have a performance of around 10^5 CPU-h/mm^2. For reference, a practical limit in mask synthesis requires a performance of the order of 100 CPU-h/mm^2, and such times include the correction of the effects being modelled, which means that the simulator needs to be run multiple times before the correction achieves convergence, thus making SCFT or molecular simulations impractical for full-chip processing.

From the point of view of application in volume manufacturing, the model needs to provide at a minimum: error placement and a signal detecting phase transition conditions.

Because of the strong runtime requirements, it is practically impossible to use first-principle models. Instead, more formal models coupled with experimental validation can be used to design, calibrate and validate compact model formulations that are sufficiently accurate but process specific. To understand this trade-off, it is important to understand the types of DSA processes along with process requirements so that the best match of process and application can be found.

Chemo-epitaxy has been found to be effective in the formation of large arrays of lines and spaces (Stoykovich et al., 2005). The pitch is defined by the natural period of the BCP, and by modifying the surface interactions between the substrate and the polymer, it is possible to create large sparse guiding patterns that result in robust assembly of narrow lines, aligned to the orientation of the guiding pattern.

This approach has been widely characterized, and it is a promising candidate to print fins (Schmid et al., 2013) or metal interconnect lines. Because of the assembly characteristics around the boundary of the guiding pattern, this approach needs to be coupled with a trim exposure, which permits the definition of discrete active devices. A detailed discussion about this subject will be presented in the next section.

In chemo-epitaxy, the main mechanism to direct orientation of the diblock-copolymer phases is related to the surface interactions between substrate and diblock-copolymer. In contrast, grapho-epitaxy exploits chain confinement to direct the orientation of the ordered phases. Grapho-epitaxy for lines and spaces has been implemented in two techniques. One that uses posts with separations less than the radius of gyration of the diblock-copolymer (Chang et al., 2014) and another that uses trenches to direct the assembly process (Wang et al., 2003).

The post-implementation of grapho-epitaxy poses mainly two challenges that limit its introduction in the manufacturing process. First, the posts need to be manufactured with a resolution that is only attainable by direct write machines. While it is clear that write times might be faster than traditional direct mask writing times because of the low density of guiding patterns, it is not clear that such approach could compete with other approaches that use traditional lithography processes. Second, the posts induce rippling in the metal lines, which can lead to increasing electric parasitic behaviour, which limits the benefit of higher density. For more details of this technique, please refer Chapter 8.

Trench-based grapho-epitaxy is different as it can use traditional lithography techniques to print the guiding pattern, and it only requires an 'end' trimming to define the fin elements.

When speaking in terms of DSA, there is a minimum of two parameters that need to be understood: diblock-copolymer phase (cylindrical or parallel lamella) and director technique (chemo-epitaxy or grapho-epitaxy). As previously stated, in the case of grapho-epitaxy, there are two possible alternatives—using either trenches or posts.

However, all these combinations are aimed for very distinct applications, with some not being considered because of the number of difficult challenges to implement in full-volume manufacturing.

It is important to note that, from the point of view of the technology to be used, memory and logic/MPU products also have very different requirements. Because of the repetitive nature of memory applications, it is possible that different combinations of processes may result in a more optimal solution. From now on, however, the assumption is that the target process is for logic/MPU applications; this means that it is difficult to rely on large-scale repetition, unlike a memory-targeted process.

If we were to arrange the techniques in order of more likely to less likely to be implemented for a given application within a logic/MPU process, the order would be:

- Approach: Cylindrical BCP, trench-style grapho-epitaxy. Application: Contact/via shrink, contact/via pitch multiplication and metal cut technique.
- Approach: Parallel lamella BCP, trench-style grapho-epitaxy. Application: Fin layer with pitch multiplication, metal gratings
- Approach: Parallel lamella BCP, chemo-epitaxy. Application: Fin layer and metal gratings.
- Approach: Parallel lamella BCP, post-style grapho-epitaxy. Application: 2D metal.

The following sections will focus primarily on how to better implement the first and second approaches in full production, but in this section we will continue to describe the necessary modelling technology.

Let us start with the second approach: Using parallel lamella BPC and a trench-style grapho-epitaxy process to print a fin layer.

From the point of view of modelling requirements for full-chip applications, there is very little left to do for one simple reason. Unlike grapho-epitaxy for contact/via layers, it is not possible to correct individual fin structures in the group. Let us consider the case of a three-line forming guiding pattern as depicted in Figure 11.1.

The walls that define the trench in which the diblock-copolymer orients itself only have a direct impact in the shape of the polymer right next to the walls of the guiding pattern. Once the polymer conformably follows the shape of the guiding pattern, the polymer adjusts its shape to achieve configurations that have the lowest energy. For that reason, the walls only indirectly affect the features towards the centre of the guiding pattern, which is solely controlled by the configuration energy of the system.

If the shape of the guiding pattern does provide a preferential axis for the self-assembly process, the block-copolymer will orient itself in one of many meta-stable

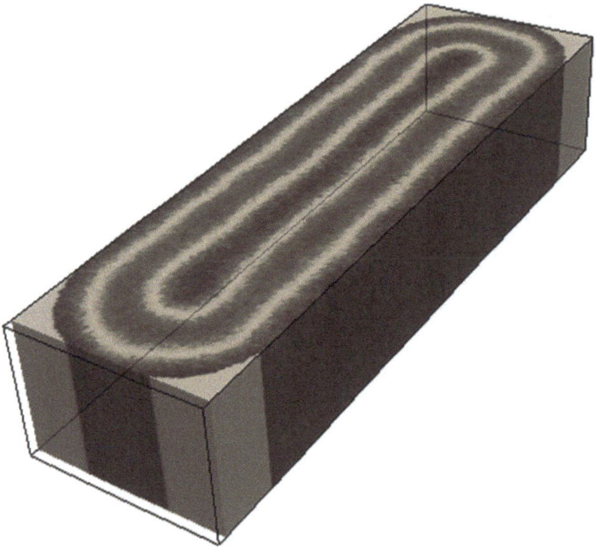

Figure 11.1 Three-line forming guiding pattern after assembly.

Figure 11.2 When the guiding pattern does not provide a preferential direction, the assembly process becomes random towards the centre of the guiding pattern.

states that minimize the energy of the system (Figure 11.2). In addition, it is not sufficient to provide a preferential axis, as the width of the guiding pattern needs to be commensurable to the period of the diblock-copolymer, otherwise, the polymer will also exhibit random behaviour (Figure 11.3).

For a given diblock-copolymer composition, there is a range of minimum aspect ratio and guiding pattern width that guarantees proper assembly for the desired number of parallel structures. What does this mean from the point of view of full-chip capable models for such technology? If it is true that one could construct a model that would predict the number of structures that a given guiding pattern would generate,

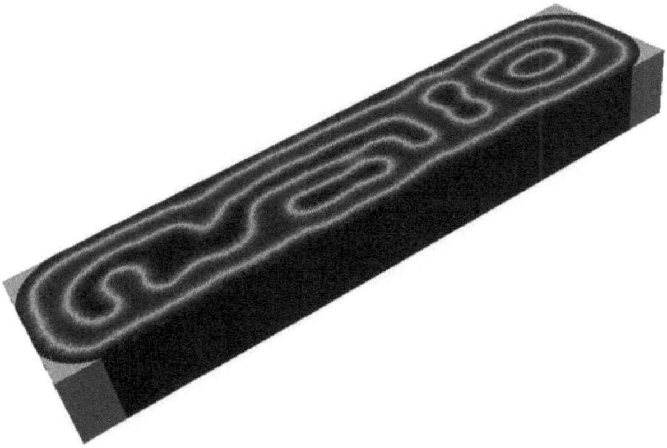

Figure 11.3 Incommensurability between the width of the guiding pattern and the natural period of the block-copolymer will result in random orientation of the diblock-copolymer chains.

the alternative, which is to pre-qualify these ranges and then simply refer to a table to implement them, is much more desirable for consistency and performance reasons. This means that models that can predict phase transitions and error placement for parallel lines using grapho-epitaxy are limited to process optimization and exploration. As once the system has been characterised, a very simple table can be used to determine which guiding pattern structures are sufficiently robust for full-volume manufacturing.

In contrast, the grapho-epitaxy process for cylinders does require a model that is able to guide the placement of the cylinders around the positions of interest.

Figure 11.4 shows a small set of guiding patterns with dimensions allowing from one to four cylinders. Figure 11.5 shows the resulting structures without showing the guiding pattern for clarity. From the modelling standpoint, it is important to consider that while robust cylindrical phase formation corresponds to thermodynamically stable

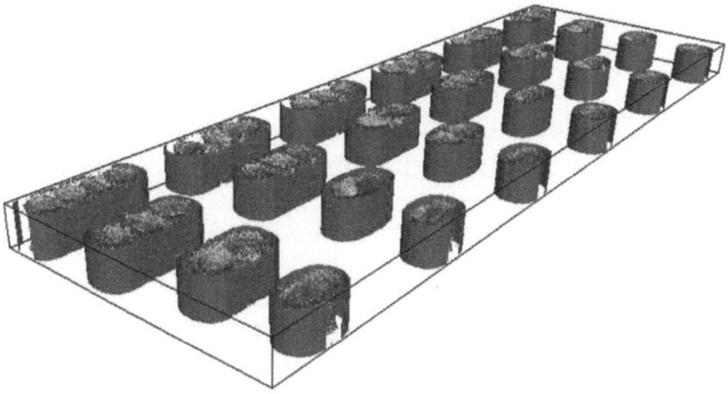

Figure 11.4 Grapho-epitaxy for directed cylindrical phase. Depending on the composition of the diblock-copolymer and the dimensions of the guiding pattern, different numbers of cylinders can be formed within the guiding pattern.

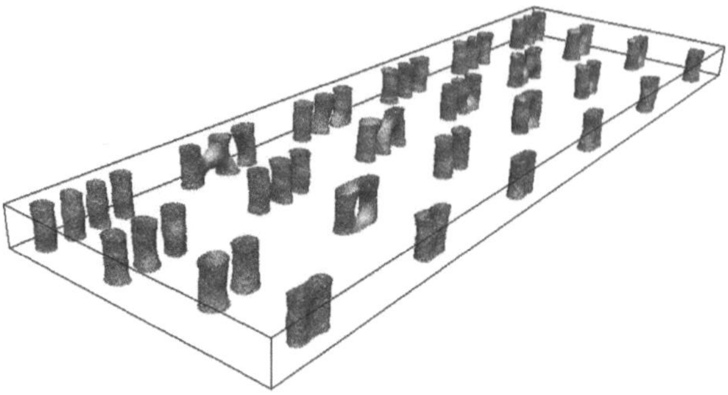

Figure 11.5 Cylinder formation without showing the guiding pattern. There are optimal regions for cylinder assembly and specific dimensions that exist in a disordered/unstable state.

configurations, there are several constructs showing merging or separating cylinders, as well as bridges or incomplete definition of the phase from top to bottom.

Those conditions are meta-stable configurations, meaning that there is a multitude of states that will exhibit the similar total energy, and depending on the dynamics of the system—small thermal fluctuations or small composition difference—they may give rise to very different configurations to be assembled.

For that reason, when considering a compact model for full-chip correction, only those guiding pattern structures that exhibit robust assembly will be considered, and as the guiding patterns get closer to regions near phase transition, the compact model's only requirement is to provide a result indicating that a phase (other than a cylinder) will be formed. Otherwise, any shape that the compact model would consistently predict will never match all the structures found in real experiments, as those results themselves vary depending on small fluctuations in the system and are random by nature (Figure 11.5).

For all these reasons, depending on the process that will use DSA as a patterning technique, these are the compact modelling requirements. For grapho-epitaxy of cylindrical structures, a model identifying feature placement, contact size and detection of phase transition within a group is required.

For grapho-epitaxy of lamella structures for fin manufacturing, a set of rules indicating the necessary size and aspect ratio for the manufacture of arrayed structures is sufficient. By making these approximations, it is expected that mask synthesis supporting either application will be of the order of 10–100 CPU-h/mm^2, which is several orders of magnitude smaller than the performance of current full physics models.

11.2 Applications of lamella systems

There has been ample research in the area of self-assembly. Different types of structures can be created in theory (Ross et al., 2014) but in practice the structures need to be reliably built and they have to be useful for constructing devices.

In order for a technology to be considered as a patterning option, it needs to meet certain criteria:

- Provide the necessary resolution.
- Minimize the number of patterning steps.
- Provide sufficient error placement and registration.
- Impose the least number of layout constraints.

When building fins, SADP (self-aligned double patterning) has been the main process to produce such structures. It has the required resolution. It requires at least two lithographic steps, one to print the mandrel pattern and one or more trim processes to remove unwanted structures. It is self-aligned since the deposition is conformal with the mandrel mask, but it does require strong design rule restrictions. Let us now compare such a process with a DSA process.

The DSA process also has the required resolution. It also requires at least two lithographic steps to remove unwanted structures. It is self-aligned because of the natural period of the polymer, but it also requires different design rules.

In other words, DSA and SADP are very similar in almost every aspect and only differ slightly in the design rules they impose and the relative cost difference between a DSA track and a deposition track.

The intent of this section is not to make a cost of ownership study for each technology, but simply to highlight that there are alternatives where there are no obvious cost savings. With this in mind, let us discuss the requirements to implement DSA as a fin-printing technology.

11.2.1 Characterising the assembly process

The first step is to characterise the process and determine the exact dimensions that lead to the desired number of fins in a group. Depending on the diblock-copolymer composition and surface treatment of the substrate, the maximum number of fins and the ranges at which phase assembly occurs will vary, Figure 11.6 shows a typical phase diagram which after experimental or numerical simulations, the ranges under which different fin counts are found.

When the width of the guiding pattern is not commensurable to the natural period of the polymer, disorder occurs, but there is certain tolerance to the exact value of width of the guiding pattern as the diblock-copolymer will preferentially order itself along the main axis of the guiding pattern.

In fact, it is a desirable property that after a certain length value, robust assembly will occur in the entirety of the guiding pattern. In a first-order approximation, the guiding pattern width is defined by the required number of periods to be assembled, but further restrictions must be considered when accounting for the required relative placement within the robust assembly region. The resulting constraints will further narrow the possible guiding pattern width tolerance.

It is also during this process optimization step that the maximum number of fins supported by the choice of diblock-copolymer, composition and surface treatment is found. This presents the first limitation of the technology compared to SADP,

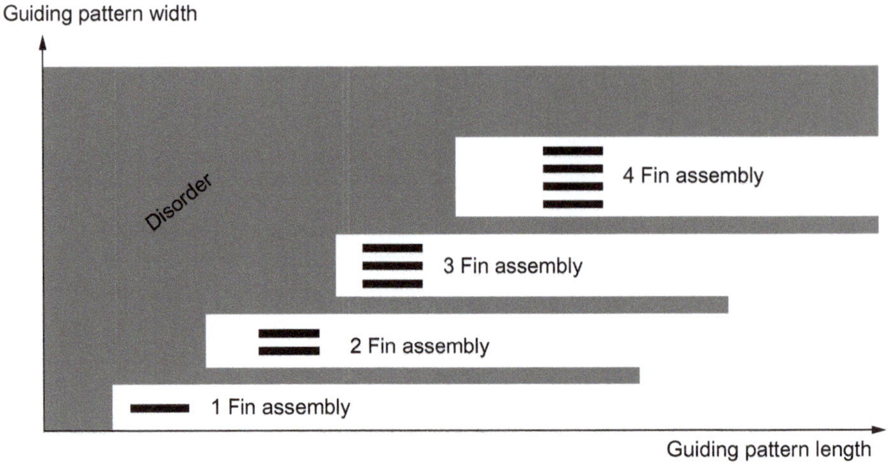

Figure 11.6 Typical phase diagram indicating disorder and ordered phases for a rectangular guiding pattern.

which does not exhibit such limitation, but which may not be relevant within the range of application of DSA for fin printing.

11.2.2 Fin patterning using DSA

Let us consider a prototypical standard cell that follows sub -7 nm best practices. Figure 11.7 shows the gate and active regions. The Fin-Fet devices are not visible, but they are captured by discrete values that the active region can take. It is not

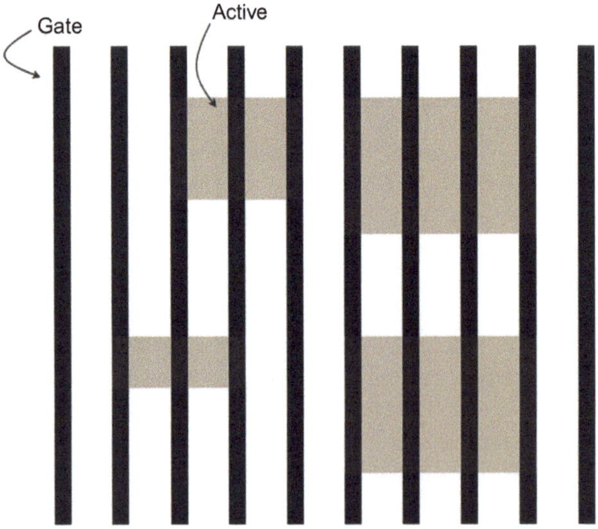

Figure 11.7 Prototypical example of a 7 nm style Fin-Fet layout.

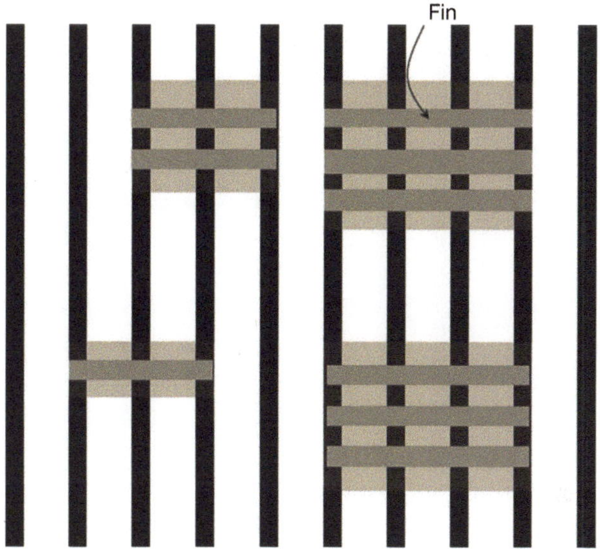

Figure 11.8 Prototypical cell indicating the location of the Fin-Fet devices along the gate; in this case, constructing devices with one, two and three fins.

possible to have any arbitrary size of the active region as only a discrete number of fins can fit within the active region.

It should be made apparent that the fin pitch is tighter than the gate pitch (Figure 11.8) and that the active area can only take incremental sizes supporting an integer number of fins.

Because of the pre-characterisation of the process, the minimum dimension of the guiding pattern required to construct such fins is now known.

Figure 11.9 shows the guiding pattern shapes. But because the fins do not align properly, several trimming steps will be required after the fin assembly. The graph previously depicted in Figure 11.6 shows that in order to guarantee robust assembly of three fins, the extensions must be longer than those for one or two fins. Also, since it is not possible to guarantee adequate assembly in non-rectangular guiding patterns, when discontinuities are found, the guiding patterns need to align to the larger number of fins. Such is the reason why the guiding pattern needs to be extended to the single- and double-fin regions.

Figure 11.10 shows the expected assembly of a lamella system forming three lines. Notice how the structures at the edges of the guiding pattern are not usable, and they define the amount of trim required in subsequent steps. The following steps must not only select those fins of interest but also must remove non-fin configurations resulting from the self-assembly process.

The next step is to remove any of the termination structures as well as the extra fins in the unwanted areas. Figure 11.11 shows a simple trim layer that would remove the unwanted structures and leave behind the required fins.

Unfortunately, unless a high resolution system such as extreme ultraviolet (EUV) or direct write is used, it is not possible to use conventional immersion lithography

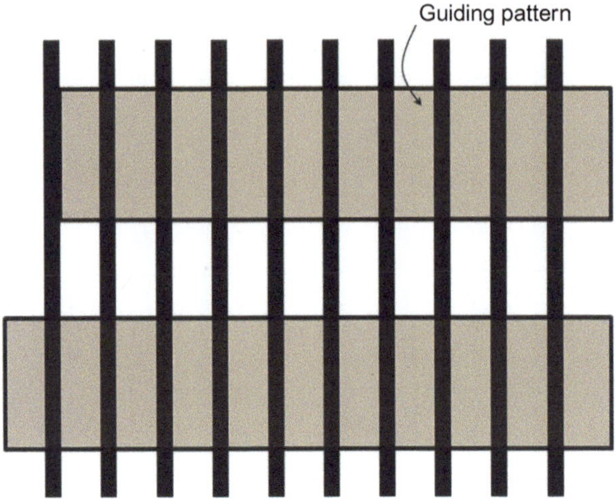

Figure 11.9 Guiding pattern defining the regions that will undergo self-assembly.

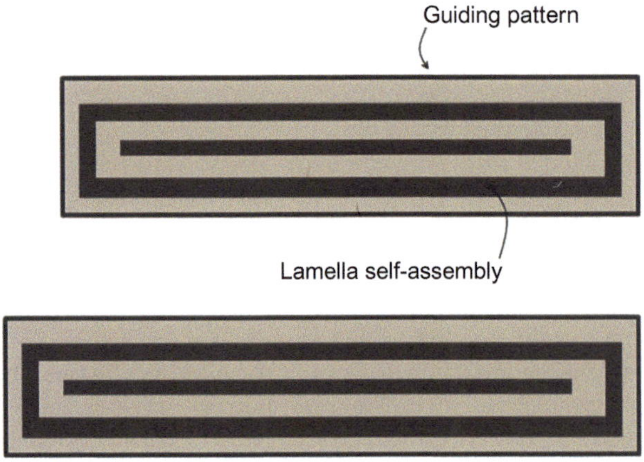

Figure 11.10 Assembly within guiding pattern.

to pattern the trim layer. Figure 11.12 shows the poor pattern fidelity and resolution, which prevents adequate trimming of the underlying fin array. It is important to mention that if EUV or direct write were available, they could also be, in principle, capable of resolving the fin structures with a smaller number of patterning steps. However, due to the unavailability of those techniques for full chip, immersion lithography is the only current technology able to be used in full-volume production.

Given the tight pitch of the fins, those structures would require a minimum of three patterning steps to properly transfer the fins onto the wafer. This is the reason to study any patterning process with the same, or preferably smaller, number of patterning steps.

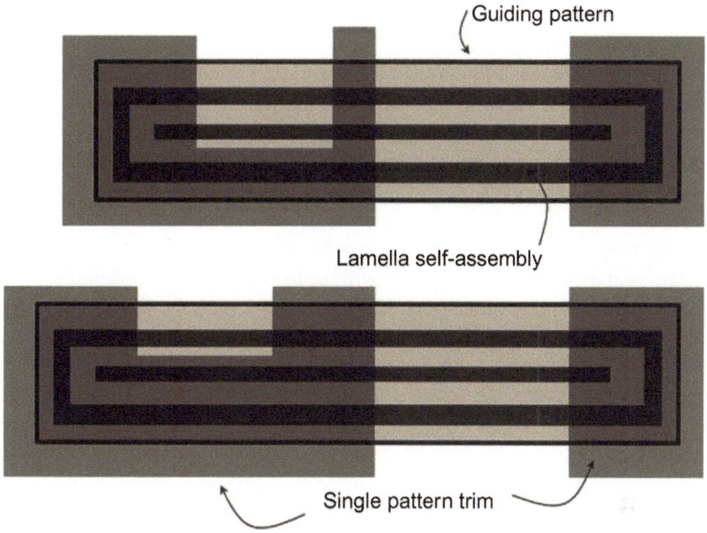

Figure 11.11 Ideal trim assuming single patterning trim.

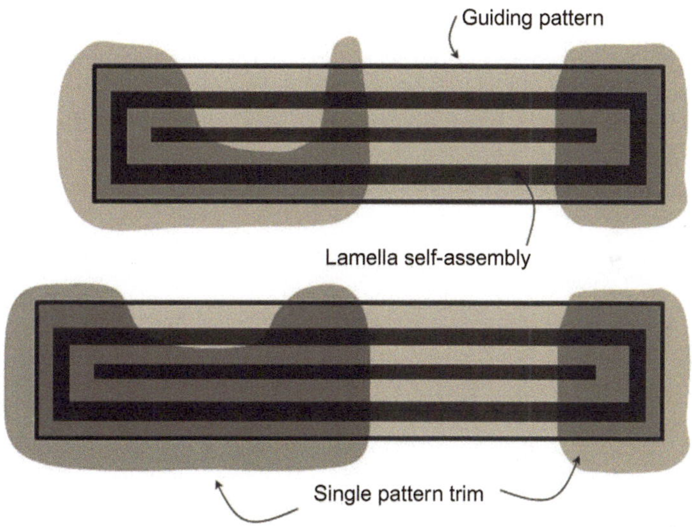

Figure 11.12 Actual trim result after using optical lithography to pattern the trim layer.

Let us consider the alternatives for fin printing.

Pitch split: In the case of using pitch split, three patterning steps are required. This means using three masks with different CD uniformity. Because the fins will not be spatially correlated, the total variability of the device will be a function of three independent fin variability distributions. In addition, the total overlay error will also be a function of three distinct patterning steps.

SADP: In the case of self-aligned double patterning, a minimum of two patterning steps are required: one defining the mandrel mask that will define the fins, and a trim patterning step to remove the unwanted fins. We have established previously that one patterning step will be insufficient given the resolution of the trimming steps, so we can consider the worst case scenario in which two patterning steps are needed to trim the unwanted fins. In such case, the fins are only a function of the local variability of the mandrel mask and the variability of the deposition process. Because the variability of the deposition process is less than that of separate patterning steps, SADP is the preferred mode of building Fin-Fet devices. However, for sub 7 nm processes, SADP is insufficient to achieve the necessary pitch requirement, which means that SAQP is needed, which will further exacerbate the variability produced by the deposition and etch non-uniformity.

DSA: DSA is very similar in many aspects to the SADP process; the difference is that the size of the fins is controlled by the ability of the polymer to self-assemble within the guiding pattern. In that respect, the question remains open as to which process (deposition or assembly) is more homogenous and provides better fin control. DSA has one advantage over SADP. SADP can only provide an even number of fins (which means that any odd number of fins will need to undergo a trimming process that can be problematic depending on the overlay control) while DSA can provide odd and even numbers of fins. However, at the same time, SADP does not have an upper limit in the number of fins that it can print within a group while DSA does.

Continuing in the implications of using a single patterning process for the trim layer, Figure 11.13 shows the final result after fin trimming. While some groups of fins can be reliably printed, those fins that were forced to be in a larger group during the assembly process may exhibit bad pattern fidelity and may be sensitive to overlay issues.

As mentioned before, these results are expected when a single patterning trimming process is used, however, if the lithography process degrades the fins to the point of making them unusable, then a double patterning trim might be required. It is important

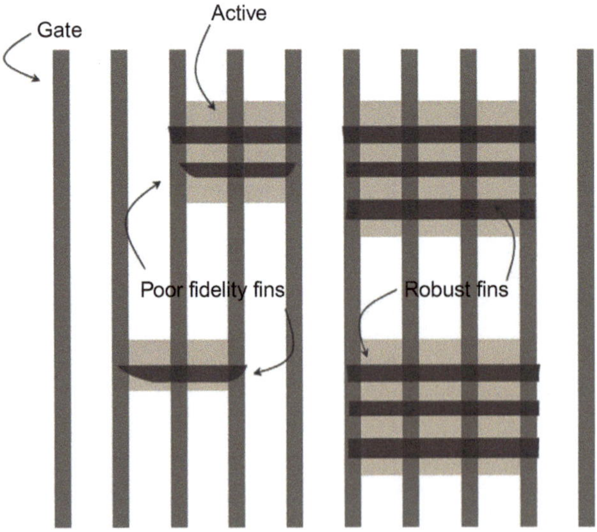

Figure 11.13 Final likely result after single patterning trim.

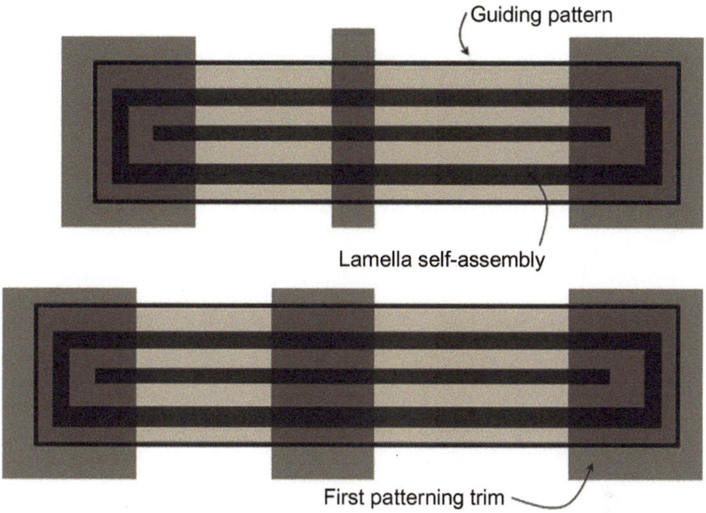

Figure 11.14 Alternative patterning trim assuming double patterning trimming.

to point out that this is a worst-case condition, these methods are still preferred over pitch split for the reasons of overlay and fin variability previously described.

If two patterning steps are allowed, then the trim mask can be simplified. Figure 11.14 shows the first patterning trim mask and the areas that it will remove from the assembled fins.

After the trim has been performed, the fins are grouped in the largest number of fins defined by the guiding pattern (Figure 11.15); this means that the next patterning steps need to have an overlay accuracy better than half a fin pitch distance in order to guarantee that the unwanted fins are removed and the desired fins are not affected during the trimming process (Figure 11.16).

Figure 11.15 Resulting fins from first patterning trimming.

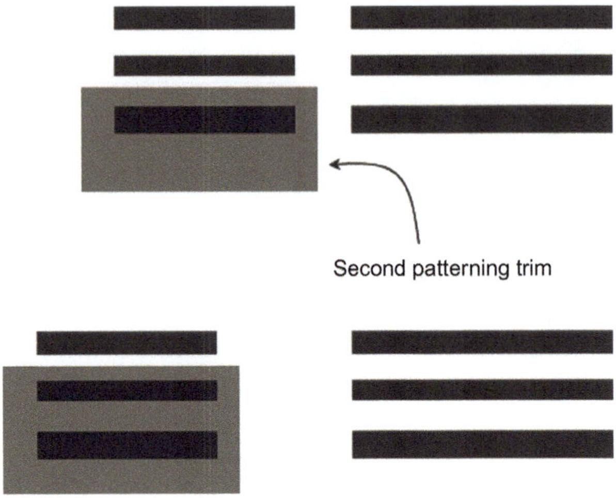

Figure 11.16 Second patterning trimming to remove unwanted fins.

After the second patterning step, the final fins have finally been transferred onto the substrate (Figure 11.17) without exhibiting fin degradation as it is in the single patterning trimming case.

However, DSA's ability to print odd and even number of fins permits one to adjust the design rules to require only a single trim step. SADP could arrive to a similar conclusion, but it would force the designer to only use an even number of fins, which is much more restrictive.

In order to achieve single trim conditions, there needs to be enough separation between groups of different number of fins. Figure 11.18 depicts the same functional

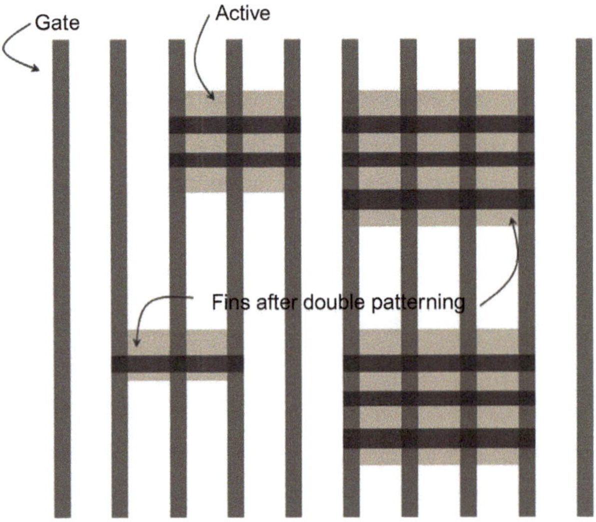

Figure 11.17 Final fins after double patterning trimming.

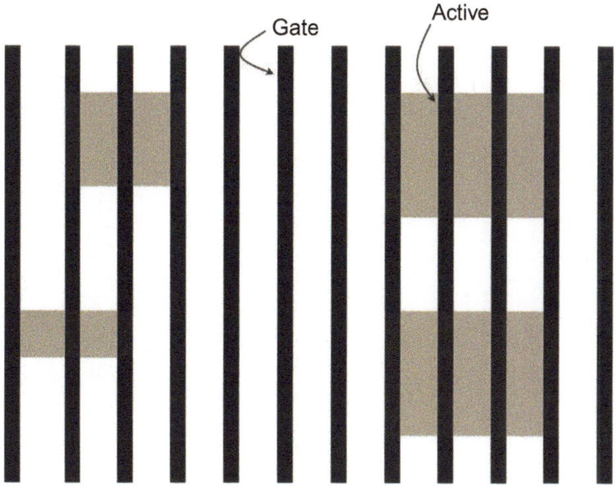

Figure 11.18 Design modification to support single patterning trim.

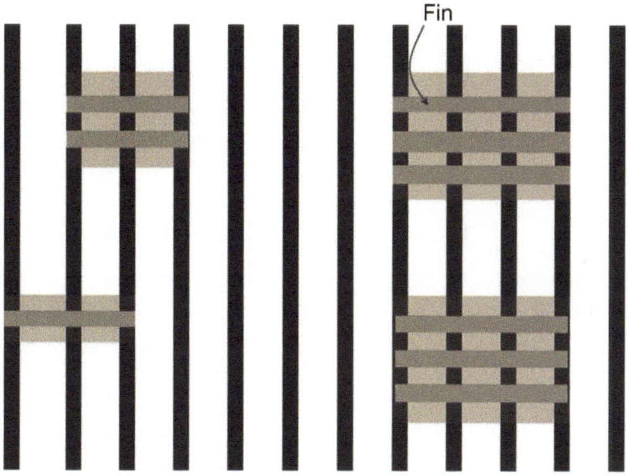

Figure 11.19 New fin locations mainly restrict layout density in one direction as additional spacing is required.

layout with additional spacing between the active layer, which defines the fin regions (Figure 11.19). It is important to highlight that such spacing only needs to be added whenever there is a different number of fins. Also, such spacing only affects density along one axis since the scaling in the vertical axis is still defined by other layers and not by limitations of the DSA process printing fins.

Because there is now sufficient space between the guiding patterns, each individual group can be properly imaged (Figure 11.20), and the assembly process can take place forming the number of fins dictated by the width of the guiding pattern (Figure 11.21).

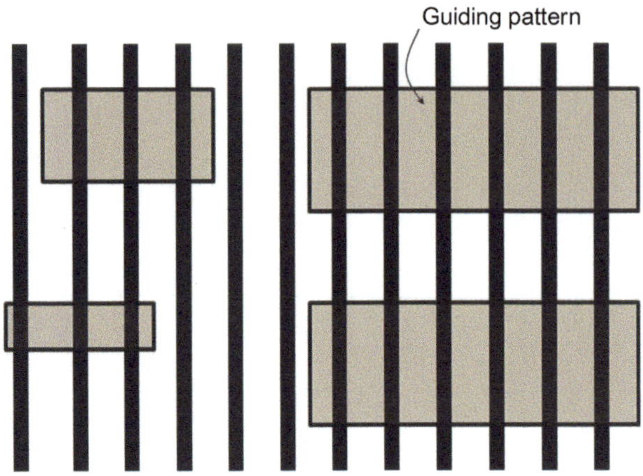

Figure 11.20 Guiding pattern capable of delivering the actual number of fins required at every location.

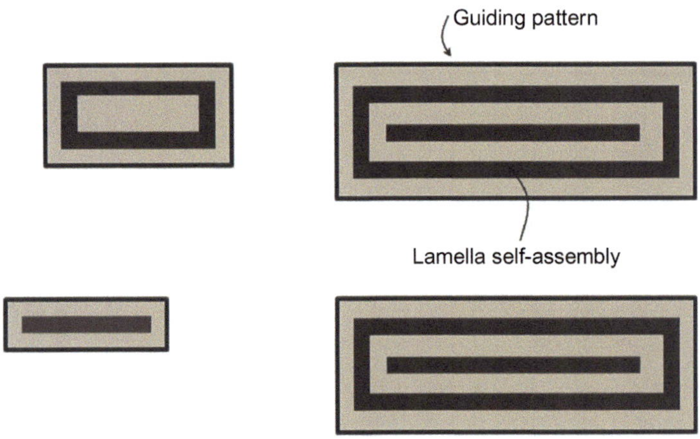

Figure 11.21 Guiding pattern and polymer assembly.

Now, the process needs to eliminate the unwanted regions, which in this case are limited to end points of the guiding pattern. It is also important to note that because a single fin can be properly imaged without the assistance of trim, it reduces the complexity of the trim layer. Figure 11.22 shows the resulting trim layer in which the fins are defined by trimming the edges of the topologies assembled inside the guiding pattern.

The resulting layout (Figure 11.23) is locally less dense than the double patterning alternative, but previous studies (Liebmann and Pileggi, 2009) have indicated that local reductions in density do not necessarily mean increase in the total area of the design, as, in many cases, there is additional space that is not traditionally used and comes from compaction and routing constraints.

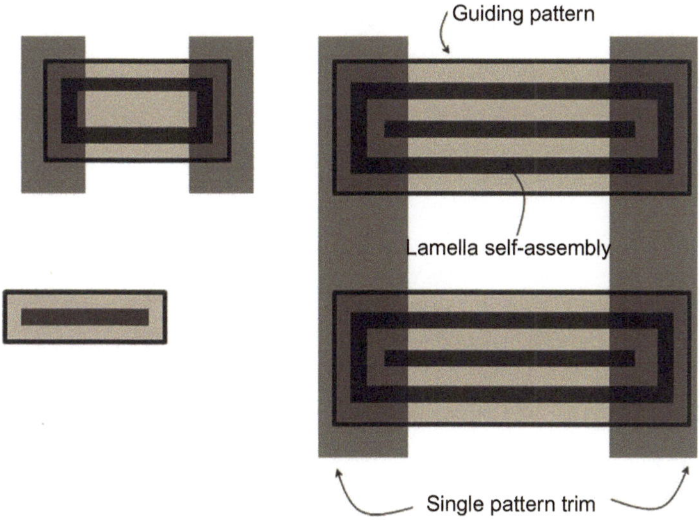

Figure 11.22 Trim layer showing how single fins may not require trim and how larger lower resolution trim blocks are required.

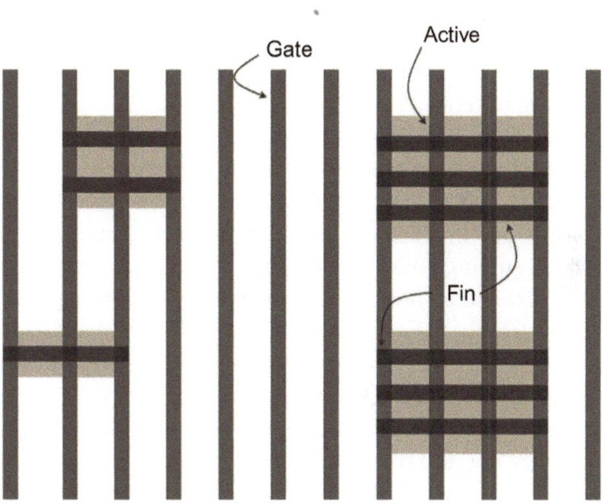

Figure 11.23 Final result after trimming and considering DSA properties during cell synthesis.

11.3 Applications of cylinder forming diblock-copolymers

In the previous section, we discussed the reason why it is not necessary to have a model that predicts the lamella structures for a fixed size guiding pattern; the necessary information can be pre-characterised and the layout can be decomposed into guiding pattern and trim geometries using only rules. This was also related to the inability to control individual fin placement by modifying the guiding pattern.

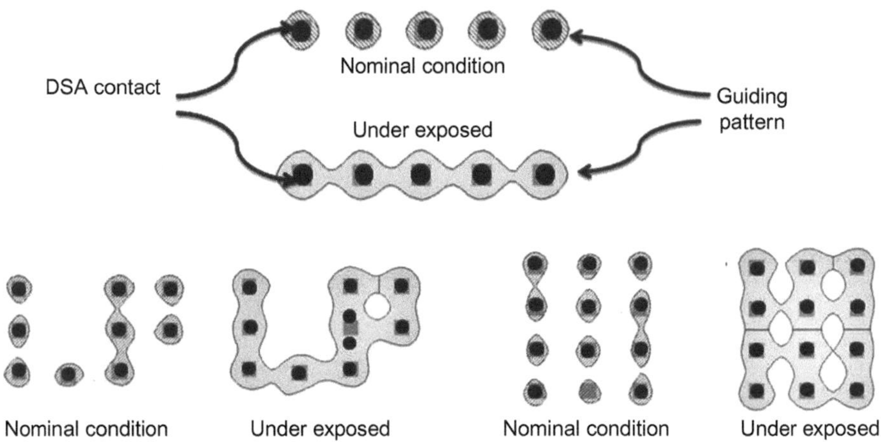

Figure 11.24 Effects of process window variations in guiding pattern. DSA is able to rectify variations in the guiding pattern but within limits, and the failure mechanism can be different at different process conditions, thus, reducing total common process window.

In the case of cylinder forming diblock-copolymers, this is not the case. First, the main mode of operation uses short strings that define the guiding pattern and not necessarily arrays. Better control in the guiding pattern has been shown to halve the error placement in grapho-epitaxy of contact holes (Wuister and Druzhinina, 2014).

Because the guiding pattern can take many different shapes, especially after considering process variations, it is impractical to try to devise complex rules to determine position and shape of the cylinders as both metrics are a function of the guiding pattern shape.

Figure 11.24 shows how different guiding patterns can lead to proper assembly of the cylinders under nominal and off-nominal conditions, but the figure also shows how the shape of the guiding pattern can vary at such conditions, leading to guiding patterns that would result in an unwanted number of cylinders being assembled.

For that reason, as long as the guiding pattern can take the shape of strings, it is still possible to control confinement, thus permitting the influence of the guiding pattern to the individual cylinder formation. In the case of arrays of cylinders, where a large guiding pattern is used to assembly a multitude of cylinders within, there is no need to use a compact model because rules implementing a pre-characterisation process (as was the fin case in the previous section) are sufficient to implement the process.

Such an approach will not be described here, and, instead, we will focus in the data flow and operations needed to implement a via/contact process using cylinder forming grapho-epitaxy.

11.4 Applying cylinder forming grapho-epitaxy to a contact/via process

Figure 11.25 shows typical low-density layout. Because of lithographic limitations and the co-optimization of the process with the layout, this particular layout has different pitches in the vertical and horizontal direction. In this particular example, we

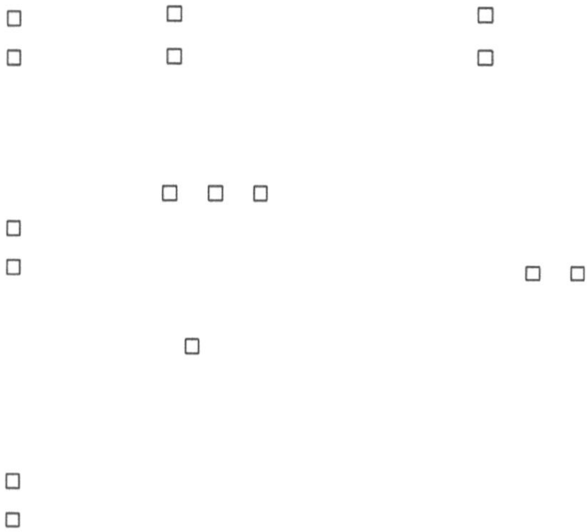

Figure 11.25 Typical layout for contact layer depicting groups of one, two and three contacts.

assume a diblock-copolymer, which has a natural period commensurable to the vertical pitch. At this moment, we know that different pitch targets will react differently to the assembly process, but, to illustrate the flow, we will assume that all these groups could be printed with a single exposure.

That being the case, before performing a guiding pattern synthesis, it is necessary to create groups of contacts. This operation group's contacts can be safely printed within the same guiding pattern. This initial step is similar in concept to the pre-colour assignment needed for multiple patterning. In multiple patterning, a colour-assignment process takes place to determine the compliance of the layout to multi-patterning. This prevents layouts that cannot be coloured to be sent to the mask synthesis process where no solution will be found.

The goal of verifying that the layout is colourable is to quickly reject configurations during physical design that would cause problems during manufacture. Similarly, grouping is performed in the layout to make sure that no long chains are present, that pitches are within the natural period of the polymer and that the groups themselves are resolvable within the number of allowed patterning steps. In the case of this example, one patterning step is allowed (Figure 11.26).

After the layout is deemed complaint by passing the previously described criteria, the groups undergo a model-based guiding patterning synthesis. The model should be aware of the natural period of the polymer as well as the volume fraction of the polymer, which ultimately defines the size of the cylinders. An ideal solution of such a synthesis process is depicted in Figure 11.27. The resulting guiding pattern after synthesis corresponds to the ideal guiding pattern needed to place the cylinders at the right positions with the necessary diameter.

At this moment, the choice of lithography needs to be made explicit (Wuister and Druzhinina, 2014). If EUV is used, it might be that assist features are not required, and the shape of the final mask will resemble the guiding pattern target. But because of delays in EUV to be used in full-volume manufacturing, in this case we will assume that a DUV process using a 193 immersion system is used. In such a case, assist features are added to

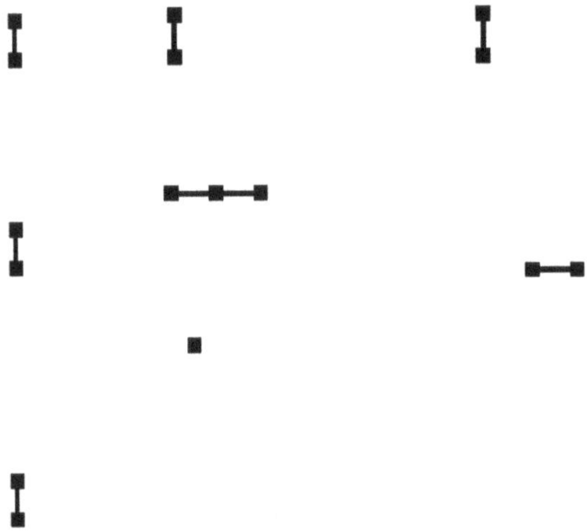

Figure 11.26 Groups of the contact layer. Shorter groups are preferred as the longer the group, the higher the probability of improper assembly.

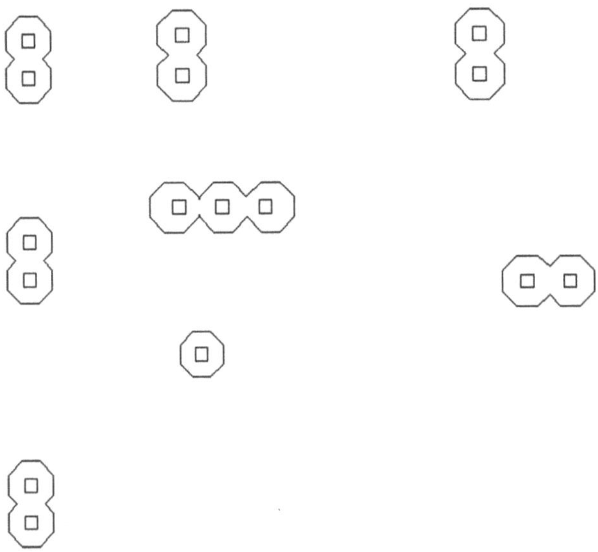

Figure 11.27 Synthesized guiding pattern using a compact model. The vertical pitch is tighter than the horizontal pitch, which is a typical side-effect of traditional lithography that requires a strong off-axis illuminator. DSA, on the other hand, benefits from single pitch dense constructs.

improve the lithographic process window (Figure 11.28). The assist features are placed with respect to the litho-target, which, in this case, is the guiding pattern for DSA.

Alternatively, the assist features can be inserted in a model-based fashion to further improve the process window of the exposure. Optical proximity correction is then used to resolve the guiding pattern, resulting in the mask depicted in Figure 11.29.

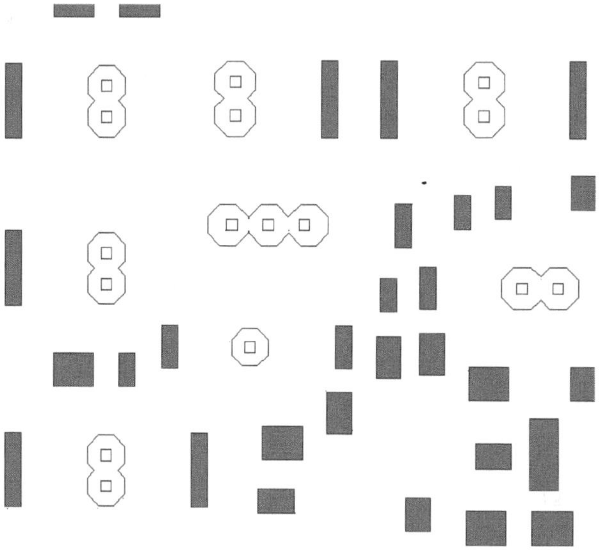

Figure 11.28 Assist features placement based on the guiding patterns to permit modulation and robustness of the mask used to generate the guiding pattern.

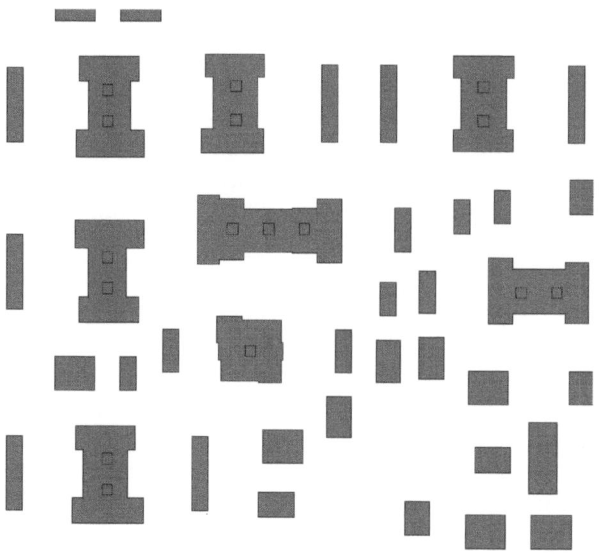

Figure 11.29 Full mask after OPC. Resolution limits and mask constraints will limit the ability of OPC to faithfully reproduce the guiding patterns.

As stated before, the lithography process imposes limitations to the types of shapes that can be printed. Figure 11.30 shows the actual guiding pattern shapes that optical proximity correction was able to deliver. The guiding pattern is clearly different from the ideal synthesized pattern because the 193i's resolution is not sufficient to provide the necessary level of detail.

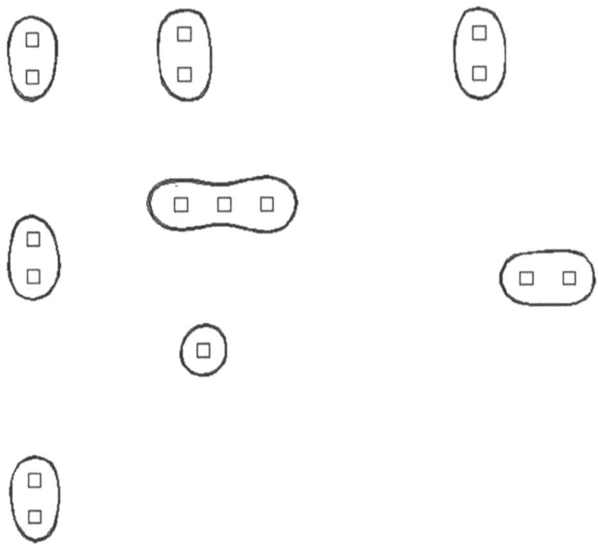

Figure 11.30 The actual guiding pattern shapes across multiple process window conditions cannot faithfully represent the synthesized guiding patterns.

In a traditional process, having this large discrepancy between the litho-target and the actual guiding pattern would mean that the guiding patterns are useless. But because the diblock self-assembles inside the guiding pattern, it is possible that several of these groups could still be valid. Having a reliable compact model becomes useful not only at generating a guiding pattern, but more importantly in the ability of the compact model to detect how highly off-target guiding patterns will affect the self-assembly process.

Figure 11.31 shows the compact model predictions for all the guiding patterns in this region. These predictions suggest that while the guiding pattern is not exactly the ideal shape computed during guiding pattern synthesis, the actual guiding pattern shapes might still be sufficiently accurate to be useful.

However, the compact model should also be able to predict whether the guiding pattern is within the range of proper assembly. A compact model is mainly a 2D model in order to be fast enough to be used in a full-chip implementation, and, for that reason, there is little hope that it can predict 3D effects that happen within the graphoepitaxy trench.

However, while the 2D model might predict a specific contact displacement, it is possible to map the total energy within the guiding pattern (Fenger and Burbine, 2014). By experimentally finding the range of energy that a given number of cylinders has while assembling properly and determining the range of energies that cylinders will have when they are under phase transition (Figure 11.32), it is possible to determine which guiding patters are not sufficiently robust.

After considering and applying those boundary conditions to the guiding patterns, it is possible to indentify that indeed there are some guiding patterns that will likely have

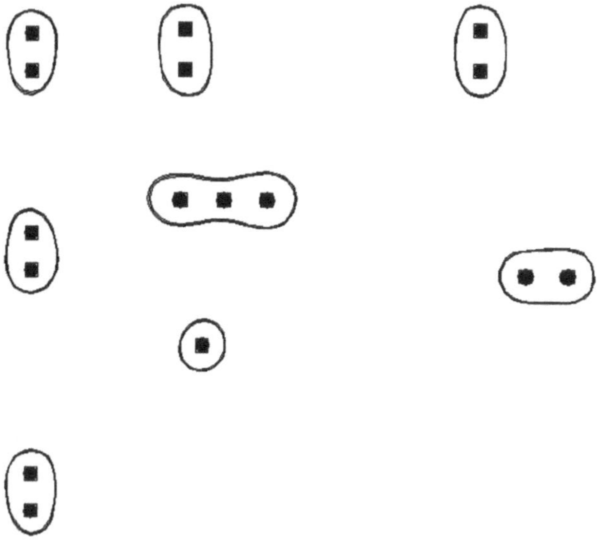

Figure 11.31 Cylinder formation inside the actual guiding pattern. When the required pitch is similar to the natural period of the guiding pattern, loss in guiding pattern quality does not strongly affect error placement.

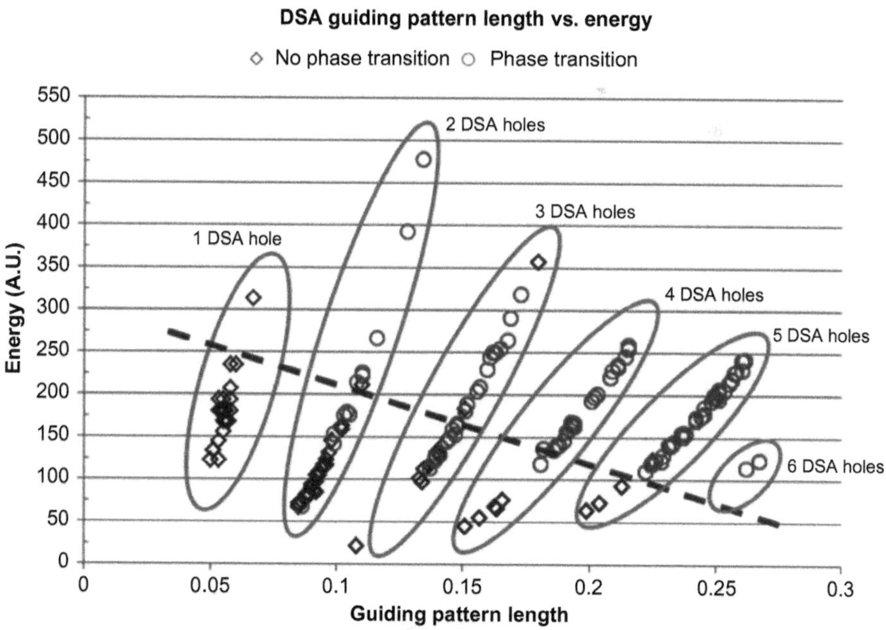

Figure 11.32 Phase transition characterisation. A compact model cannot predict 3D behaviour, but it can predict cylinder configurations whose internal energy is outside a pre-characterised calibration range.

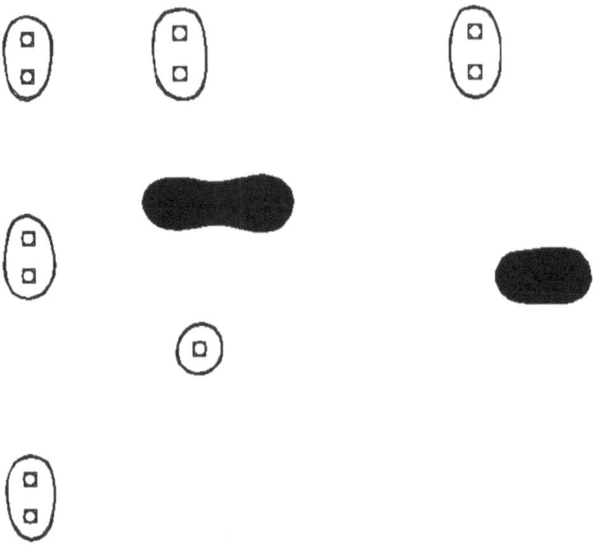

Figure 11.33 Typical output of a compact model considering phase transition regions. The dark objects correspond to an output of the simulator, indicating that such guiding patterns will not reliably assemble in the desired configurations.

problems in assembling in the right configurations. In this case, it is interesting to note that the horizontal pitch that deviates further from the natural period of the diblock-copolymer are the structures most likely to undergo phase transition (Figure 11.33). This is why it is important to consider the natural period of the polymer in any design rules that define the required contact placement.

While DSA used in fin manufacturing imposes asymmetrical shrink constraints, DSA for contacts prefers to have all the contacts within a single pitch; however, the single pitch restriction is limited to interactions within a group, which means that any relative position of contacts is allowed as long as it belongs to different guiding patterns.

At of the end of 2014, there is still work needed to implement DSA in a full-volume process, but the likely applications that will adopt DSA first as well as the necessary steps to implement DSA in the mask synthesis flow are likely to remain constant.

References

Chang, J.-B., Choi, H.K., Hannon, A.F., Alexander-Katz, A., Ross, C.A., Berggren, K.K., 2014. Design rules for self-assembled block copolymer patterns using tiled templates. Nat. Commun. 5, 3305.

Fenger, G., Burbine, A., 2014. Calibration and application of a DSA compact model for grapho-epitaxy hole processes using contour-based metrology. In: Proc. SPIE 9235. Photomask Technology, pp. 92351.

Gronheid, R., et al., 2014. Defect reduction and defect stability in IMEC's 14 nm half-pitch chemo-epitaxy DSA flow. In: Proc. SPIE 9049, Alternative Lithographic Technologies VI, pp. 904905.

In, I., et al., 2006. Side-chain-grafted random copolymer brushes as neutral surfaces for controlling the orientation of block copolymer microdomains in thin films. Langmuir 22, 7855–7860.

Jarnagin, N.D., 2013. High χ Block Copolymers for Sub 20 nm Pitch Patterning: Synthesis, Solvent Annealing, Directed Self Assembly, and Selective Block Removal. Georgia Tech, Georgia.

Kim, S.O., et al., 2003. Epitaxial self-assembly of block copolymers on lithographically defined nanopatterned substrates. Nature 424, 411–414.

Liebmann, L., Pileggi, L., 2009. Simplify to survive: prescriptive layouts ensure profitable scaling to 32 nm and beyond. In: Proc. SPIE 7275.

Peters, A.J., et al., 2013. Effects of block copolymer polydispersity and chi-N on pattern line edge roughness and line width roughness from directed self-assembly of diblock copolymers. In: Proc. of SPIE Alternative Lithographic Technologies V, p. 868020.

Reilly, M., et al., 2013. Combining physical resist modeling and self-consistent field theory for pattern simulation in directed self-assembly. In: Proc. SPIE 8682, Advances in Resist Materials and Processing Technology XXX, p. 86820G.

Ross, C.A., Berggren, K.K., Cheng, J.Y., Jung, Y.S., Chang, J.-B., 2014. Three-dimensional nanofabrication by block copolymer self-assembly. Adv. Mater. 26, 4386–4396.

Schmid, G., Farrell, R., Xu, J., Park, C., Preil, M., Chakrapani, V., Mohanty, N., Ko, A., Cicoria, M., Hetzer, D., Somervell, M., Rathsack, B., 2013. Fabrication of 28 nm pitch Si fins with DSA lithography. In: Proc. SPIE.

Stoykovich, M.P., Müller, M., Kim, S.O., Solak, H.H., Edwards, E.W., de Pablo, J.J., Nealey, P.F., 2005. Directed assembly of block copolymer blends into nonregular device-oriented structures. Science 308, 1442–1446.

Wang, Q., Yan, Q., Nealey, P.F., de Pablo, J.J., 2000. Monte Carlo simulations of diblock copolymer thin films confined between two homogeneous surfaces. J. Chem. Phys. 112, 450–464.

Wang, Q., Nealey, P.F., de Pablo, J.J., 2003. Simulations of the morphology of cylinder-forming asymmetric diblock copolymer thin films on nanopatterned substrates. Macromolecules 36 (5), 1731.

Wuister, S., Druzhinina, T., 2014. Influence of litho patterning on DSA placement errors. In: Proc. SPIE 9049.

Index

Note: Page numbers followed by *f* indicate figures.

E

F

G

FSC
www.fsc.org
MIX
Papier | Fördert gute Waldnutzung
FSC® C083411

Kolibri 360 GmbH
Ferdinand-Jühlke-Straße 7
99095 Erfurt, Deutschland
produktsicherheit@kolibri360.de